PERSPECTIVES ON THE FACE

M. MICHAEL COHEN JR.
Professor Emeritus of Pediatrics
Faculty of Medicine
Dalhousie University
Halifax, Nova Scotia

OXFORD
UNIVERSITY PRESS
2006

OXFORD
UNIVERSITY PRESS

Oxford University Press, Inc., publishes works that further
Oxford University's objective of excellence
in research, scholarship, and education.

Oxford New York
Auckland Cape Town Dar es Salaam Hong Kong Karachi
Kuala Lumpur Madrid Melbourne Mexico City Nairobi
New Delhi Shanghai Taipei Toronto

With offices in
Argentina Austria Brazil Chile Czech Republic France Greece
Guatemala Hungary Italy Japan Poland Portugal Singapore
South Korea Switzerland Thailand Turkey Ukraine Vietnam

Published by Oxford University Press, Inc.
198 Madison Avenue, New York, New York 10016

www.oup.com

Library of Congress Cataloging-in-Publication Data
Cohen, M. Michael (Meyer Michael), 1937–
Perspectives on the face / M. Michael Cohen Jr.
p. ; cm.
Includes bibliographical references and index.
ISBN 13: 978-0-19-530040-6
ISBN 0-19-530040-8
1. Face.
[DNLM: 1. Face—physiology. 2. Body Image. 3. Esthetics.
4. Face—surgery. 5. Human Body. WE 705 C678p 2006] I. Title.
QM535.C59 2006
611'.92—dc22

2006003078

9 8 7 6 5 4 3 2 1

Printed in the United States of America
on acid-free paper

Foreword

Biomedical texts usually require the efforts of several authors in order to be comprehensive. The solitary author must possess an encyclopedic knowledge of the subject. Dr. M. Michael Cohen Jr. displays such intellectual breadth in *Perspectives on the Face*.

Michael is a polymath: anthropologist, art historian, dysmorphologist, embryologist, geneticist, ichthyologist, paleontologist, and pathologist. In his latest book, he describes the face from eight perspectives: evolutionary, developmental, anatomical, genetic, surgical, psychological, sociocultural, and artistic. These chapters are "classic Cohen"—written in a sprightly and informative style. As a collector of scientific esoterica, he effortlessly integrates data from diverse disciplines. The text is lavishly illustrated with strange and wonderful images, including photographs of his friends, his wife and children, and even his own young visage (find him if you can). Michael is a master table-maker—for fun look at Tables 7–6, 7–10, and 8–2. The heavy subjects are lightened by clever juxtapositions and incongruities. One hears his impish chuckle when reading such headings as "Put a Head on It: Neural Crest and the Origin of Vertebrates" (Chapter 2); "Noses Nobody Knows: The Pathology of Sniffing" (Chapter 4); and "Sociable Robots" (Chapter 7).

The many faces of Dr. Cohen will delight the reader with kaleidoscopic reflections on that wondrous feature that defines our species.

John B. Mulliken, M.D.
Director, Craniofacial Centre
Children's Hospital
Professor of Surgery
Harvard Medical School
Boston, Massachusetts

Preface

Faces, like images seen in a kaleidoscope, are infinitely complex and beautiful. My main purpose in writing this book is to provide eight facial perspectives that can be differentiated from one another as separate yet related fields of knowledge. What constitutes a face depends on how the question is framed. In Chapter 1, I give two anatomic–evolutionary definitions, two behavioral definitions, and a symbolic definition.

The face can be studied from so many different perspectives–evolutionary, embryonic, developmental, postnatal, anatomical, physiological, dysmorphic, genetic, surgical, psychological, sociocultural, symbolic, and artistic–that no one discipline can lay claim to the face as its exclusive domain. Yet within each discipline, many authoritative articles and some authoritative books have been written. However, no well-documented books covering the face *in toto* have ever been written, although several picture books with very modest texts suitable for coffee tables have appeared.

Of several medical books that I have written, two of them contain portions with detailed information about the face and craniofacial regions (1, 2). In writing this volume, I have attempted to produce a comprehensive, well-referenced text on the face by overviewing each area in a separate chapter with depth provided by selecting a few ideas for detailed consideration.

Readers who should find this volume of interest include biologists, pediatricians, plastic surgeons, craniofacial surgeons, oral and maxillofacial surgeons, orthodontists, craniofacial biologists, psychologists, sociologists, anthropologists, art historians, students, and others with a special interest in the face.

M. Michael Cohen Jr.
Halifax, Nova Scotia, Canada,
2005

REFERENCES

1. Cohen MM Jr: The Child with Multiple Birth Defects. Second Edition. New York: Oxford University Press, 1997.
2. Cohen MM Jr, MacLean RE: Craniosynostosis: Diagnosis, Evaluation, and Management. Second Edition. New York: Oxford University Press, 2000.

Acknowledgments

I am grateful to John B. Mulliken (Boston) for writing the Foreword. I wish to thank the following colleagues and friends for proofreading various chapters and offering valuable suggestions: Brian K. Hall (Halifax), Chapter 2, Evolutionary Perspective; Joy Richman (Vancouver), Chapter 3, Embryonic, Developmental and Postnatal Perspectives; Peter Camfield (Halifax), Chapter 4, Anatomical and Physiological Perspectives; John B. Mulliken (Boston), Jeffrey C. Posnick (Chevy Chase), and S. Anthony Wolfe (Miami), Chapter 6, Surgical Perspective; and Fay G. Cohen (Halifax), Chapter 10, Conclusion. However, the responsibility for these chapters is mine. Although I have made some modifications based on their suggestions, I have not taken their advice in every instance. Therefore, they should not be blamed for any mistakes—real or perceived—in the final versions of these five chapters.

I want to acknowledge the following individuals who provided information or help on specific topics, references, websites, or pictures: Gaurav Aggarwal (Baltimore), John H. Barker (Louisville), Nancy Burson (New York), Peter Camfield (Halifax), Jaroslav Cervenka (Minneapolis), Rama Chellappa (Baltimore), Fay G. Cohen (Halifax), Aaron M. Cohen (New York), Justin D. Cohen (Malden), Lionel Egger (Geneva), Donald H. Enlow (formerly Cleveland), Reginald Goodday (Halifax), Robert J. Gorlin (Minneapolis), Raoul C.M. Hennekam (London), Susan W. Herring (Seattle), Thomas H. Hohl (Seattle), Robin Howell (Halifax), Leonard B. Kaban (Boston), Sven Kreiborg (Copenhagen), Takayuki Kuroda (Yokohama), Jan E. Jirasek (Prague), the late Jan Langman (Charlottesville), Richard Le Grand (Hamilton), James A. Levine (Rochester), Rhonda Miller (Halifax), Susan Milmoe (Cambridge), Benjamin C. Moffett, Jr. (formerly Seattle), Archie Morrison (Halifax), Kimie Ohyama (Tokyo), Barry Pass (Arlington), Nelson Powell (Palo Alto), David Precious (Halifax), the late Samuel Pruzansky (Chicago), Stuart Richardson (Halifax), Antonio Richieri–Costa (Bauru), Kathleen A. Russell (Halifax), Kohei Shiota (Kyoto), Paul J.W. Stoelinga (Nijmegen), Kathleen K. Sulik (Chapel Hill), Paul Tessier (Boulogne), Nadia Magnenat–Thalmann (Geneva), Michael W. Vannier (Chicago), Richard J. Wassersug (Halifax), and Roger A. West (Seattle)

I am grateful to Medical Computing & Media Services of the Faculty of Medicine at Dalhousie University for their support and dedication to excellence. Particularly helpful were James R. Clark, Alan J. Dorey, Judelyn R. Ferguson, Norene Reinhardt, and Jordan Urquhart. I want to thank Marilyn Klein (Instructional Resources, Faculty of Dentistry) for the wonderful diagrams she provided.

Finally, I wish to thank my administrator Ruth E. MacLean (Halifax) for her able executive skills. This book would not have been possible without her expertise.

Contents

Perspectives on the Face

1

Introduction

Faces, like images seen in a kaleidoscope, are infinitely complex and beautiful. Together with their associated structures, they "allow us to speak and smile; sigh and kiss; smell, touch, chew, and swallow; cry out in pain; and convey a world of feelings and emotions through facial expression" (16). However, faces are also found in a variety of different animals (20) (Fig. 1-1).

WHAT IS A FACE?

What constitutes a face depends on how the question is framed. There are three general ways to look at faces—structurally, behaviorally, and symbolically.

Figure 1-1 Face of an adult male proboscis monkey. From Schultz, 1969 (20).

Anatomic–Evolutionary Definitions

If two eyes, nostrils,[a] and a mouth define a face, then fish have faces (10). They may have one or two nostrils, which are not connected to the mouth. However, water entering the nostrils does bathe the olfactory mucosa. Fish lack an ear canal, but the inner ear is present. In some fish, Weber's bones connect the swim bladder to the inner ear, and sound is transmitted.

Another structural definition of the face requires evolutionary transformation of the skull, in which a face is recognized in mammals, but not in fish, amphibians, or reptiles. The jaw is suspended from the braincase in reptiles, but mammals have (a) three ear ossicles, (b) a secondary palate separating the airway passage from the mouth,[b] and (c) vertical positioning of the dentary, with alterations of the jaw muscles (see Chapter 2).

Behavioral Definitions

All animals communicate with each other in various ways by tactile, chemical, visual, and acoustical signaling. Insects may have heads, but do they have faces? Most say no. However, in paper wasps (*Polistes dominulus*), the number of black

[a]Most jawed fishes have two external nostrils on each side of the snout: an anterior nostril permitting water to enter the nasal cavity and a posterior nostril for water to exit. In contrast, tetrapods have a single pair of external nostrils, but their nasal cavity communicates with the mouth via the choanae.
[b]A secondary palate evolved independently in crocodiles and in mammal-like reptiles called therapsids.

splotches on their yellow faces signal their strength and status. The more facial splotches, the higher the status (21).

It is sometimes suggested that communication by facial expression is necessary to define a face. Facial muscles do not exist in fish, amphibians, reptiles, or birds, who can neither smile nor frown. In contrast, mammals can suckle and later chew, supported by a muscular tongue and movable lips and cheeks. Mammals also have external ears with muscles to change their positions, a flexible and often movable nose for use in smelling and touching, and facial hair—the vibrissae—on either side of the muzzle, chin, cheeks, and above the eyes to serve as tactile organs, which are associated with musculature that allows them to move voluntarily (18). The muscles of facial expression in primates, including humans, allow direct emotional communication (19) (see Chapter 7), and this provides a behavioral definition of the face.

Symbolic Definition

The term *symbol*, which has many different meanings, is defined in Chapter 8 as *the ambiguous representation of two things by association or resemblance*. Symbolic transformation of the face can be shown in Oscar Wilde's *Picture of Dorian Gray*. Faces are also suggested in various visual illusions, in vegetable carvings, and even in an inverted coffee percolator. Multicolored pansies too can suggest faces (see Chapter 8).

THE FACE IN THE LIFE CYCLE

The face shows remarkable alterations during different stages of the life cycle (Table 1-1). A painting, *The Stages of Human Life*, by Hans Baldung Grien[c] is illustrated in Figure 1-2. In this allegory, a young woman, an old woman, and a dead woman are linked by their hands and arms (2). The dead woman holds an hourglass timer, indicating that life is over.

The stages of human prenatal development are summarized in Table 1-2. In a 34-day-old embryo, the prospective face consists of a cephalic region and pharyngeal arches (Fig. 1-3). In a 42–44 day embryo, the developing face has a frontal area with bulging cerebral hemispheres; a nasodorsal center with nasoseptal and nasozygomatic portions; nasal pits delineated by nasal ridges with premaxillary, medionasal, and lateronasal portions; maxillary primordia; inter-

Table 1-1 Stages in the Life Cycle

Intrauterine development
Birth
Infancy
Childhood
Adolescence
Adulthood
Old age
Death

Figure 1-2 The Stages of Human Life (1530–1545), by Hans Baldung Grien, oil on wood, 61 × 151 cm, Prado Museum, Madrid.

[c]The artist has also been known as Hans Baldung. Alternative titles for his painting have included *The Stages of Life, With Death* and *The Stages of Man and Death*.

Table 1-2 Timetable and Staging of Human Prenatal Development

Periods	Age[a] (days)	Length (mm)	External Characteristics	Jirásek's Stages	Streeter's Horizons (Roman Numerals) or Carnegie Stages (Arabic Numbers)
Blastogenesis	0–2	0.2	Unicellular (fertilized oocyte)	1	1
	2–4	0.2	Blastomeric (16–20 blastomeres, morula)	2	2
	4–6	0.4	Blastodermic (blastocyst)	3	3
			Bilaminar embryo stage (round embryonic disk)		4
	6–15	0.1	Bilaminar plate	4-1	
			Primary yolk sac	4-2	5
		0.2–0.4	Secondary yolk sac	4–3	6a
			Trilaminar embryo stage (pear-shaped embryonic disk)		6b
	15–17	0.4–1.0	With primitive streak	5-1	
	17–20	1.0–2.0	With notochordal process	5-2	7–8
Organogenesis			Early somite stage (shoe sole-shaped embryo)		IX
	20–21	1.5–2.0	Completely open neural groove	6-1	X
			Neural tube closing		
	21–26	1.5–4.0	Both ends open	6-2	XI
	26–30	3–5	One or both neuropores closed	6-3	XII
			Stage of limb development (C-shaped embryo)		
	28–32	4–6	Buds of proximal extremity	7-1	
	31–35	5–8	Buds of proximal and distal extremities	7-2	XIII
	35–38	7–10	Proximal extremities: two segments	7-3	XIV–XV
	37–42	8–12	Proximal and distal extremities: two segments	7-4	XVI–XVII
	42–44	10–14	Digital rays, foot plates	7-5	XVIII
	44–51	13–21	Digital tubercles	7-6	XIX
	51–53	19–24	Digits, toe tubercles	7-7	XX
			Late embryonic stage (embryo with differentiated extremities, including fingers and toes)		XXI
	52–56	22–25	Open eyes	8-1	XXII
	56–60	27–35	Fusing eyelids	8-2	XXIII
Fetal period	60–182+	31–200	Fused eyelids	9	
Perinatal period	170–266+	201–350	Third trimester fetus and newborn (open eyes)	10	

[a]Age from conception.
From Jirásek, 1983 (13).

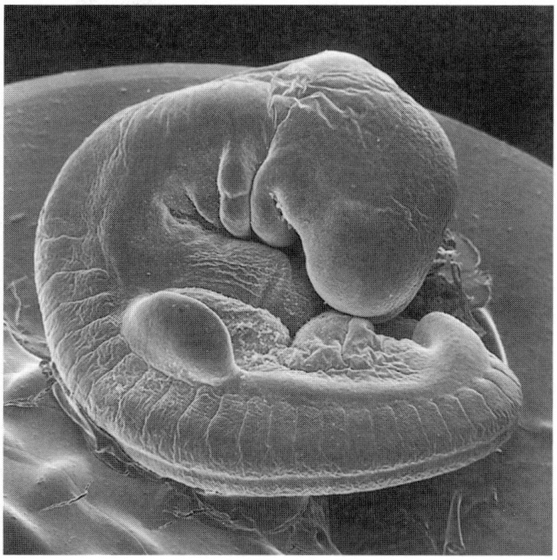

Figure 1-3 Embryo of 34 days with 34 paired somites and buds of proximal and distal extremities. Note cephalic region and pharyngeal arches. From Jirásek, 1983 (13).

premaxillary depression; premaxillary–maxillary junction; and lower jaw (Fig. 1-4). In a 52–55 day embryo, the face is recognizable and the eyes are open (13) (Fig. 1-5).

Infant faces are shown in Figure 1-6. Molding of the head is caused by compression during passage through the birth canal and is made possible by the underlying sutures that permit adjustive overlap of the calvarial bones (see also Fig. 3-21). During infancy, the head is relatively large with prominent eyes and a diminutive face. Figure 1-7 illustrates faces during childhood and adolescence. Figure 1-8 compares the same person at 1 year and at 20 years. Adult faces of different ages are shown in Figure 1-9. With age, loss of elasticity in the skin results in facial wrinkling. The first to appear are transverse lines of the forehead, nasolabial furrows from the nose to the outer corners of the lips, and "crow's feet" at the outer corner of each eye. The oldest woman shown is a 90-year-old Yahgan Indian

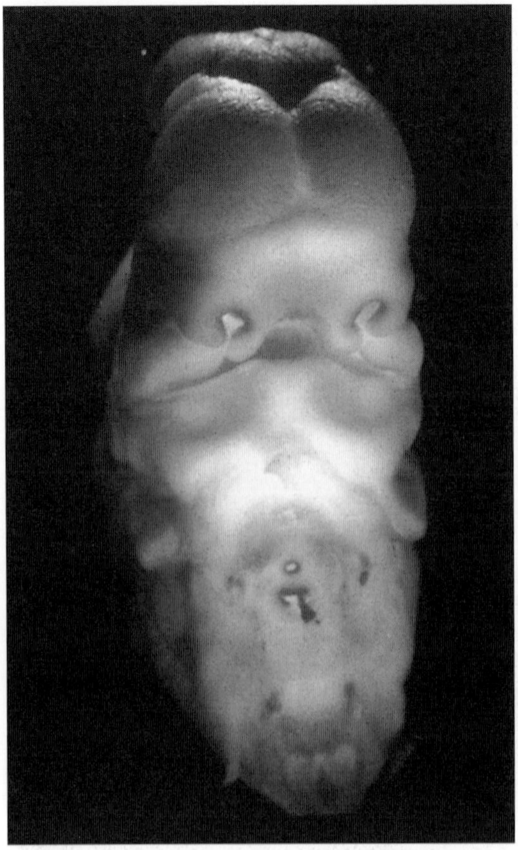

Figure 1-4 Embryo of about 42–44 days. The developing face has a frontal area with bulging cerebral hemispheres; a nasodorsal center with nasoseptal and nasozygomatic portion; nasal pits delineated by nasal ridges with premaxillary, medionasal, and lateronasal portions; maxillary primordia; interpremaxillary depression; premaxillary-maxillary junction; and lower jaw. From Jirásek, 1983 (13).

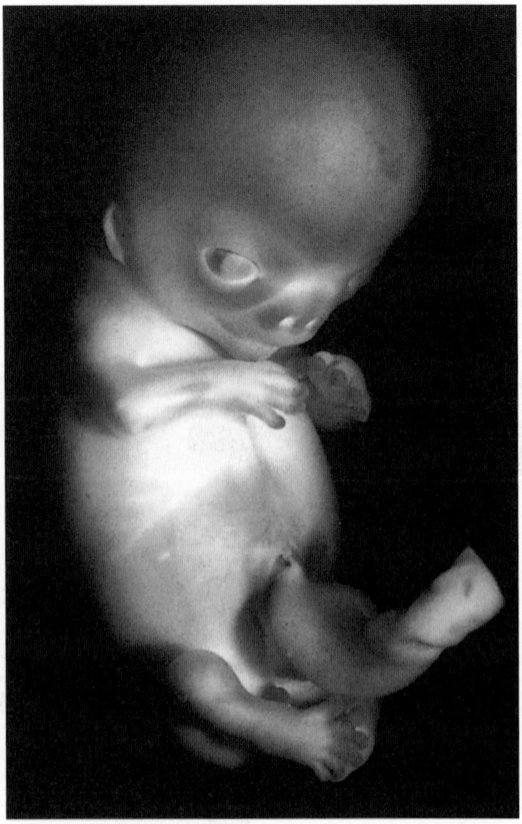

Figure 1-5 Embryo of about 52–55 days. The face is recognizable and the eyes are open. From Jirásek, 1983 (13).

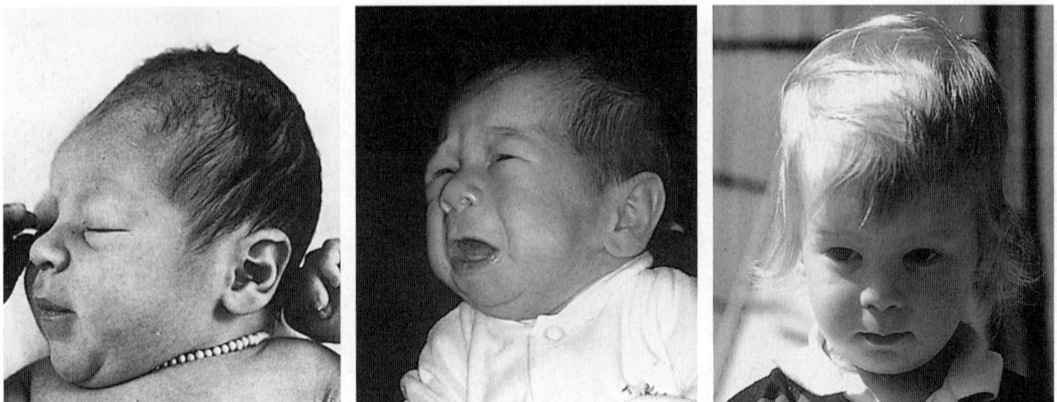

Figure 1-6 Infant faces. Left, Molding of the head caused by compression during passage through the birth canal. See also Figure 3-17. Center, Crying infant. Right, Note the relatively large head with prominent eyes and a diminutive face.

Figure 1-7 Faces during childhood and adolescence. Left, Age 9. Center, Age 16. Right, Age 18.

Figure 1-8 Note similarity in facial features and expression in the same person at 1 year and at 21 years.

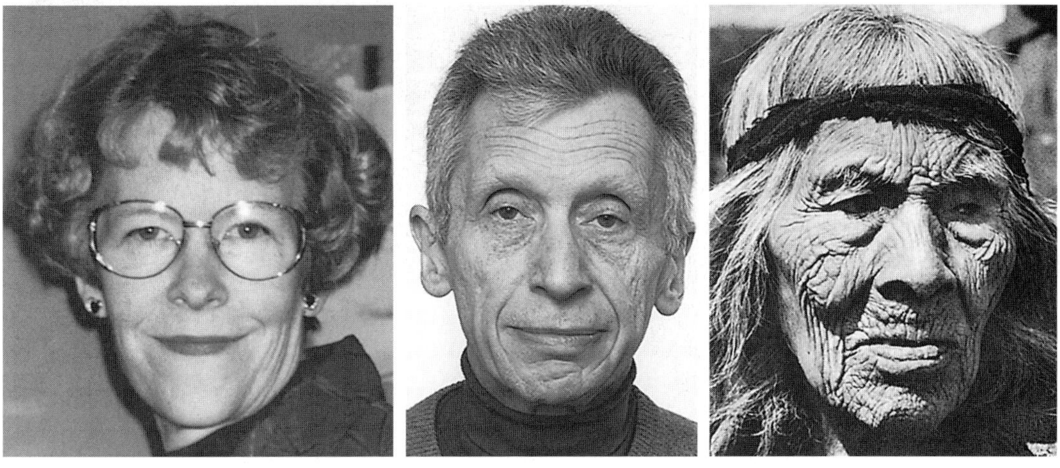

Figure 1-9 Adult faces of different ages. Right, 90-year-old Yahgan Indian woman. Right, From Moore, 1971 (15).

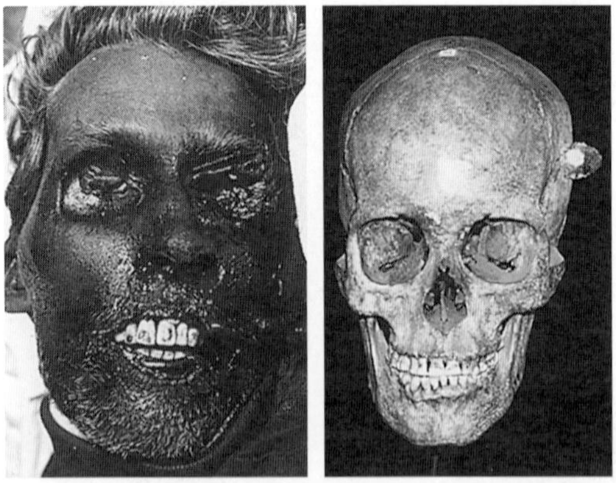

Figure 1-10 Face in death. Left, Natural death and decomposition for one week at room temperature before he was discovered. Right, Skull with projectile point lodged in the cranium—the cause of death. Right, From Bernice P. Bishop Museum, Honolulu.

from Tierra del Fuego at the southern tip of South America (15). Faces in death are illustrated in Figure 1-10. One shows a natural death and decomposition for a week at room temperature before he was discovered. The other is a skull with a projectile point lodged in the cranium—the cause of death.

PERSPECTIVES ON THE FACE

The face can be studied from so many different perspectives—evolutionary, embryonic, developmental, postnatal, anatomical, physiological, dysmorphic, genetic, surgical, psychological, sociocultural, symbolic, and artistic—that no one discipline can lay claim to the face as its exclusive domain. Yet the lack of communication between the various disciplines is striking. For example, it is not commonly thought that artistic approaches can teach the scientifically inclined about the face, much less teach it scientifically. However, in addition to aesthetic value and intrinsic interest, artistic approaches can provide research and experimental dimensions (4, 5) (see Chapter 9). Perspectives on the face are listed in Table 1-3. One chapter is devoted to each, but examples are introduced here.

Table 1-3 Perspectives on the Face

Evolutionary perspective
Embryonic, developmental, and postnatal perspectives
Anatomical and physiological perspectives
Dysmorphic and genetic perspectives
Surgical perspective
Psychological perspective
Sociocultural and symbolic perspectives
Artistic perspective

Evolutionary Perspective

Figure 1-11 compares the facial profiles of various animals. In the shark, the "nasal" area is pointed and slopes downward with the lower jaw recessed. Progressing through reptiles, marsupials, and primates, including humans, the jaw moves downward, the eyes move forward, and the head becomes more spherical (15). The brain is compared in a fish, reptile, rabbit, and human in Figure 1-12. The size of the cerebral hemispheres increases in relation to the older parts of the brain. With increased size, the cranium becomes larger and more rounded (3). Within the nervous system, rates of protein evolution are significantly higher in primates than in rodents, and within the primate order itself, protein evolution is most prominent in the human lineage. Many candidate genes have been identified that may have evolutionary implications in regulating both brain size and behavior (9).

Embryonic, Developmental, and Postnatal Perspectives

At the beginning of the third week of embryonic life, the neural plate—an ectodermal thickening overlying the notochord—develops. During the next few days, the neural folds form the neural groove (Fig. 1-13). Fusion begins in the region of the fourth somite and proceeds in cephalic and caudal directions to form the neural tube (Fig. 1-14). At the cranial and caudal ends of the tube, the anterior and posterior neuropores are temporary openings (Fig. 1-14; see also Figs. 3-3, 3-4) that eventually close. The broader cephalic portion of the neural tube becomes the future brain and the cylindrical caudal portion becomes the future spinal cord (14).

Figure 1-11 Facial profiles of various animals. Modified from Moore, 1971 (15).

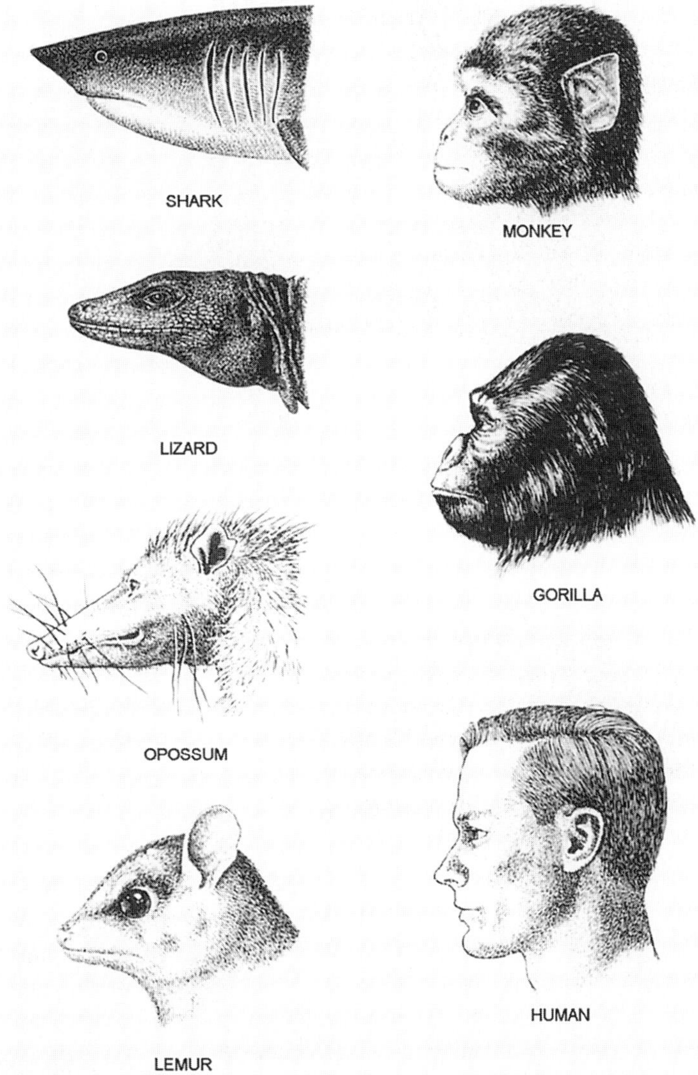

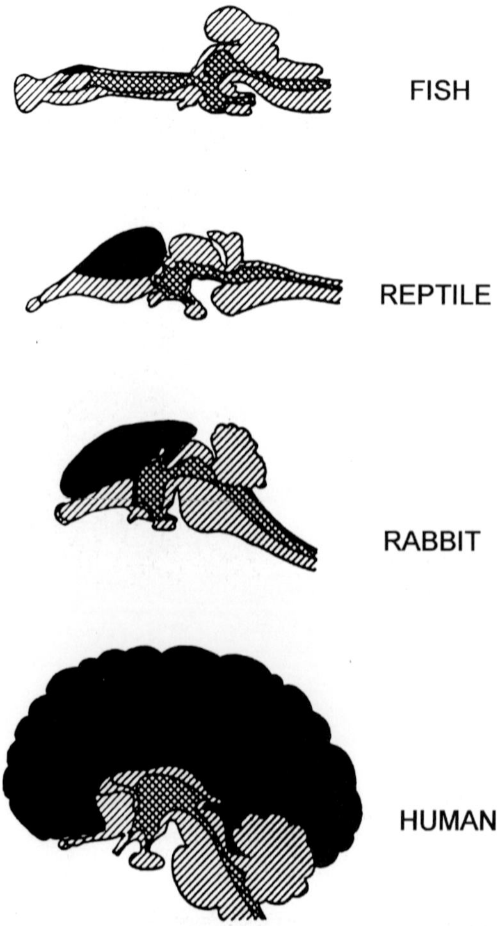

Figure 1-12 Comparison of brain size in fish, reptile, rabbit, and human. The size of the cerebral hemisphere increases in relation to the older parts of the brain. As the cerebral cortex increases in size, the cranium becomes larger and more rounded. Modified from Campbell, 1970 (3).

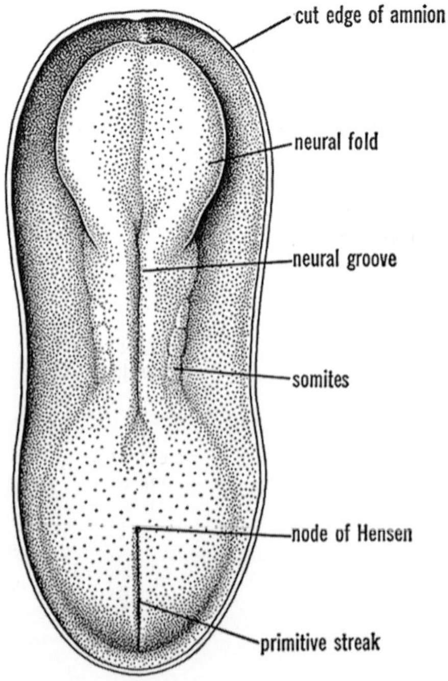

Figure 1-13 Human embryo at the third week of development. Note the neural plate and neural folds on each side of the neural groove. From Langman, 1963 (14).

Anatomical and Physiological Perspectives

The ability to identify human faces is remarkable. Although the face ages with time, it is possible to instantly identify the face of someone who has not been seen for 25 or 30 years. An area in the brain that governs facial recognition is found at the base of the cerebral cortex (Fig. 1-15). Any lesion that destroys this area impairs the ability to identity faces, but has almost no other effects, although dimming of vision may occur. The patient can read, name objects, and match a full-face picture with a profile picture of the same person. Only the ability to recognize specific faces is lost—a neurological condition known as prosopagnosia. An affected man who cannot recognize his own wife can see her, but can only recognize who she is when he hears her speak (11).

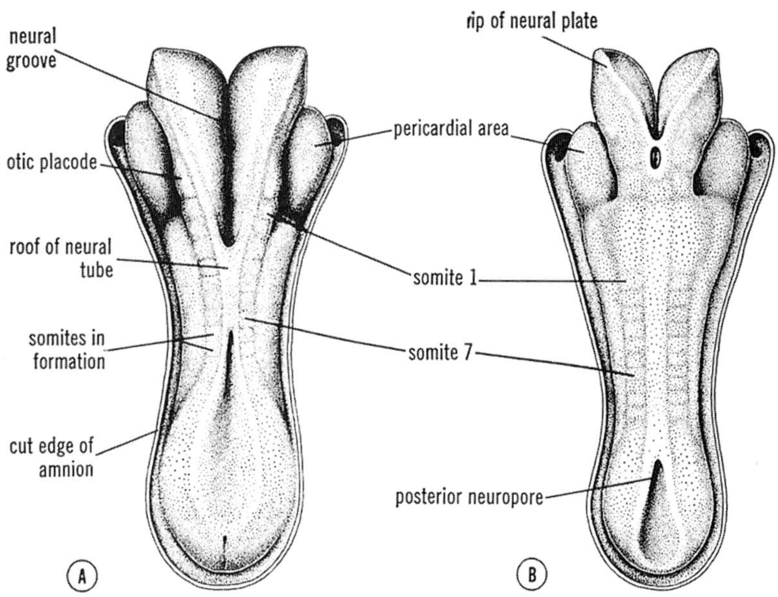

Figure 1-14 A, Human embryo of about 22 days with seven somites. The neural folds have fused at the level of the fourth to seventh somites. B, Human embryo of about 23 days. Note the neural tube with anterior and posterior neuropores. See also Figures 3-3, 3-4. From Langman, 1963 (14).

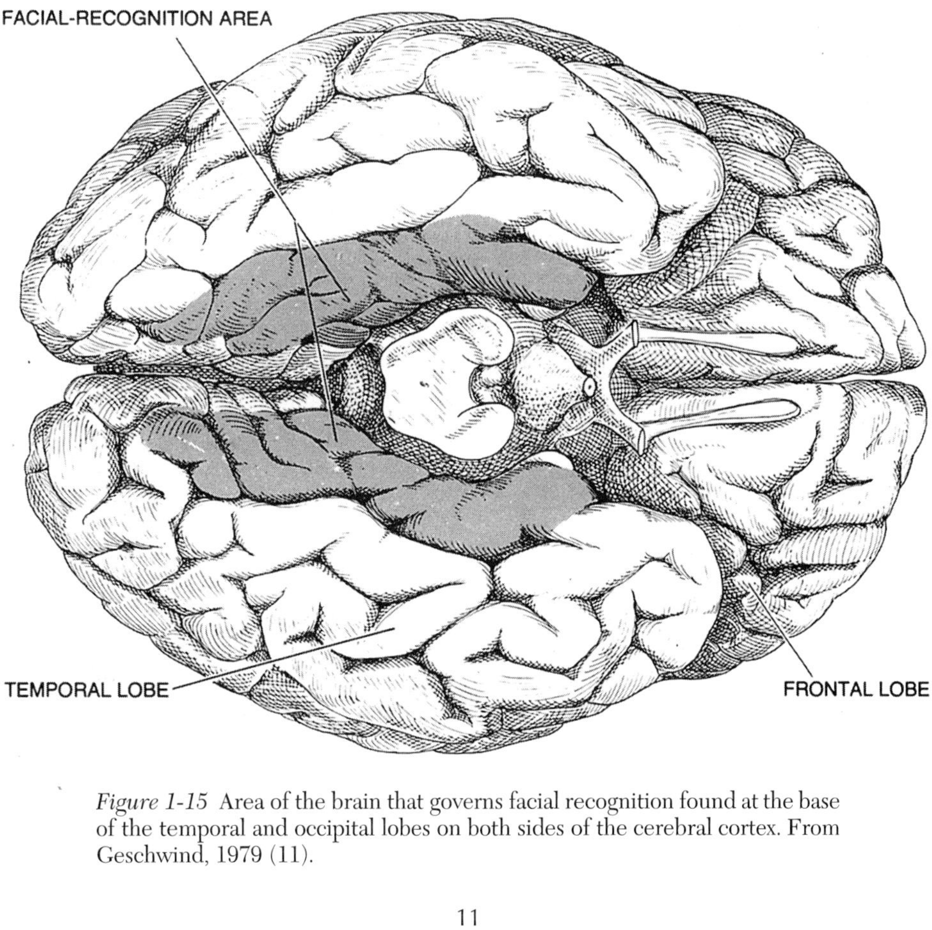

Figure 1-15 Area of the brain that governs facial recognition found at the base of the temporal and occipital lobes on both sides of the cerebral cortex. From Geschwind, 1979 (11).

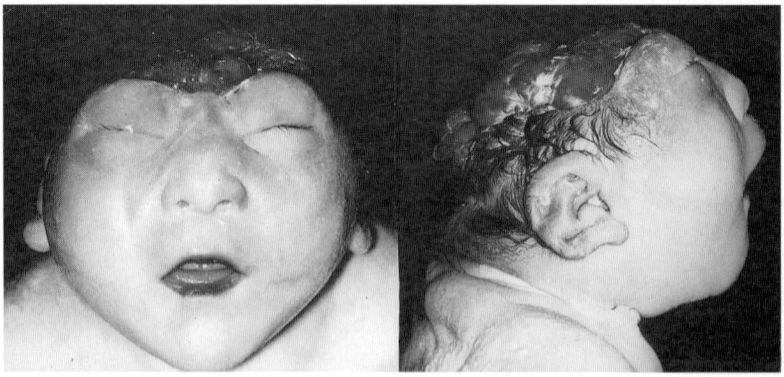

Figure 1-16 Anencephaly, resulting from persistence of the anterior neuropore. From Cohen, 2002 (7).

Dysmorphic and Genetic Perspectives

If the anterior neuropore persists, anencephaly results, characterized by an open neural tube in the cephalic region with an exposed mass of degenerating neural tissue on the skull floor. The cranial vault is absent, producing bulging of the eyes and absence of the neck (Fig. 1-16). Both the membranous neurocranium and the chondrocranium are grossly malformed (7).

Surgical Perspective

A girl born with a large venous malformation is shown at 6 months and after 1 year following debulking with preservation of the facial nerve carried out through a preauricular and a coronal incision in the scalp. Posnick (17) used microvascular and craniofacial techniques to accomplish the results (Fig. 1-17).

Psychological Perspective

Figure 1-18 illustrates a hospitalized infant with Baller–Gerold syndrome and a rubber Tweety Bird toy purchased by the infant's mother at the hospital gift shop and placed in her baby's crib. Craniosynostosis is one feature of the Baller–Gerold syndrome (6) that may result in an elongated head similar to Tweety's head shape. It suggests that the mother's choice of a purchase might have been a subconscious desire: "If Tweety is cute, so is my son."

Sociocultural and Symbolic Perspectives

The ancient Greek anatomist and physician Galen thought that personality was a reflection of four humors or fluids that make up the human body. If one of these fluids was dominant, the associated personality emerged. Yellow bile indicated bad temper and irritability; such a person was said to be choleric. Black bile indicated gloom and pessimism; such a person was said to be melancholic. Phlegm indicated sluggishness and lack of excitability; such a person was said to be phlegmatic. Blood indicated cheerfulness and passion; such a person was said to be sanguine (12) (Fig. 1-19).

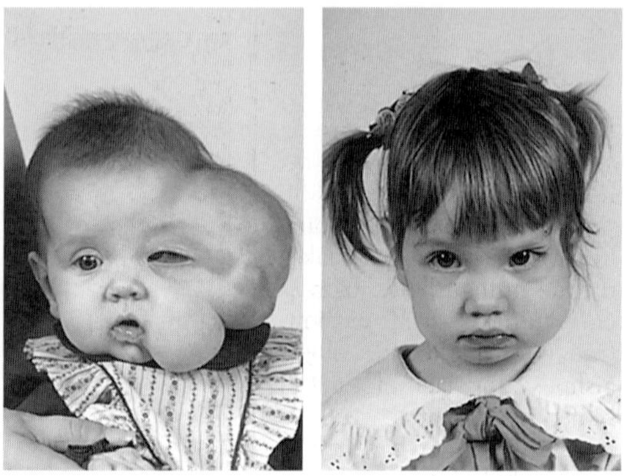

Figure 1-17 A girl before and after surgical removal of a massive venous malformation. From Posnick, 1993 (17).

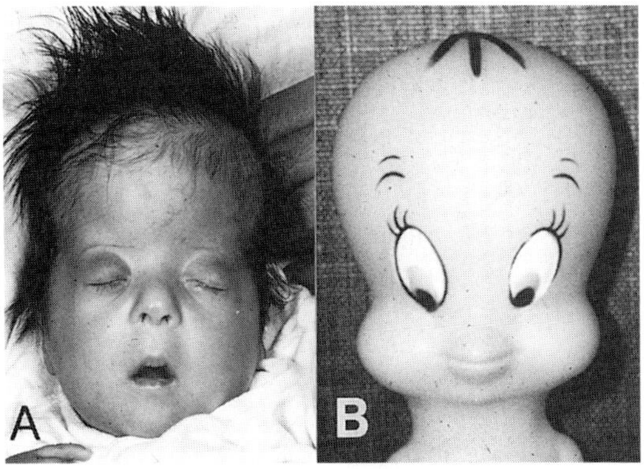

Figure 1-18 A, Infant with Baller-Gerold syndrome. Craniosynostosis and an elongated head. B, Rubber Tweety Bird toy that the infant's mother purchased at the hospital gift shop and placed in the baby's crib. Note the similarity in the shape of the infant's head and Tweety's head.

Artistic Perspective

Figure 1-20 compares two Renaissance paintings: a Florentine *Portrait of a Youth* by Botticelli and a Venetian *Portrait of Jacopo Soranzo* by Tintoretto. The difference in painting style is striking. In Florentine painting, the outlines of the face are sharply drawn with a three-dimensional sculptural quality together with a sense of palpability. In contrast, in Venetian painting, the lines of the face are soft and less distinct, but a sense of atmosphere surrounds the figure (1).

Figure 1-19 Ancient Greek anatomist and physician Galen thought that personality was a reflection of the four humors or fluids that make up the human body: blood, cheerfulness and passion (sanguine); phlegm, sluggishness and lack of excitability (phlegmatic); yellow bile, bad temper and irritability (choleric); and black bile, gloom and pessimism (melancholic). From Comb, 1834 (8).

Figure 1-20 Left, Close-up of a Florentine portrait of a youth with a Medal of Cosimo de Medici, by Botticelli. From Uffizi Gallery, Florence. Right, Close-up of a Venetian portrait of Jacopo Soranzo, by Tintoretto. From Gallerie del' Accademia, Venice.

REFERENCES

1. Berenson B: Italian Painters of the Renaissance. New York: Meridian, 1976.
2. Buendia JR: A Basic Guide to the Prado. Madrid: Silex, 1989.
3. Campbell B: Human Evolution. Chicago: Aldine, 1970.
4. Cohen MM Jr: Picasso and the art of distortion and dislocation: The artist as researcher and experimentalist. J Craniofac Genet Dev Biol 11:61–95, 1991.
5. Cohen MM Jr: Toomey Memorial Lecture, University of Maryland, Baltimore, April 15, 1999.
6. Cohen MM Jr: Other syndromes with craniosynostosis. In Craniosynostosis: Diagnosis, Evaluation, and Management. Second Edition. M.M. Cohen, Jr. and R.E. MacLean, Editors. New York: Oxford University Press. Chapter 31, pp. 385–440, 2000.
7. Cohen MM Jr: Malformations of the craniofacial region: Evolutionary, embryonic, genetic, and clinical perspectives. Am J Med Genet (Semin Med Genet) 115:245–268, 2002.
8. Comb G: A System of Phrenology. Boston, 1834.
9. Dorus S, Vallender EJ, Evans PD, Anderson JR, Gilbert SL, Mahowald M, Wyckoff GJ, Malcom CM, Lahn BT: Accelerated evolution of nervous system genes in the origin of *Homo sapiens*. Cell 119:1027–1040, 2004.
10. Doubilet D: FishFace. New York: Phaidon Press, 2003.
11. Geschwind N: Specailization of the human brain. In The Brain: A Scientific American Book. San Francisco: W.H. Freeman and Co., 1979, pp. 108–117.
12. Hill T: The Contemplation of Mankind, 1588.
13. Jirásek JE: Atlas of Human Prenatal Morphogenesis. Boston: Martinus Nijhoff, 1983.
14. Langman J: Medical Embryology. Baltimore: Williams & Wilkins, 1963.
15. Moore R: Evolution. New York: Time-Life Books, 1971.
16. Oral Health in America: A Report of the Surgeon General. Department of Health and Human Services, National Institute of Dental and Craniofacial Research, Rockville, Maryland, 2000.
17. Posnick JC, Al-Qattan MM, Zuker RM: Large vascular malformation of the face undergoing resection with facial nerve preservation in infancy. Ann Plast Surg 30:67–70, 1993.
18. Preuschoft S: Primate faces and facial expressions. Soc Res 67:245–271, 2000.
19. Romer AS: Man and the Vertebrates. Baltimore: Penguin Books Ltd., 1963.
20. Schultz AH: The Life of the Primates. New York: Universe Books, 1969.
21. Tibbetts EA, Dale J: A socially enforced signal of quality in a paper wasp. Nature 432:218–222, 2004.

2

Evolutionary Perspective

HOW OLD IS OUR HEAD?

Animals can be characterized as being composed of two or three cell layers: the former are diploblastic; the latter, triploblastic. Examples of diploblastic animals include hydra, jellyfish, sea anemone, and coral. Most animals are triploblastic—for example, nematode worms, fruit flies, mice, and humans.

Triploblastic animals have a conserved set of genes responsible for head-organizing activity during embryogenesis. Structural equivalents to Spemann's head organizer in amphibian embryos include the embryonic shield in fish, Hensen's node in avian embryos, and the node in mouse and human embryos (48). Head-organizing genes include several classes of homeobox genes: paired class (*Oxt*, *Goosecoid*, and *Ant/Hesx1*), antennapedia class (*Emx* and *Hex/Prh*), Lim class (*Lim1*), and winged class (*Forkhead* and *Hnf3*). Later during development, these genes have many functions, including, among others, brain patterning and neurogenesis (1, 4, 7, 28).

In evolutionary terms, developmental processes that lead to head development have been considered to be relatively recent, but evidence is accumulating that the origin may trace back to diploblastic animals. If a head is minimally defined as feeding behavior through an apical mouth opening associated with tentacles coordinated by nerve cells, the hydra (Fig. 2-1) qualifies. When the hydra's nerve net is disrupted, it is unable to catch food. Homologs of head-specific genes have been cloned for coral, jellyfish, and hydra (Table 2-1). In some instances, these diploblastic genes are expressed in an apical-specific manner (20).

DISRUPTED HEAD DEVELOPMENT

Disruption of head-organizing genes can have a detrimental effect on craniofacial morphogenesis. Three examples follow.

Shisa normally antagonizes Wnt and Fgf signaling and is expressed in the prospective head ectoderm and in Spemann's organizer. Loss of Shisa function sensitizes the neuroectoderm to Wnt signaling and suppresses head formation (64).

Epidermal IκB kinase-α controls skeletal and craniofacial morphogenesis. IκB kinase-α-deficient mice lack epidermal differentiation and have abnormal skeletal development with disrupted craniofacial morphogenesis. Faulty epidermal differentiation interferes with proper epithelial–mesenchymal interaction (54) (see Chapter 3).

Fgf signaling is important in the patterning of the anteroposterior axis during gastrulation. Overexpression of embryonic Fgf in *Xenopus* results in suppression of head development (49). The creation of headless tadpoles was sensationalized in the media. On October 19, 1997, Britain's *Sunday Times* carried a dramatic headline: "Headless Frog Opens Way for Human Organ Factory." Needless to say, the headline set off a spirited debate about the ethics of creating brainless humans for medical purposes (14, 47).

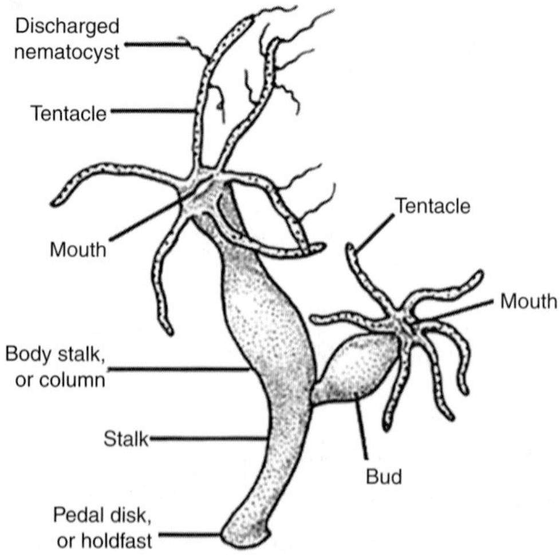

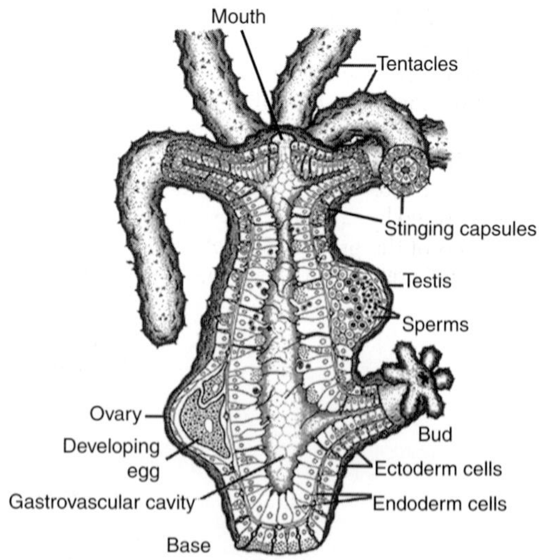

Figure 2-1 Hydra (*Chlorohydra viridissima*), a two-layered radially symmetrical carnivorous animal. Left, Tentacles have nematocysts, specialized stinging cells. Nerve net controls the tentacles which bring the prey to its mouth. Note new budding hydra. Right, Sections through hydra showing ectoderm and endoderm, and lack of mesoderm. Note hydra bud. Note also sexual reproduction with sperm and a developing egg. Modified from http://www.Ideo.columbia.edu/ . . . /life/slides/phyla/cnidaria.html and http://www.amicros.org/img/hydra.jpg

Table 2-1 Diploblastic Animal Homologues of Head-Organizer Genes in Triploblastic Animals

Genes in Diploblastic Animals[a]	Related Genes in Triploblastic Animals
Paired-class	
Paired-like	
Pdla	*Arx*
Pdlb	*Arx*
Otx	*Otx*
Pax-type	
PaxA	*Paxneuro*
PaxB	*Pax2*, *Pax5*, *Pax8*
PaxC	*Eyegone*, *Pax4*, *Pax6*
Antp-class	
Cnemx	*Empty spiracle*, *emx*
cnox1	Hox-related (paraHox group 1)
Winged-helix	
Budhead	*Forkhead*, *axial*, *Hnf3*, *HFH4*, *XFKH5*, *Hyfkh2*, *FD5*, *lin31*, *FKH4*, *Hyfkh3*
Tbox	
HyBra1	*Brachyury*

[a]Hydra, jellyfish, sea anemones, coral. These diploblastic animals are cnidarians.

Modified from Galliot and Miller, 2000 (20).

MOVING TOWARD EXTINCTION

Extinction has many causes. The Tasmanian devil (*Sarcophilus harrisii*), a carnivorous marsupial, has exhibited facial cancer with 100% mortality for the past decade. Tumors grow over the face and muzzle, so that eventually devils cannot feed. Death occurs within six months of acquiring the facial tumor. Thus, instead of breeding four-to-five times in a lifetime, Tasmanian devils may breed only once before they succumb to facial cancer (6b).

Ten years ago, the population size was estimated to be about 150,000 and now it may be about one half that number. Although there is no threat of immediate extinction, "functional extinction" may be a possibility because illegally-introduced foxes have invaded the ecological niche occupied by the Tasmanian devils (6b).

Careful search has not revealed a viral cause for the deadly facial tumors, but a specific karyo-

Figure 2-2 Highly conserved *Hox* genes for determining the anterior-posterior axis. In contrast to a single *Hox* cluster in fruit flies and most other invertebrates, humans and other vertebrates have four clusters of *Hox* genes that most likely evolved by two successive duplications of the primary *Hox* cluster. From Veraksa et al., 2000 (59).

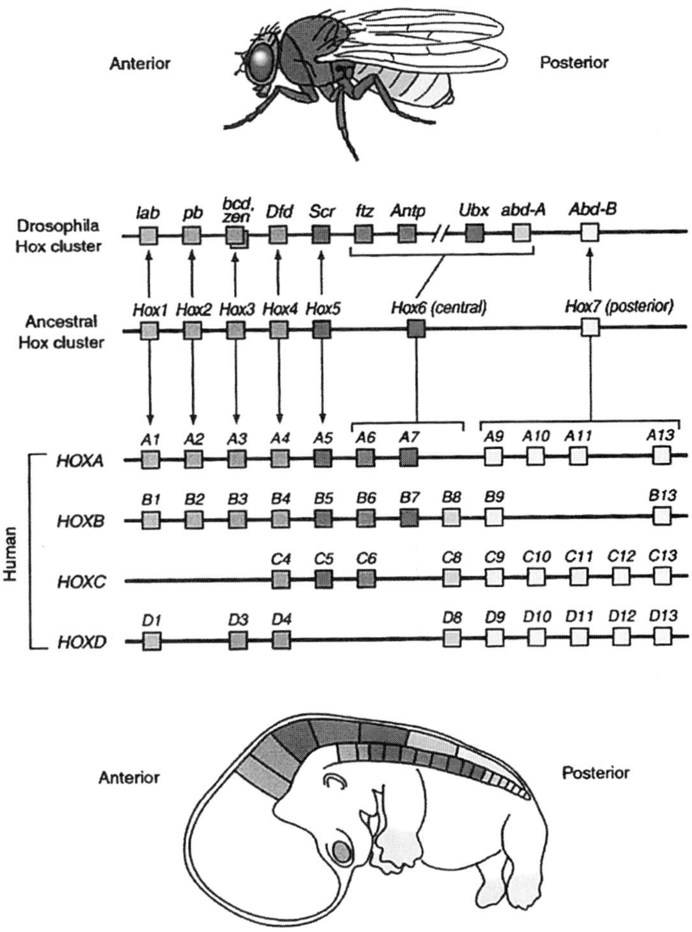

type characterizes all the tumors. Because Tasmanian devils engage in fighting and facial biting during mating,[a] inoculation of tumor cells by implantation is a likely mode of transmission from one animal to another.

HEAD END VS. REAR END

In the hydra *Chlorohydra viridissima* (Fig. 2-1), the "head end" and the "rear end" are the same; the mouth that admits food also expels waste. However, when the hydra's head regenerates, *cnox* genes (short for cnidarian *homeobox*) are involved in this nonsegmented animal that lacks mesoderm (Table 2-1). Thus, *homeobox* genes in bilateral animals have orthologs in the hydra.

Bilateral animals such as nematode worms, fruit flies, mice, and humans share a common genetic mechanism for patterning their head and rear ends (anterior–posterior axis), involving multiple *Hox* (*Homeobox*) genes (Fig. 2-2). The proteins made from these genes all contain a similar homeodomain motif made up of 60 amino acids that are highly conserved in evolution. For example, the homeodomain in fruit flies and humans differs by only one amino acid. In contrast to the single Hox cluster in fruit flies and most other invertebrates, humans and other vertebrates have four clusters of Hox genes (*HOXA*, *HOXB*, *HOXC*, and *HOXD*) that most likely evolved by two successive duplications of the primary Hox cluster (10, 13, 17, 42, 45, 59).

[a]Facial injuries from biting occur when males fight for access to females and especially when females fight off males at the end of the mating period.

HIGHLY CONSERVED GENES IN EVOLUTION

Comparative genome analysis shows that genes are highly conserved in evolution. On the average, vertebrate genomes have about twice the number of genes as invertebrate genomes. About one-third of human genes are related to those found in plants. About two-thirds are shared with fruit flies. Mice are at least 90% genetically similar to us, and gorillas and chimpanzees are 99% similar. Worldwide, the genomes of all human populations are 99.9% identical (24, 29, 32, 36).

Despite genetic similarity between various organisms, chromosome size and number can vary widely (Table 2-2). Humans have 46 chromosomes; sheep, 54; Indian muntjac deer, 6; and *Amoeba dubia*, several hundred. However, comparisons of sequences from different species suggest that bursts of activity of *jumping genes*, genetic duplications, and chromosome fusions play a key role in evolution (32).

Protein coding sequences account for only a small fraction of metazoan genomes. In humans, less than 2% of the genome codes for proteins (29, 36). Of 1,308 *groups* of proteins, humans have 3,129 *specific* proteins; fruit flies, 1,445; worms, 1,503; and yeast, 1,441. The 1,308 conserved core of proteins are, for the most part, responsible for basic "housekeeping" functions of the cell, such as metabolism, DNA replication and repair, and translation (29).

Given that the increase in the number of genes during evolution arises by duplication rather than by the invention of new genes, how can evolutionary diversity and increased complexity arise? Two mechanisms have been proposed to account for increasing complexity at the genetic level: (a) greater elaboration of *cis*-regulatory DNA sequences to control the expression of nearby genes and (b) increased intricacy of multiprotein transcription complexes that regulate gene expression (36).

Table 2-2 Genome Size and Chromosome Number

	Genome Size (base pairs)	Chromosome Number (n)[a]
Amoeba dubia	670,000,000,000	Several hundred
Trumpet lily (*Lilium longiflorum*)	90,000,000,000	12
Mouse (*Mus musculus*)	3,454,200,000	20
Human (*Homo sapiens*)	3,200,000,000	23
Carp (*Cyprinus carpio*)	1,700,000,000	49
Chicken (*Gallus gallus*)	1,200,000,000	39
Housefly (*Musca domestica*)	900,000,000	6
Tomato (*Lycopersicon esculentum*)	655,000,000	12

[a]Haploid number.

From Knight, 2002 (32).

PUT A HEAD ON IT: NEURAL CREST AND THE ORIGIN OF VERTEBRATES

The emphasis on similarities between deuterostomes, particularly the protochordates, and vertebrates has sometimes obscured some fundamental differences between them (Table 2-3). The transition from protochordates to vertebrates was associated with a shift from passive filter feeding to active predation, which required a new head. Vertebrates are mobile predators, jawless or jawed, with modified skeletal elements and pharyngeal muscles. Active preda-

Table 2-3 Differences between Vertebrates and Protochordates

Features	Differences
Anterior brain	Occurs only in vertebrates; non-neural in cephalochordates
Sense organs	Complex paired peripheral sense organs in vertebrates; simple peripheral and central receptors in protochordates
Pharyngeal skeleton	Cartilaginous and bony in vertebrates; collagenous and chitinous in protochordates
Water transport	Pumped through pharynx by muscle in vertebrates; driven by ciliary movement in protochordates
Gas exchange	Capillary gill expansion in vertebrates; lack of pharyngeal specialization in protochordates
Aortic arches	Muscle-walled in vertebrates; covered with collagen in protochordates
Food transport	Muscular peristalsis in vertebrates; ciliary motion in protochordates

Modified from Gans and Northcutt, 1983 (21).

Table 2-4 Some Characteristics of Vertebrates

Features	Embryonic Origins	Associated Functions
Nervous system		
Cranial nerves with sensory ganglia	Neural crest	Food detection, gas exchange
Trunk nerves with sensory ganglia	Neural crest	Food detection
Peripheral motor ganglia	Neural crest	Gas exchange, food processing
Paired special sense organs		
Nose	Epidermal placodes	Food detection
Eyes	Epidermal placodes	Food detection
Ears	Epidermal placodes	Food detection
Gustatory organs	Epidermal placodes	Food detection
Pharyngeal and alimentary modifications		
Cartilaginous bars	Neural crest	Food detection, food processing, gas exchange
Branchiomeric muscle	Muscular hypomere	Food detection, food processing, gas exchange
Smooth muscle of gut	Muscular hypomere	Food processing
Circulatory system		
Gill capillaries	Hypomere	Gas exchange
Muscularized aortic arches	Neural crest	Gas exchange
Skeletal system		
Anterior neurocranium and sensory capsules	Neural crest	Food detection, food processing
Cephalic exoskeleton and derivatives	Neural crest	Food detection

Only some of the features are listed here. Others include lateral line mechanoreceptors, lateral line electroreceptors, calcitonin cells, chromaffin cells, and a muscular heart.

Modified from Gans and Northcutt, 1983 (21).

tion, supported by a special exchange of respiratory gases, is directed by a complex array of sense organs together with their integrated circuitry (21).

In a classic article on the origin of vertebrates, Gans and Northcutt (21) discussed shared characteristics (Table 2-4) that developed from three embryonic tissues: hypomere, neural crest, and epidermal placodes. The hypomere is capable of producing well-developed muscular layers (21). The sensory placodes develop as ectodermal thickenings either adjacent to the neural crest or from the neural folds. The precise relationship between the placodes and the neural crest is unresolved (26).

Gans and Northcutt (21) postulated several structural and functional transitions during evolution of the vertebrates (Fig. 2-3). Before the protochordates evolved, filter feeding involved a collagenous ciliated pharynx, an integumentary exchanger, and an epidermal nerve net. Then, segmented muscle and a notochord followed. The transition from the protochordates to the protovertebrates involved the beginning of a new head characterized by pharyngeal breathing with branchiomeric muscle, gill capillaries, and pharyngeal cartilage. Predation involved paired external sense organs and an anterior brain. Then external bone evolved—an exoskeleton—for electroreception. In the evolution from agnathans to gnathostomes, jaws, vertebrae, and a post-otic skull developed.

JAWS

In primitive vertebrates, the pharyngeal arches resemble one another without specialization. The mandibular arch was then subdivided into upper and lower jaw cartilages (Fig. 2-4). The lamprey is a jawless vertebrate (agnathan) with well-differentiated upper and lower lips, although morphologic homologies with the jaws are uncertain. Embryonic patterning is very similar to that of jawed vertebrates (gnathostomes), such as the chick. Similarities include the deployment of neural crest cells and regulatory

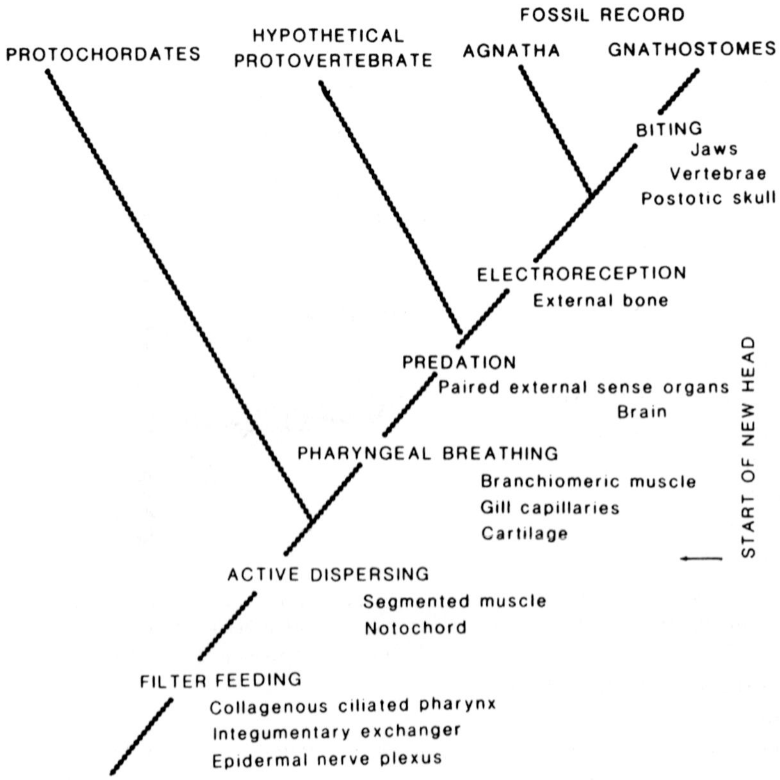

Figure 2-3 Hypothesized structural and functional transitions in vertebrate evolution. The postulated functional states (upper case letters) precede the modified structures (lowercase letters). From Gans and Northcutt, 1983 (21).

genes such as *Otx* (53). The lamprey is the only known vertebrate that demonstrates *HoxL6* expression in the mandibular arch. Loss of Hox expression in the mandibular arch of gnathostomes may have facilitated evolution of the jaws (15). *Dlx1* and *Msx1* overlap in expression in neural crest-derived tissue. Homology and gene expression became uncoupled during the transformation from agnathan to gnathostome. Thus, a heterotopic shift in mesenchymal interactions occurred during the evolution of the jaws (53) (Fig. 2-4).

The *Dlx* genes became essential for the proximal–distal identity of neural crest cells in the pharyngeal arches. The first pharyngeal arch has two principle subdivisions: a proximal maxillary division and a distal mandibular division. *Dlx5* and *Dlx6* specify these divisions. If both are inactivated ($Dlx5^{-/-}$ + $Dlx6^{-/-}$), the lower jaw is transformed into an upper jaw (16).

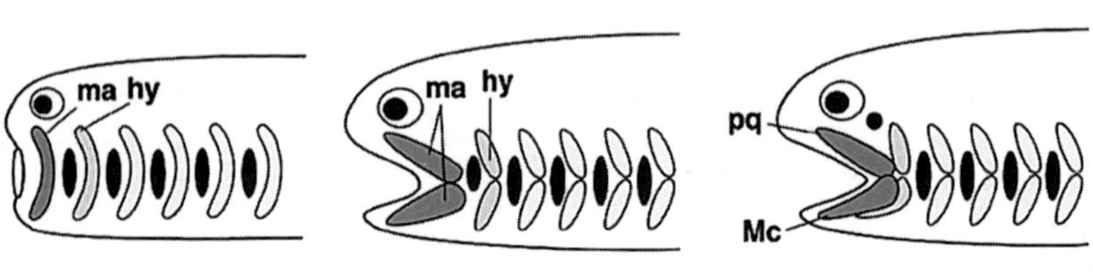

Figure 2-4 Hypothetical evolutionary transition of the jaw. Left, Pharyngeal arches resemble each other in the ancestral condition. Mandibular arch (ma); hyoid arch (hy). Center, Alterations in mandibular and hyoid arches. Right, Mandibular arch is enlarged in protognathostomes. Finally, the palatoquadrate (pq) cartilage and Meckel's cartilage (Mc) differentiate into the upper and lower jaw respectively. From Shigetani et al., 2002 (53).

THE SKELETON IN THE CLOSET

Five skeletal tissues are known: cartilage, bone, dentine, cementum, and enamel. All except enamel, which is derived from epithelium, originate from mesenchyme. Progressive complexity is usually considered to characterize general evolution. However, skeletal evolution is characterized by a *reduction* in types of tissue, in the number of skeletal elements, and in the location of skeletal elements (25).

The earliest fossilized vertebrate skeletal tissues known were dentine with a continuous layer of supportive basal bone underneath—the dermal exoskeleton of neural crest origin—which covered the animal. This external shield consisted of individual units variously termed tubercles, tesserae (Fig. 2-5), denticles, or odontodes (25, 56). The dermal exoskeleton persists in shark and crocodile scales, but has been lost in birds and mammals (25, 41, 55).

Figure 2-5 Dorsal armor of *Cardipeltis* (Heterostraci), an early Devonian jawless fish. The armor is composed of individual tesserae. From Halstead, 1973 (27).

It cannot be proven that external dentine and bone were the first vertebrate skeletal tissues, but they were the first to be preserved as fossils. Because endochondral bone formation begins with a cartilaginous model, whether cartilage or bone came first has been debated. If unmineralized cartilage appeared first in evolution, it would not have been fossilized except under rare and unusual circumstances (25, 56).

It has sometimes been assumed that cartilage preceded bone because sharks have a cartilaginous skeleton devoid of bone. However, ontogenetic sequences need not parallel phylogenetic sequences; the ancestors of sharks had two independent skeletal systems—a dermal exoskeleton *and* an internal endoskeleton (Fig. 2-6). Thus, sharks lost their bony endoskeleton *secondarily* (25).

The endoskeleton may or may not have been present at the time the exoskeleton arose. The primary endoskeleton may have arisen first as unmineralized cartilage followed by endochondral bone formation. It cannot be determined if the endoskeleton arose as cartilaginous support

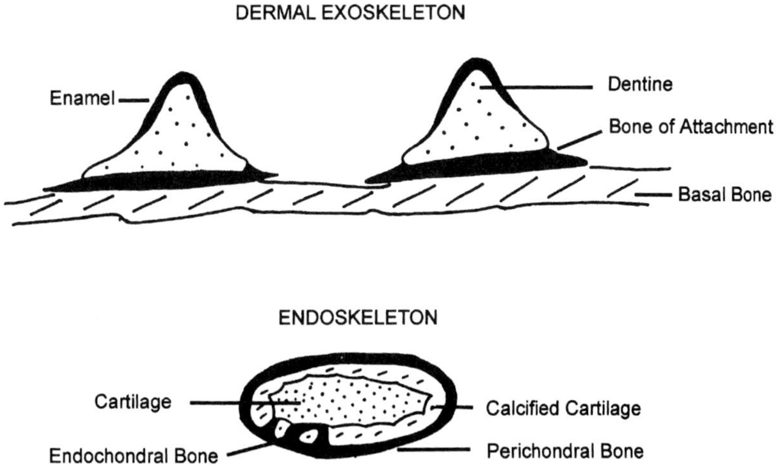

Figure 2-6 The dermal exoskeleton and endoskeleton of early vertebrates. The exoskeleton is based on the sequence: dentine (± enamel), bone of attachment, and basal bone. The endoskeleton is based on the sequence: cartilage, calcified cartilage, perichondral bone, and endochondral bone. Modified from Hall, 1992 (25).

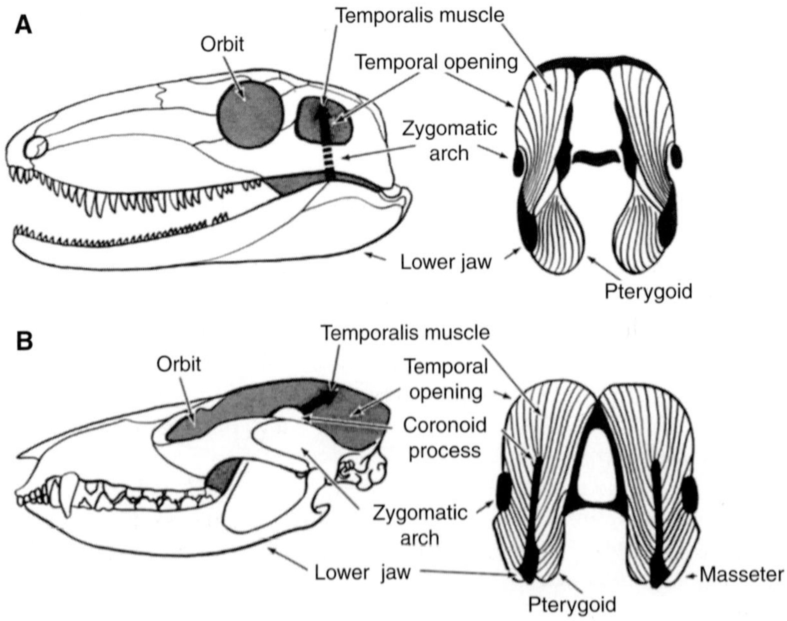

Figure 2-7 Transformation of the mammal-like reptilian skull to the mammalian skull. A, Lateral and cross-sectional views of the skull of a primitive mammal-like reptile. B, Lateral and cross-sectional views of the skull of a mammal. See text for details. From Radinsky, 1987 (51).

for the gills or as an axial skeleton for the notochord and neural tube. If it arose from the primitive pharyngeal arches as support for active gill respiration, then the vertebrate exoskeleton and the endoskeleton (Fig. 2-6) both arose from neural crest (25, 56).

In extant mammals, cranial neural crest originating from the forebrain, midbrain, and hindbrain neural tube differentiates into skeletal tissues that form the major parts of the skull (12). However, it has been shown experimentally that trunk neural crest does have both skeletogenic and odontogenic potential (40, 46).

TRANSFORMATION OF THE MAMMALIAN SKULL

About 190 million years ago, the first mammals evolved from one line of small cynodont therapsids.[b] The lower jaw in therapsids was composed largely of a single bone, the dentary, the remaining original bones having been reduced to small remnants at the back of the jaw (Fig. 2-7). In mammals, the dentary developed a condyle that became articulated with the squamosal bone. The remnants of the reptilian jaw joint became the mammalian ear ossicles. Modification in feeding and hearing may have led to the development of the definitive middle ear. This evolutionary transformation is reflected in mammalian embryological development (51). Evidence is accumulating to suggest that the separation of the middle ear bones from the mandible evolved independently in monotreme mammals and therians (marsupials, placentals) (43).

In non-mammalian vertebrate jaws, the craniomandibular joint is between the quadrate region of the palatoquadrate and the articular region of Meckel's cartilage. In mammalian vertebrate jaws, the joint is between the squamosal bone and the dentary. The middle ear is formed by the transfer of the post-dentary elements—articular, prearticular, and quadrate—as auditory ossicles. An ossified Meckel's cartilage has been reported in some Cretaceous mammals. It has been hypothesized that Meckel's cartilage extended from the dentary to the ear region. The dentary became tilted in position because of its close relationship to Meckel's cartilage. Reduction in the post-dentary unit gradually weakened its tie to the dentary until a critical point was reached at which the dentary assumed a

[b]Cynodont (dog-toothed) therapsids (mammal-like reptiles).

more vertical position during ontogeny; it was no longer attached to the post-dentary unit, which became anchored to the basicranium by connective tissue (60).

A horizontal flange of bone developed, forming a secondary palate that separated the food and air passageways. A secondary palate evolved independently in crocodiles and therapsids. In mammals, the secondary palate extended backward to form the soft palate (51, 52). The development of choanae permitted breathing with the mouth closed. The lungfish, which bridge the gap between fish and tetrapods, have "posterior nostrils" that superficially resemble choanae, and an intermediate fossil form strongly suggests that the choanae evolved from posterior nostrils (30).

The transformed skull had special functional significance for mammals. First, the jaw was strengthened because it was composed of one bone instead of several. Second, the three ear ossicles produced better amplification of sound. Third, horizontal flanges grew from each side of the upper jaw, forming the secondary palate, which separated the air passageway from the mouth. This secondary palate also strengthened the upper jaw by acting as a brace. Fourth, alterations in the musculature of the jaw involved an increase in both size and differentiation. The temporalis changed its direction from straight upward to upward and backward (Fig. 2-7). The lower portion of the temporalis extended downward to form the masseter, which attached to the angle of the mandible and was unique to mammals. The old reptilian pterygoid muscle became the medial pterygoid in mammals, forming a sling with the masseter. The mammalian jaw, dentition, and musculature permitted stronger and more extended chewing and a more efficient digestive system (51).

COMPARISON OF SOME MAMMALIAN SKULLS

Figure 2-8 compares the lateral view of three mammalian skulls: dog (A), chimpanzee (B), and human (C). The cranium shows a proportionate increase in size accompanied by a change in the position of the face.

In the dog, the face is situated in front of the cranium. A median bony crest on the surface of the cranium serves as an attachment for powerful muscles of mastication. A posterior bony ridge serves to attach the muscles of the neck. The bony orbits are not sharply defined and are widely open behind. The zygomatic arches are prominent and support strong masticatory muscles (35) (Fig. 2-8).

In the chimpanzee, the cranium is enlarged with a tendency for the eyes to rotate forward, facilitating stereoscopic vision. The orbits are surrounded by a complete ring of bone. The zygomatic arch has become progressively attenuated with a reduction in the relative size of the jaws (35) (Fig. 2-8).

In humans, the cranium has become even larger. The expanded frontal lobes have resulted in an upright forehead above the eyes. The face

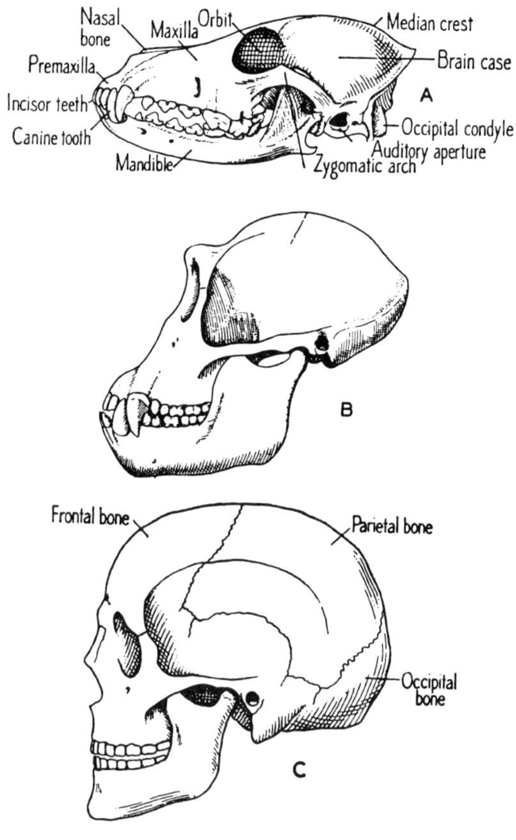

Figure 2-8 Lateral view of the skulls of a dog (A), chimpanzee (B), and human (C). For comparison, the skulls are of the same length. From Le Gros Clark, 1966 (35).

is reduced and has a downward and backward position relative to the expanded forebrain. The zygomatic arches have become even more attenuated than in other higher primates (35) (Fig. 2-8).

Figure 2-9 shows forward and lateral expansion of the anterior and middle cranial fossae, resulting in rotation of the orbits (F,F), facilitating stereoscopic vision in humans. In contrast, vision is divergent (G,G) in a mammal such as a deer (18). Expansion of the brain and a reduction in the facial prominence of the human result in a flat facial angle (Fig. 2-10) compared with a smaller brain and more protrusive face in the gorilla, resulting in a sloping facial angle. Figure 2-11 compares the morphology of chimp and human skulls. Unlike the chimp, the adult human skull more closely resembles its fetal counterpart. Neoteny, a slowdown in the growth rate with a delay in maturation, has occurred in hu-

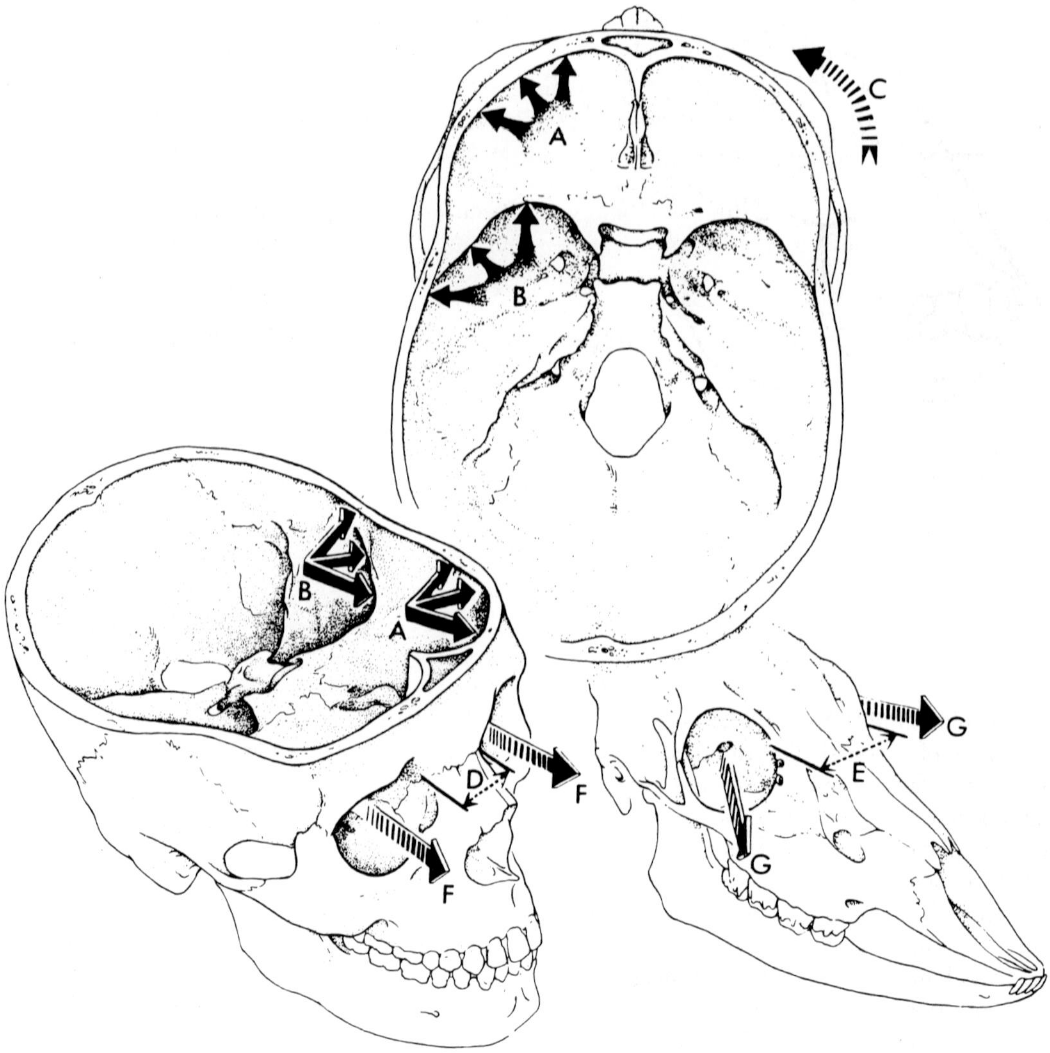

Figure 2-9 Marked forward and lateral expansion of the anterior and middle cranial fossae with altered positioning of the orbits, permitting stereoscopic vision (A, B, C, arrows). Note forward facing orbits (F, arrows) in humans. In a mammal such as a deer, the orbits diverge (G, arrows). Also note the difference in the breadth of the nasal bridge and interorbital region in the human (D) and the deer (E). From Enlow, 1968 (18).

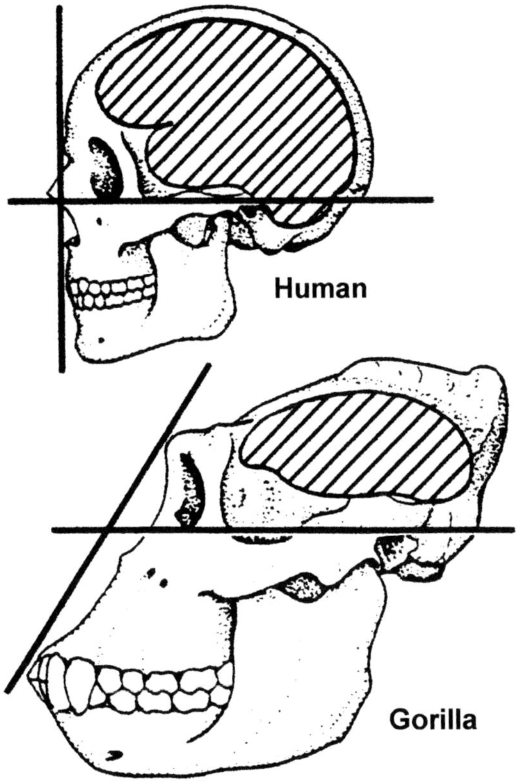

Figure 2-10 Expansion of the brain and a reduction in the facial prominence of humans result in a flat facial angle compared with a smaller brain and more protrusive face in the gorilla, resulting in a sloping facial angle.

man evolution (23). Figure 2-12 compares a lateral view of a dog skull with a Chevrolet Corvette; the face is anterior to the cranium. A lateral view of a human skull is compared with a camper; the expanded forebrain results in an upright forehead above the face (19).

Figure 2-13 compares the inferior surface of three mammalian skulls: dog (A), chimpanzee (B), and human (C). The foramen magnum is at the posterior end of the dog skull, but is positioned inferiorly in the chimpanzee, and in a more balanced position in humans (35). Figure 2-14 contrasts the orientation of the foramen magnum and the anterior cranial base in a human and a rodent. The foramen magnum is vertically placed in the human (A) and posteriorly placed in the rodent (C). The anterior cranial floor and cribriform plate face downward in the human (B) and forward in the rodent (D) (18). The muzzle of a typical mammal points obliquely downward with the orbital axis approximately parallel to the ground (Fig. 2-15). With enlargement of the brain and bending of the cranial base in the human, the foramen magnum faces downward—consistent with upright stance—and the orbital axis faces forward (19).

Two general trends in the evolution of the human skull consist of (a) enlargement of the brain over the fore- and rear-skull base and (b) reduction in the size of the face by alterations in the nasal chamber and dentition (6, 19, 61) (Table 2-5). Changes in the methods of food preparation, particularly cooking, softening, and grinding, have contributed to smaller face size in humans. Although comparative evidence has been used to support this notion, experimental evidence is more recent. In the retrognathic mammal, the rock hyrax (*Procavia capensis*), facial growth was examined in response to raw/dried foods vs. cooked foods. Hyraxes raised on cooked foods had significantly decreased facial growth (~10%) (38).

Primitive mammals had 44 teeth, which can be expressed by a dental formula for one half the jaw: $\frac{3.1.4.3}{3.1.4.3}$ three incisors, one canine, four premolars, and three molars in the upper (above the line) and lower (below the line) jaws. Primates, like most mammals, are both diphydont, having both deciduous and permanent dentitions, and heterodont, having teeth of different shapes that perform different functions. Humans have 32 teeth: $\frac{2.1.2.3}{2.1.2.3}$ (61). Dental formulas for various animals are summarized in Table 2-6.

EVOLUTION OF THE HOMINIMS

Fossil and molecular evidence indicates that the evolution of the hominims[c] diverged from that of the chimpanzee about 6-8 million years ago (5). Many evolutionary changes preceded the origin of the genus *Homo*. The history of our own

[c]Hominim refers to humans and our evolutionary ancestors back to the separation of humans and apes. Hominids include hominims *and* African apes.

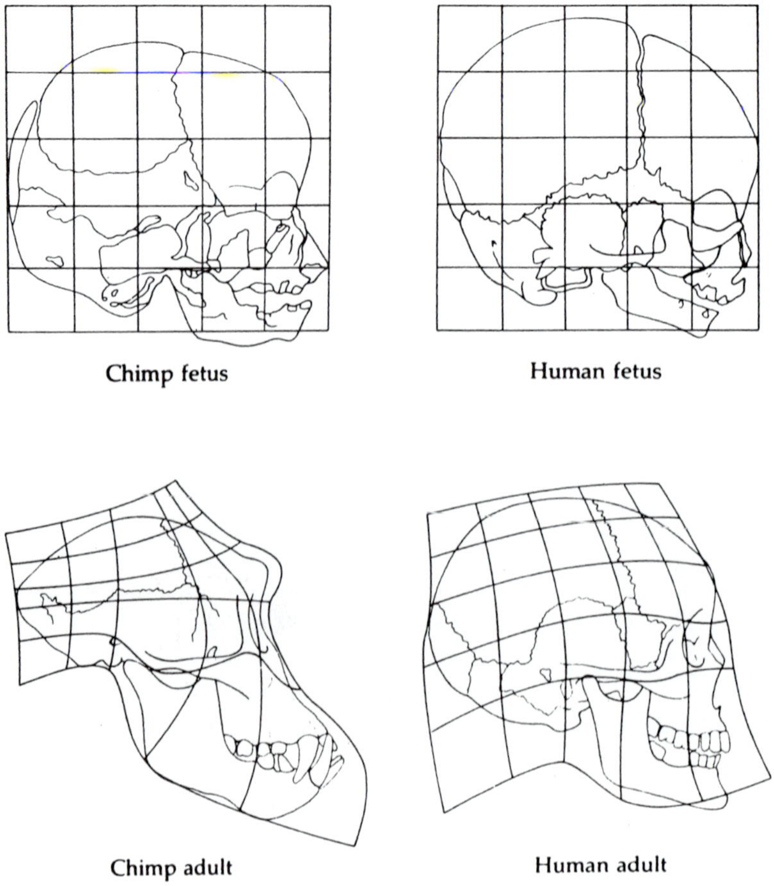

Figure 2-11 Morphology of chimpanzee and human skulls. Compared to the chimp, the adult human skull more closely resembles its fetal counterpart. Neoteny, a slowdown in the growth rate with a delay in maturation, has occurred in human evolution. Modified from Gould, 1977 (23).

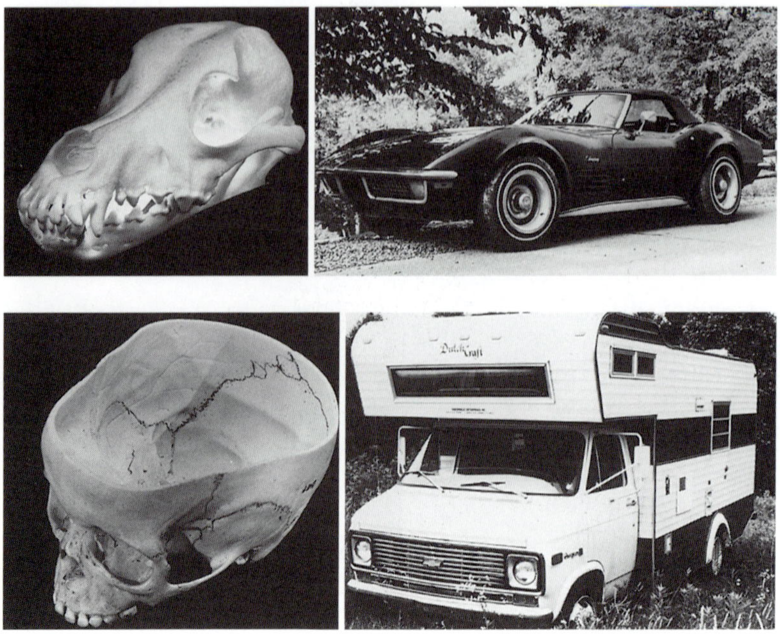

Figure 2-12 Top, Lateral view of a dog skull compared to a Chevrolet Corvette; the face is anterior to the cranium. Bottom, Lateral view of a human skull is compared to a camper; the expanded forebrain results in an upright forehead above the face. From Enlow, 1990 (19).

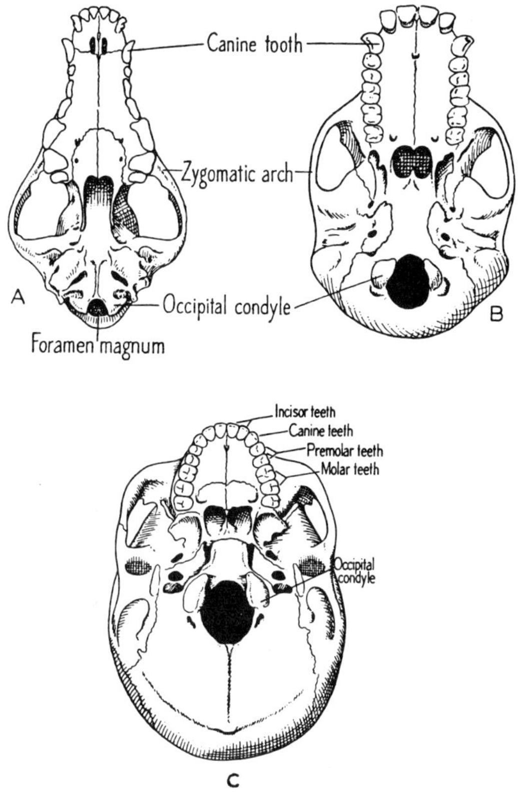

Figure 2-13 Inferior view of the skulls of a dog (A), chimpanzee (B), and human (C). From Le Gros Clark, 1966 (35).

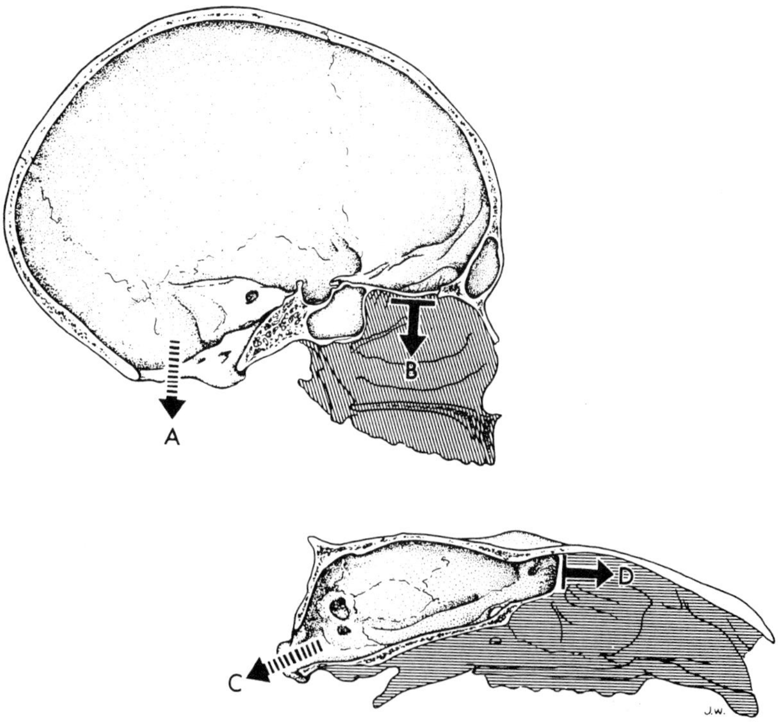

Figure 2-14 Orientation of the foramen magnum and the anterior cranial base in a human and a rodent. The foramen magnum is vertically placed in the human (A) and posteriorly placed in the rodent (C). The anterior cranial floor and cribriform plate face downward in the human (B) and forward in the rodent (D). From Enlow, 1968 (18).

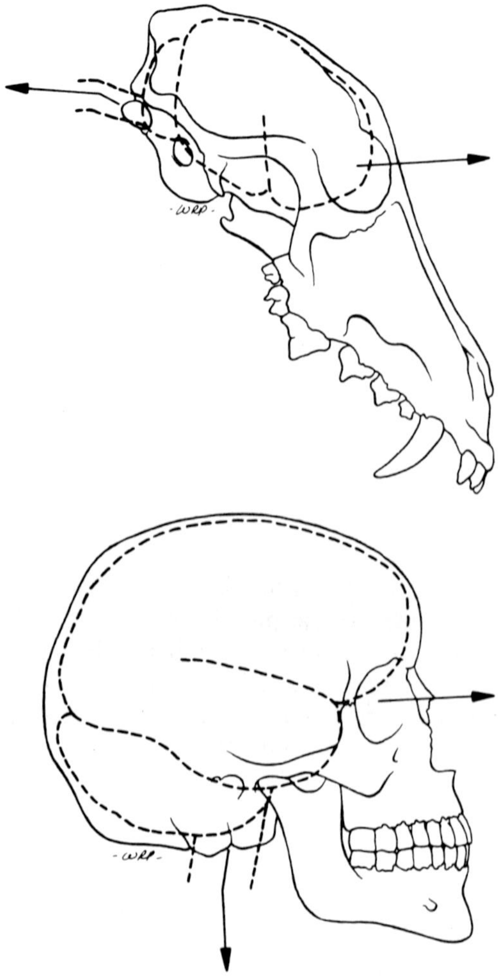

Figure 2-15 The muzzle of a typical mammal points obliquely downward with the orbital axis approximately parallel to the ground. With enlargement of the brain and bending of the cranial base in the human, the foramen magnum faces downward—consistent with upright stance—and the orbital axis faces forward. From Enlow, 1990 (19).

Table 2-5 Trends in Evolution of the Human Skull

Trends	Consequences
Enlargement of the brain over the fore- and rear-skull base	Enlargement of the skull Rotation of eyes forward, facilitating stereoscopic vision Expansion of frontal lobes of the brain, resulting in upright forehead above the eyes Bending of skull base associated with shift in foramen magnum position from back of skull to inferior part of skull
Reduction in size of nasal chamber and reduction of dentition	Reduction of the face Downward and backward position relative to the expanded frontal lobes with upright forehead

Data from Biegert, 1969 (6), Watts, 1975 (61), and Enlow, 1990 (19).

species, *Homo sapiens*, represents a mere 3% of the time span of hominim evolution. The cranium of the 6-7 million-year-old fossil *Sahelanthropus tchadensis* had a chimpanzee-sized brain but hominim-like facial and dental features. Figure 2-16 shows hominim evolution as a series of adaptations in which different branches of the hominim lineage were formed, with some dying out. Selected craniofacial traits that distinguish humans from the apes are summarized in Table 2-7. The evolution of brain size in hominids is shown in Table 2-8 (9, 11).

Issues about classification and chronological order of hominim fossils remain unsettled (Fig. 2-16), in part because of the pace of new fossil finds in recent times (11, 22, 57, 62, 63). In "archaic" *Homo sapiens* and "anatomically modern" *Homo sapiens*, a combination of shifts has been found in the cranial base angle, length and width of the cranial fossae, and facial length. These alterations probably reflect the relative increase in the sizes of the temporal and frontal lobes, and their effects on facial retraction and neurocranial globularity; they discriminate between archaic and modern *Homo sapiens* (39).

Shortening of the sphenoid bone has had an effect on human cranial shape by altering the spacial relationships between the face, cranial base, and neurocranium, which, in concert, determine the degree of facial projection. Anatomically modern *Homo sapiens* differs from other *Homo* taxa, including Neanderthals, by a globular brain case, vertical forehead, canine fossa, and pronounced chin (37).

Mitochondrial DNA from Neanderthal sites is radically different from that of present-day humans as well as that found in early human sites. The DNA evidence suggests that the evolutionary separation of the Neanderthals must have occurred at least 300,000 years ago; they disappeared about 30,000 to 40,000 years ago (33).

Neanderthal specializations include forward projection of the face, lateral bulging of the braincase, depressed elliptical area of roughened

Table 2-6 Dental Formulas and Numbers of Teeth in Various Mammals

Group	Dental Formulas (Max) Incisors.Canine.Premolars.Molars / (Mand) Incisors.Canine.Premolars.Molars	Total Number of Permanent Teeth (× 2)[a]
Primitive mammals	3.1.4.3 / 3.1.4.3	44
Tree shrews	2.1.3.3 / 3.1.3.3	38
New World monkeys	2.1.3.3 / 2.1.3.3	36
Old World monkeys	2.1.2.3 / 2.1.2.3	32
Gorillas and chimpanzees	2.1.2.3 / 2.1.2.3	32
Humans	2.1.2.3 / 2.1.2.3	32

[a]The dental formula is multiplied by 2 to arrive at the total number of teeth because the formula itself only refers to one side of the maxillary and mandibular teeth. Humans, for example, with a dental formula of $\frac{2.1.2.3}{2.1.2.3}$ have 16 maxillary teeth (4 incisors, 2 canines, 4 premolars, and 6 molars) and 16 mandibular teeth (4 incisors, 2 canines, 4 premolars, and 6 molars), making a total of 32 teeth.

Adapted from Watts, 1975 (61).

Table 2-7 Selected Craniofacial Traits That Distinguish Humans from Apes

Craniofacial proportions
Relative brain size
Brain topology
Skull balanced upright on vertebral column
Small canine size
Presence of a chin[a]

Other traits that distinguish humans from apes include body shape and thorax, relative limb length, long ontogeny and life span, reduced hair covering, elongated thumbs and shortened fingers, pelvic alterations, S-shaped spine, language, and advanced tool making.
[a]Except in *Homo floresiensis*.

Data from Carroll, 2003 (11).

Table 2-8 Evolution of Brain Size in Hominids

Hominid	Estimated Age (Millions of Years Ago)	Average Brain Size (Cubic Centimeters)
Sahelanthropus tchadensis	6–7	~320–380
Australopithecus africanus	2.6–3	457
Paranthropus boisei	1.2–2.2	510
Homo habilis	1.6–2.3	552
Homo rudolfensis	1.8–2.4	752
Homo ergaster	1.5–1.9	854
Homo erectus	0.2–1.9	1,016
Homo heidelbergensis	0.3–1	1,198
Homo neanderthalensis	0.03–0.25	1,512
Homo sapiens	0–0.2	1,355
Homo floresiensis	?[a]	380

This table does not include all recognized hominids.
[a]Known range to date. Lived as recently as 18,000 years ago and as early as 90,000 years ago.

Adapted from Carroll, 2003 (11) and Lahr and Foley, 2004 (34).

bone at the back of the skull, bumps and crannies of the mastoid process (Fig. 2-17), and a singular configuration of the bony labyrinth of the inner ear. Modern humans lack the Neanderthal labyrinth configuration (31). The average brain size in Neanderthals, however, exceeded, on the average, the brain size of *Homo sapiens* (11) (Table 2-8).

Homo floresiensis is characterized by a diminutive body (~1 m), long arms in relation to the legs, unusual pelvis, small brain size (380 cm^3), highly convoluted frontal lobes (CT imaging), and a chinless mandible. It is newly recognized hominim species living as recently as 18,000 years ago on an island in Indonesia. Also found were stone tools, dwarf elephants, and Komodo dragons. Some skeptics have suggested that the skull represents human microcephaly, but this is highly unlikely (2, 3, 34, 37a).

MOLECULAR CORRELATES OF HUMAN EVOLUTION

A *FOXP2* mutation that affects both language and the ability to articulate (34a) was thought to be the human speech gene, which had undergone natural selection, reaching its final form between 100,000 to 200,000 years ago (28b). However, *FOXP2* is highly conserved in evolution with only 4 of 700 amino acid differences between humans and mice. Furthermore, the gene is also expressed in several other tissues besides

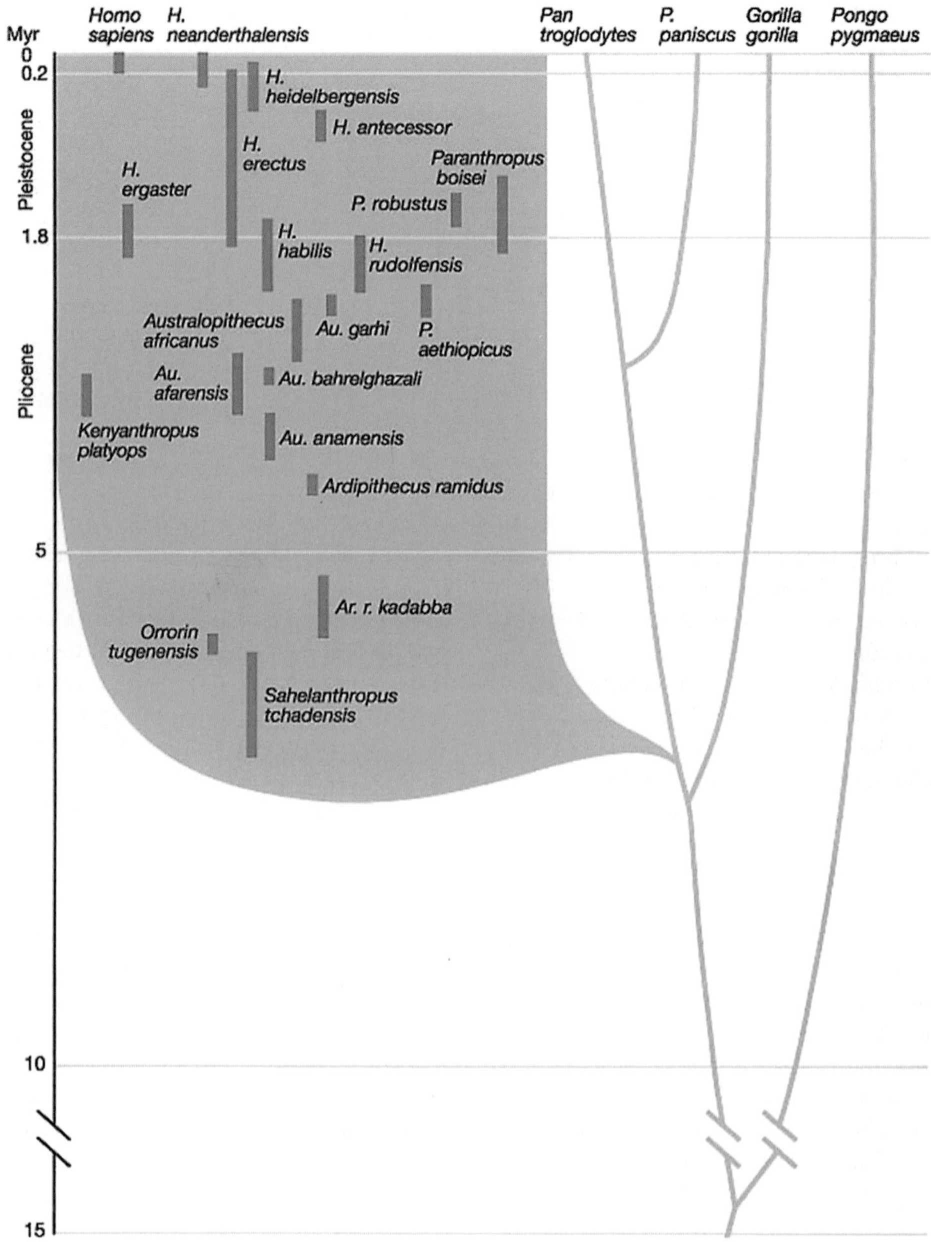

Figure 2-16 Timescale and phylogeny of the hominids. Note relationships of the apes: chimpanzee (*Pan troglodytes*), bonobo (*P. paniscus*), gorilla (*Gorilla gorilla*), and orangutan (*Pongo pygmaeus*). The phylogenetic relationships among hominids (shaded) are uncertain. Note *Homo sapiens* and *H. neanderthalensis*. Note also that the estimated age of *Sahelanthropus tchadensis* predates the chimpanzee-human divergence. *Homo sapiens* represents only the last 3% of the time span of hominid evolution. Myr, millions of years ago. From Carroll, 2003 (11).

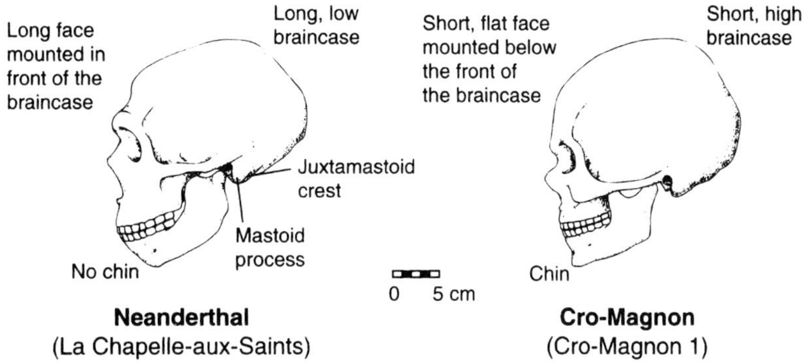

Figure 2-17 Lateral view of Neanderthal and Cro-Magnon skulls. See text. From Klein, 2003 (31).

the brain. The gene product belongs to the forkhead-domain family of transcription factors, suggesting that it plays a key role in upregulating and downregulating other genes. *FOXP2* is not responsible for language per se but rather influences other genes involved in crucial neural pathways (6a).

Microcephalin (MCPH1) mutations cause microcephaly with normal brain structure but with severe reduction in brain size. The normal *microcephalin* gene may control proliferation and/or differentiation of neuroblasts during neurogenesis and thereby brain size itself. *Microcephalin* has been subject to strong natural selection beginning about 37,000 years ago, with a confidence interval ranging from 14,000 to 60,000 years (19a).

Homozygous ASPM mutations also cause microcephaly. The gene may regulate neural stem-cell proliferation and/or differentiation, possibly by mediating spindle-cell assembly during cell division. This gene too is under strong positive selection. The allele arose about 5,800 years ago, with a possible range of 500 to 14,100 years (46a).

SIGLEC11, a gene involved in sialic acid biology, is expressed in high concentration in microglial cells of the human brain, but only occasionally in the cells of chimpanzees. This gene has also been subject to selection during evolution (28a).

EVOLUTION OF FACIAL MUSCLES

In reptiles, which are cold-blooded, the skin of the head is not movable, and musculature is restricted to opening and closing the mouth, eyes, and nose. "Facial expression" is limited to the threat of biting. Unlike reptiles, birds and mammals are warm-blooded, requiring a rearrangement of the layer that insulates the body. Emerging features include a soft, pliable leathery skin and growth of feathers and hairs together with the development of muscles that regulate their movement. One mammalian feature—suckling of the young—is associated with mobile lips and cheeks; another—extensive chewing—is associated with specialization of the dentition and a muscular tongue (58).

Primitive mammals with divergent eyesight developed tactile vibrissae of the face controlled by muscles attached to their underlying connective tissues. Binocular vision, found in primates and carnivores, reduces the necessity for tactile vibrissae, allowing the muscles to become associated with facial expression. Although all mammals have some pattern of facial musculature, elaborate facial musculature has evolved in primates, but also in carnivores and ungulates (58).

In the vertebrate ground plan, the facial nerve innervates the superficial constrictor muscle, which is attached to the branchial skeleton and serves as a respiratory mechanism. In amphibians, the muscle is split in two, with the lower portion giving rise to the mandibular depressor muscle. In reptiles, the superficial layer forms the sphincter colli muscle. In mammals, the sphincter colli splits into the superficial platysma and the deep sphincter colli (58).

In mammals, particularly primates, facial muscle systems expand, enabling blinking, squinting, frowning, eyebrow raising, ear movements, and alterations of the nose, lips, and corners of the mouth. In lemurs and lorises, the auriculolabial muscles draw back the corners of the mouth, moving the ears at the same time. The zygomatic muscles, evolving from the platysma, have lost their connection with the ear in New World monkeys. In Old World monkeys and apes, the zygomatic muscles have two branches: the major zygomatic, which lifts the corners of the mouth, and the minor zygomatic, which elevates the upper lip. In lemurs and lorises, the maxillonasolabial muscle dilates the nostril, probably aided by the orbicularis oris, with which it is closely associated. In New World monkeys and gibbons, the development of the triangularis permits depression of the corners of the mouth. In Old World monkeys and apes, the maxillonasolabial muscle becomes the levator labii superioris proprius. In humans, the risorius, a branch of the orbicularis oris, retracts the corners of the mouth (50).

NONHUMAN PRIMATE FACES

Many nonhuman primates have specific facial hair patterns, although sexual dimorphism is less pronounced than in humans. However, differences in facial skin color in males and females are much more pronounced in some instances, such as the brilliant red nose, shiny blue lateral nasal corrugations, and yellowish-white beard of the mandrill. Most nonhuman primates have brown or amber irises, but very few, such as the pigtailed macaque, show white sclerae. Primates such as lemurs, lorises, and tarsiers, are *strepsirhines*, characterized by a moist rhinarium on the tip of the nose. The other primates are *haplorhines*, characterized by dry noses. The *catarrhines*, characterized by narrow noses, include Old World monkeys and apes from Africa and Asia. The *platyrrhines*, characterized by wide noses, include New World monkeys of South America and Central America (50). The long nose of the male proboscis monkey is shown in Figure 1-1.

VARIATION OF SOME HUMAN FACIAL FEATURES

Facial features vary greatly in different populations. In many Asian countries, a flat face, epicanthal folds, and low nasal bridge are distinctive characteristics. In contrast, in many European countries, a projecting nose and receding zygomatic arches are distinctive characteristics (Figs. 2-18, 2-19). The brow ridges and mandible are

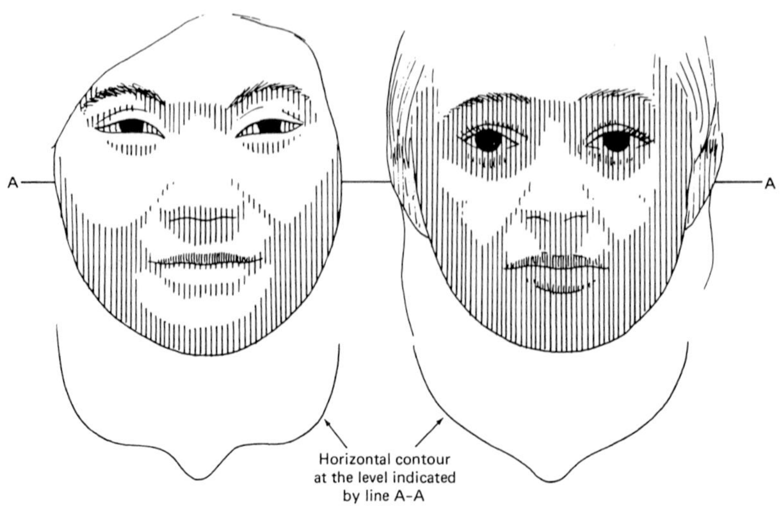

Figure 2-18 Comparison of Asian (left) and European (right) faces. From Brues, 1977 (8).

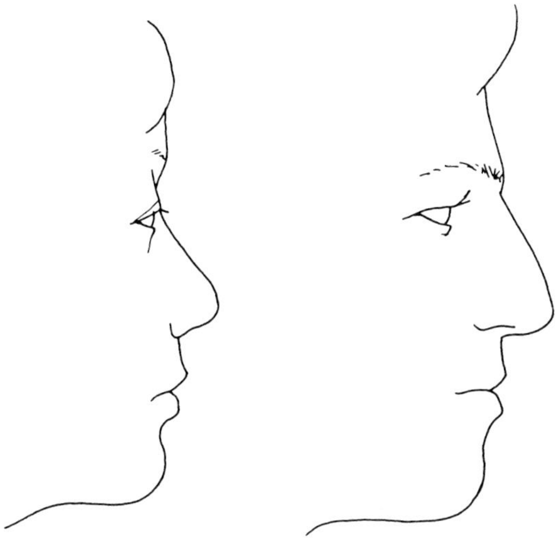

Figure 2-19 Comparison of Asian (left) and European (right) facial profiles. From Brues, 1977 (8).

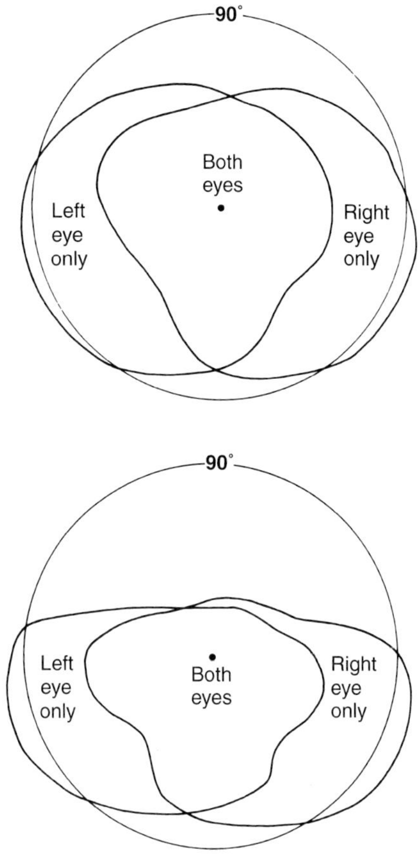

Figure 2-21 Mapping the field of vision has shown that heavy brow ridges have a limiting effect on upper vision. See text. From Brues, 1977 (8).

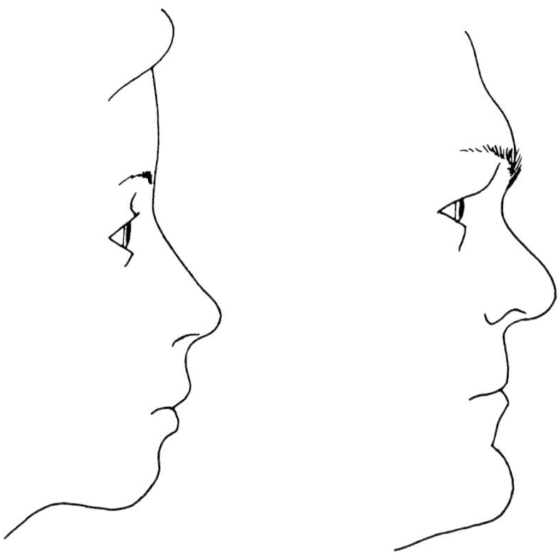

Figure 2-20 Brow ridges and mandible are more prominent in men (right) than in women (left). From Brues, 1977 (8).

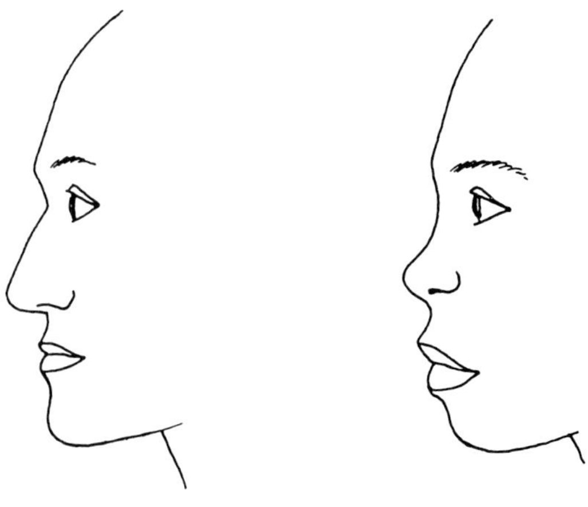

Figure 2-22 In contrast to many European faces (left), African faces (right) tend to have forward projection of the lower maxillary region, altering the shape of the nose and lips. From Brues, 1977 (8).

more prominent in men than in women (Fig. 2-20). Heavy brow ridges are characteristic of Melanesian and Australian aboriginal people. Mapping the field of vision has shown that heavy brow ridges have a limiting effect on upper vision (Fig. 2-21). Prominent zygomas have a limiting effect on side vision, whereas a prominent nose slightly limits the visual field of each eye. In contrast to many European faces, African faces tend to have forward projection of the lower maxillary region, altering the shape of the nose and lips (8, 44) (Fig. 2-22).

REFERENCES

1. Bally–Cuif L, Boncinelli E: Transcription factors and head formation in vertebrates. BioEssays 19: 127–135, 1997.
2. Balter M: Skeptics question whether Flores hominid is a new species. Science 306:1116, 2004.
3. Balter M: Small but smart? Flores hominid shows signs of advanced brain. Science 307:1386–1389, 2005.
4. Beddington RS, Robertson EJ: Anterior patterning in mouse. Trends Genet 14:277–284, 1998.
5. Begun DR: The earliest hominins–Is less more? Science 303:1478–1480, 2004.
6. Biegert J: The evaluation of characteristics of the skull, hands, and feet for primate taxonomy. In Classification and Human Evolution, S.L. Washburn, Editor. Chicago: Aldine. pp. 116–145, 1969.

6a. Bishop DVM: Putting language genes in perspective. Trends Genet 18:57–59, 2002.

6b. Bostanci A: A devil of a disease. Science 307:1035, 2005.

7. Bouwmeester T, Leyns L: Vertebrate head induction by anterior primitive endoderm, BioEssays 19:855–863, 1997.
8. Brues AM: People and Races. New York: MacMillan, 1977.
9. Brunet M, Guy F, Pilbeam D, Mackaye HT, Likius A, Ahounta D, Beauvilain A, Blondel C, Bocherens H, Boisserie J-R, De Bonis L, Coppens Y, Dejax J, Denys C, Duringer P, Eisenmann V, Fanone G, Fronty P, Geraads D, Lehmann T, Lihoreau F, Louchart A, Mahamat A, Merceron G, Mouchelin G, Otero O, Campomanes PP, Ponce De Leon M, Rage J-C, Sapanet M, Schuster M, Sudre J, Tassy P, Valentin X, Vignaud P, Viriot L, Zazzo A, Zollikofer C: A new hominid from the Upper Miocene of Chad, Central Africa. Nature 418:145–151, 2002.
10. Carroll SB: Homeotic genes and the evolution of arthropods and chordates. Nature 376:479–485, 1995.
11. Carroll SB: Genetics and the making of *Homo sapiens*. Nature 422:849–857, 2003.
12. Chambers D, McGonnell IM: Neural crest: Facing the facts of head development. Trends Genet 296:381–384, 2002.
13. Cohen MM Jr: Craniofacial disorders caused by mutations in homeobox genes *MSX1* and *MSX2*. J Craniofac Genet Dev Biol 20:19–25, 2000.
14. Cohen MM Jr: Is Kermit the frog in trouble? Am J Med Genet 104:98–100, 2001.
15. Cohn MJ: Lamprey *Hox* genes and the origin of jaws. Nature 416:386–387, 2002.
16. Depew MJ, Lufkin T, Rubenstein JLR: Specification of jaw subdivisions by *Dlx* genes. Science 298:381–385, 2002.
17. de Rosa R, Grenier JK, Andreeva T, Cook CE, Adoutte A, Akam M, Carroll SB, Balovoine G: Hox genes in brachiopods and priapulids and protostome evolution. Nature 399:772–776, 1999.
18. Enlow DH: The Human Face. New York: Harper & Row, 1968.
19. Enlow D: Facial Growth. Third Edition. Philadelphia: Saunders, 1990.

19a. Evans PD, Gilbert SL, Mekel-Bobrov N, Vallender EJ, Anderson JR, Vaez-Azizi LM, Tishkoff SA, Hudson RR, Lahn BT: *Microcephalin*, a gene regulating brain size, continues to evolve adaptively in humans. Science 309:1717–1720, 2005.

20. Galliot B, Miller D: Origin of anterior patterning: How old is our head? Trends Genet 16:1–4, 2000.
21. Gans C, Northcutt RG: Neural crest and the origin of vertebrates: A new head. Science 220:268–274, 1983.
22. Gibbons A: African skull points to one human ancestor. Science 295:2192–2193, 2002.
23. Gould SJ: Ontogeny and Phylogeny. Cambridge: Harvard University Press, 1977.
24. Hacia JG: Genome of the apes. Trends Genet 17: 637–645, 2001.
25. Hall BK: The origin and evolution of vertebrate skeletal tissues. In Chemistry and Biology of Mineralized Tissues, H. Slavkin and P. Price, Editors. Oxford: Elsevier Science Publishers. pp. 103–111, 1992.
26. Hall BK, Hörstadius S: The Neural Crest. New York: Oxford University Press, 1998.
27. Halstead LB: The heterostracan fishes. Biol Rev 48:279–332, 1973.
28. Hartmann B, Reichert H: The genetics of embryonic brain development in *Drosophila*. Mol Cell Neurosci 12:194–205, 1998.

28a. Hayakawa T, Angata T, Lewis AL, Mikkelsen TS, Varki NM, Varki A: A human-specific gene in microglia. Science 309:1693, 2005.

28b. Holden C: The origin of speech. Science 303: 1316–1319, 2004.

29. International Human Genome Sequencing Consortium: Initial sequencing and analysis of the human genome. Nature 409:860–921, 2001.
30. Janvier P: Wandering nostrils. Nature 432:23–24, 2004.

31. Klein RG: Whither the Neanderthals? Science 299:1525–1527, 2003.
32. Knight J: All genomes great and small. Nature 417:374–376, 2002.
33. Krings M, Capelli C, Tschentscher F, Geisert H, Meyer S, von Haeseler A, Grossschmidt K, Possnert G, Paunovic M, Pääbo S: A view of Neandertal genetic diversity. Nat Genet 26:144–146, 2004,
34. Lahr MM, Foley R: Human evolution writ small. Nature 431:1043–1044, 2004.
34a. Lai CS, Fisher SF, Hurst JA, Vargha-Khadem F, Monaco AP: A forkhead-domain gene is mutated in a severe speech and language disorder. Nature 413:519–523, 2001.
35. Le Gros Clark WE: History of the Primates. Fifth Edition. Chicago: University of Chicago Press, 1966.
36. Levine M, Tijan R: Transcription regulation and animal diversity. Nature 424:147–151, 2003.
37. Lieberman DE: Sphenoid shortening and the evolution of modern human cranial shape. Nature 393:158–162, 1998.
37a. Lieberman DE: Further fossil finds from Flores. Nature 437:957–958, 2005.
38. Lieberman DE, Krovitz GE, Yates FW, Devlin M, St. Claire M: Effects of food processing on masticatory strain and craniofacial growth in a retrognathic face. J Hum Evol 46:655–677, 2004.
39. Lieberman DE, McBratney BM, Krovitz G: The evolution and development of cranial form in *Homo sapiens*. Proc Natl Acad Sci USA 99:1134–1129, 2002.
40. Lumsden AG: Spatial organization of the epithelium and the role of neural crest cells in the initiation of the mammalian tooth germ. Development 103:155–169, 1988.
41. Maisey JG: Phylogeny of early vertebrate skeletal induction and ossification patterns. Evol Biol 22: 1–36, 1988.
42. Mark M, Rijli FM, Chambon P: Homeobox genes in embryogenesis and pathogenesis. Pediatr Res 42:421–429, 1997.
43. Martin T, Luo Z-X: Homoplasy in the mammalian ear. Science 307:861–862, 2005.
44. McCullough JH: The visual field in man and its evolution as related to facial morphology. Doctoral Dissertation, University of Colorado, 1970.
45. McGinnis W, Krumlauf R: Homeobox genes and axial patterning. Cell 68:283–302, 1992.
46. McGonnell IM, Graham A: Trunk neural crest has skeletogenic potential. Curr Biol 12:767–771, 2002.
46a. Mekel-Bobrov N, Gilbert SL, Evans PD, Vallender EJ, Anderson JR, Hudson RR, Tishkoff SA, Lahn BT: Ongoing adaptive evolution of *ASPM*, a brain size determinant in *Homo sapiens*. Science 309:1720, 2005.
47. Morton O: First Dolly, now headless tadpoles. Science 278:729, 1997.
48. Niehr C: Head in the WNT: The molecular nature of Spemann's head organizer. Trends Genet 15: 314–319, 1999.
49. Pownall ME, Tucker AS, Slack JMW, Isaacs HV: *eFGF*, *Xcad3* and *Hox* genes form a molecular pathway that establishes the anteroposterior axis in *Xenopus*. Development 122:3881–3892, 1996.
50. Preuschoft S: Primate faces and facial expressions. Soc Res 67:245–271, 2000.
51. Radinsky LB: The Evolution of Vertebrate Design. Chicago: University of Chicago Press, 1987.
52. Romer AS: Man and the Vertebrates. Baltimore: Penguin Books Ltd., 1963.
53. Shigetani Y, Sugahara F, Kawakami Y, Murakami Y, Hirano S, Kuratani S: Heterotopic shift of epithelial–mesenchymal interactions in vertebrate jaw evolution. Science 296:1316–1319, 2002.
54. Sil AK, Maeda S, Sano Y, Roop DR, Karin M: IκB kinase-α acts in the epidermis to control skeletal and craniofacial morphogenesis. Nature 428:660–664, 2004.
55. Smith M, Hickman A, Amanze D, Lumsden A, Thorogood P: Trunk neural crest origin of caudal fin mesenchyme in the zebrafish *Brachydanio rerio*. Proc R Soc Lond 256:137–145, 1994.
56. Smith MM, Hall BK: Development and evolutionary origins of vertebrate skeletogenic and odontogenic tissues. Biol Rev Camb Philos Soc 65:277–373, 1990.
57. Stringer C: Out of Ethiopia. Nature 423:692–695, 2003.
58. van Hooff JARAM: The facial displays of the catarrhine monkeys and apes. In Primate Ethology. D. Morris, Editor. New York: Doubleday & Company. Chapter 2, pp. 9–88, 1969.
59. Veraksa A, Del Campo M, McGinnis W: Developmental patterning genes and their conserved functions: From model organisms to humans. Molec Genet Metab 69:85–100, 2000.
60. Wang Y, Hu Y, Meng J, Li C: An ossified Meckel's cartilage in two Cretaceous mammals and origin of the mammalian middle ear. Science 294:357–361, 2001.
61. Watts ES: Biology of Living Primates. Dubuque: W.C. Brown Co. Publishers, 1975.
62. White T: Early hominids–Diversity or distortion? Science 299:1994–1997, 2003.
63. Wolpoff MH, Senut B, Pickford M, Hawks J: *Sahelanthropus* or *Sahelpithecus*? Nature 419:581–582, 2002.
64. Yamamoto A, Nagano T, Takehara S, Hibi M, Aizawa S: Shisa promotes head formation through the inhibition of receptor protein maturation for the caudalizing factors, Wnt and FGF. Cell 120: 223–235, 2005.

3

Embryonic, Developmental, and Postnatal Perspectives

NEURULATION

The neural tube is formed by fusion of the neural folds (Fig. 3-1). The process of neurulation is complex and involves proliferation of neuroblasts or matrix cells, development of the neuroepithelium and surface ectoderm, formation of the neural plate median hinge point, apical constriction of neuroepithelial cells, and expansion of the mesoderm and extracellular matrices. Four stages can be discerned: (a) induction of the neural plate by the notochord; (b) rostrocaudal lengthening and mediolateral narrowing; (c) lateral folding of the neural plate; and (d) apposition and fusion of the apical surfaces of the neural folds. There are three initiation sites of neural tube closure (81, 82) (Fig. 3-2).

At the beginning of the third week, the neural plate appears, followed by elevation of the neural folds (see Fig. 1-13). With further development, fusion of the neural folds forms the neural tube (see Fig. 1-14) with anterior and posterior neuropores (Figs. 3-3, 3-4). Closure of the anterior neuropore occurs at the 18–20 somite stage and closure of the posterior neuropore at the 25 somite stage. During the fourth week, the cephalic end of the neural tube develops three primary brain vesicles: the forebrain (prosencephalon), midbrain (mesencephalon), and hindbrain (rhombencephalon). As these vesicles appear, two flexures form: a cephalic flexure located in the midbrain region and a cervical flexure located at the junction of the hindbrain and spinal cord (Fig. 3-5). At the beginning of the sixth week, the prosencephalon has become subdivided into an anterior part (telencephalon), two lateral bulges (cerebral hemispheres), and a posterior part (diencephalon with outgrowth of the optic stalks). The rhombencephalon becomes subdivided into an anterior part (metencephalon) and a posterior part (myelencephalon) (73) (Fig. 3-6).

EMBRYOLOGY OF THE FACE

Neural crest cells play an integral part in facial morphogenesis. Just before the neural folds fuse to form the neural tube, neuroectodermal cells adjacent to the neural plate migrate into the facial region, where they form the skeletal and connective tissue of the face: bone cartilage, fibrous connective tissue, and all dental tissues except enamel (66, 97, 99, 105). Thus, facial mesenchyme is of neural crest origin. Vascular endothelium and skeletal muscle, however, are of mesodermal origin. Table 3-1 summarizes the origin of some craniofacial components.

By the end of the fourth week, the maxillary swellings appear lateral to, and mandibular swellings caudal to, the stomodeum. Above, the frontal prominence forms ventral to the prosencephalon. Nasal placodes arise as thickenings on either side of the frontal prominence. During the fifth week, a horseshoe-shaped ridge, consisting of medial and lateral nasal swellings, sur-

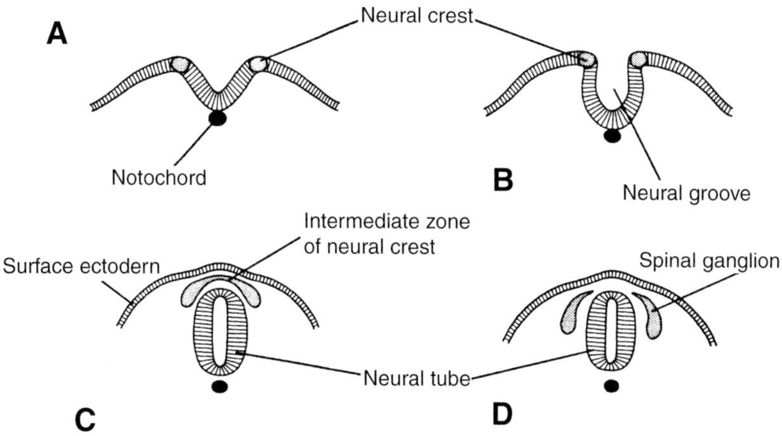

Figure 3-1 Transverse sections through successively older embryos, showing the formation of the neural folds, neural groove, and neural tube (A–D). The cells of the neural crest initially form an intermediate zone between the neural tube and surface ectoderm and later form the spinal and cranial sensory ganglia (C, D). From Langman, 1963 (73).

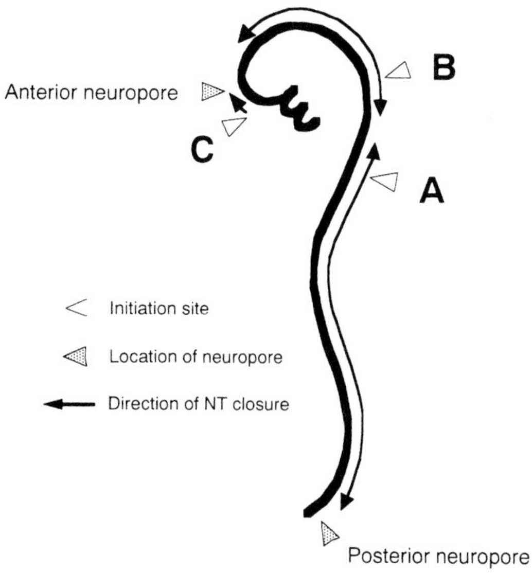

Figure 3-2 Model for multi-site neural tube closure in the human embryo. Open triangles show three initiation sites: cervical level (A), mesencephalic-rhombencephalic boundary (B), and rostral tip of the neural groove (C). Arrows indicate the directions of neural tube closure. Dotted triangles show sites of closure of anterior and posterior neuropores. From Nakatsu et al., 2000 (82).

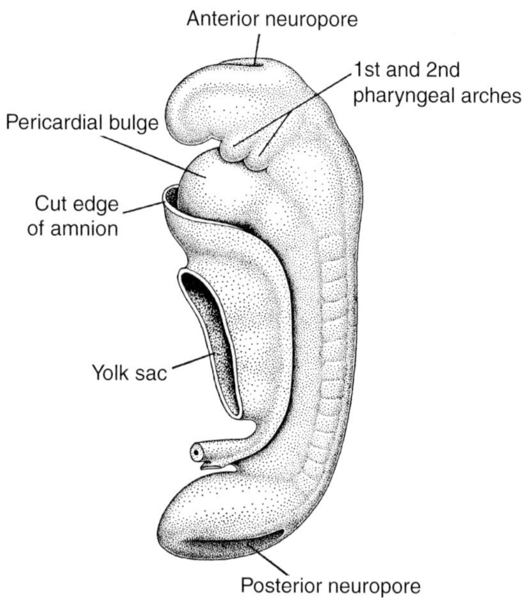

Figure 3-3 Left lateral view of 14 somite embryo (approximately 25 days). Note anterior and posterior neuropore. From Langman, 1963 (73).

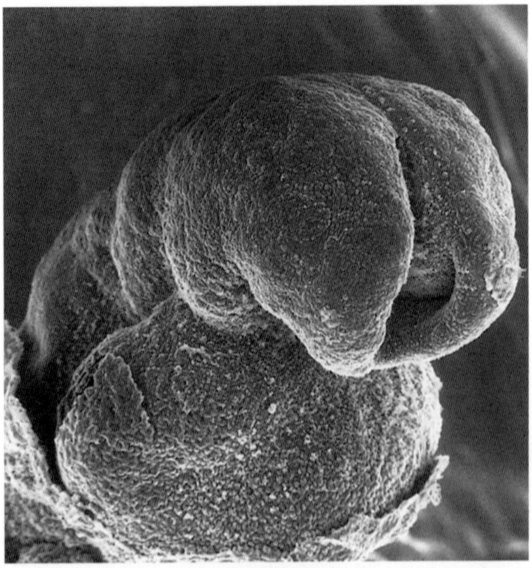

Figure 3-4 Scanning electron microscopy image of human with open anterior neuropore. From Jirásek, 1983 (64).

Table 3-1 Origin of Some Craniofacial Components

Neural crest origin
Facial connective tissue
Facial bones
Facial sutures
Frontal bone
Metopic suture
Coronal suture
Nasal capsule
Chondrocranium
Temporal squama
Mesodermal origin
Facial muscles
Endothelial lining of blood vessels
Parietal bones
Sagittal suture
Lambdoid suture
Temporal bones except squama
Occipital bone (somites plus squama)

rounds each nasal placode. As mesenchyme elevates the ridges, the nasal pits form (Fig. 3-7). By the sixth week, with continued growth of the maxillary swellings, the nasal and maxillary swellings become separated by deep furrows (49, 50, 66, 73, 99).

During the sixth and seventh weeks, the maxillary swellings fuse with the medial nasal swellings and both medial nasal swellings merge with each other (Fig. 3-8). The deep furrow separating the lateral nasal swelling and the maxillary swelling, the nasolacrimal groove, develops a solid epithelial cord that detaches from the overlying ectoderm in the furrow. After canalization of the cord, the nasolacrimal duct forms and widens at its upper end to create the lacrimal sac. After detachment of the cord, the lateral nasal and maxillary swellings merge to form the alae of the nose (49, 50, 66, 73, 99).

The fusion of the medial nasal swellings with the maxillary swellings and the merging of the medial nasal swellings with each other produce the primary palate or intermaxillary segment, which is composed of three parts: (a) a labial component that later forms the philtrum; (b) a medial section, the future dental arch containing the four maxillary incisor teeth; and (c) a triangular palatal component extending posteriorly to the incisive foramen (49, 50, 66, 73, 99).

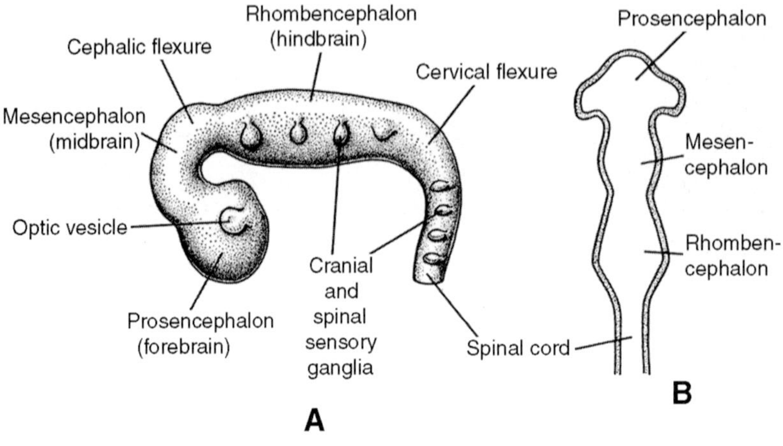

Figure 3-5 A, Lateral view of the brain vesicles and part of the spinal cord in a 4-week embryo. Note the ganglia formed by the neural crest on each side of the rhombencephalon and spinal cord. B, Diagram to show the lumina of the three brain vesicles and spinal cord. From Langman, 1963 (73).

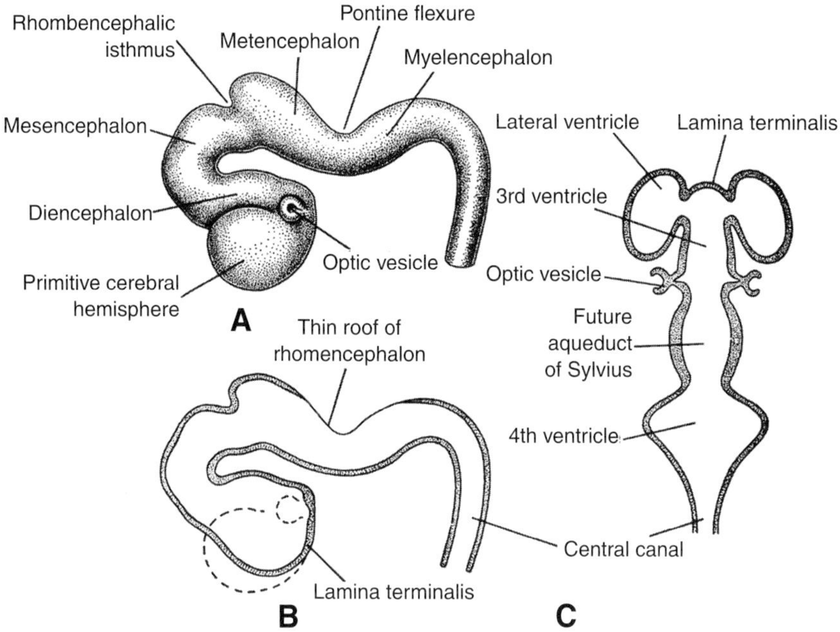

Figure 3-6 A, Lateral view of the brain vesicles of the human embryo at the beginning of the sixth week. B, Midline section through the brain vesicles and spinal cord of an embryo of the same age as shown in A. Note the thin roof of the rhombencephalon. C, Lumina of the spinal cord and brain vesicles. From Langman, 1963 (73).

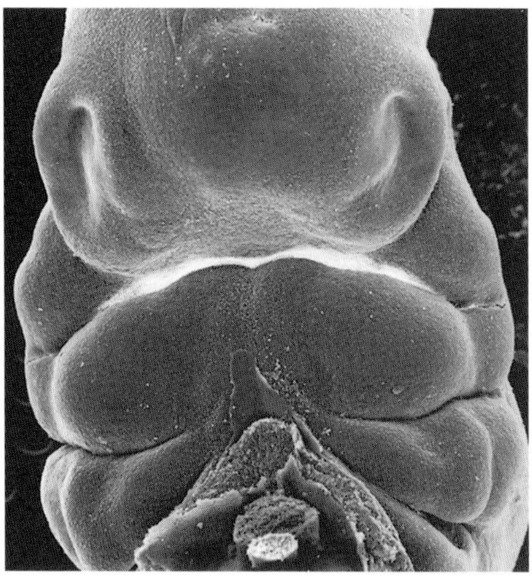

Figure 3-7 Scanning electron microscopy of the face of a 5-week-old embryo. Note the olfactory pits developing on the lateral portion of the frontonasal prominence. Courtesy of K. Sulik, Chapel Hill, North Carolina.

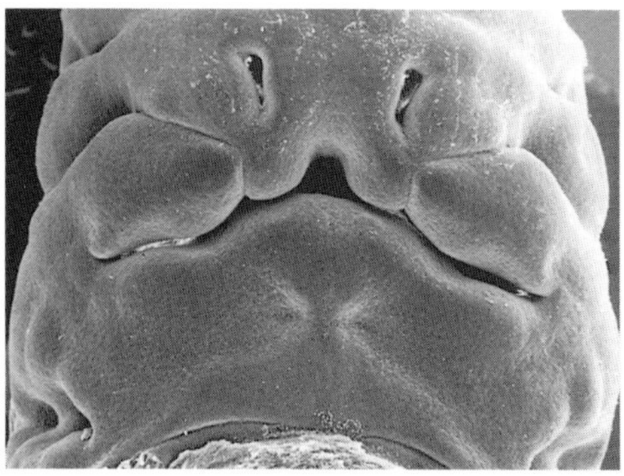

Figure 3-8 Scanning electron microscopy of a 6-week-old embryo. During the sixth and seventh weeks, maxillary swellings fuse with medial nasal swellings and both medial nasal swellings merge with each other. The upper lip is still incompletely formed at the sixth week. Courtesy of K. Sulik, Chapel Hill, North Carolina.

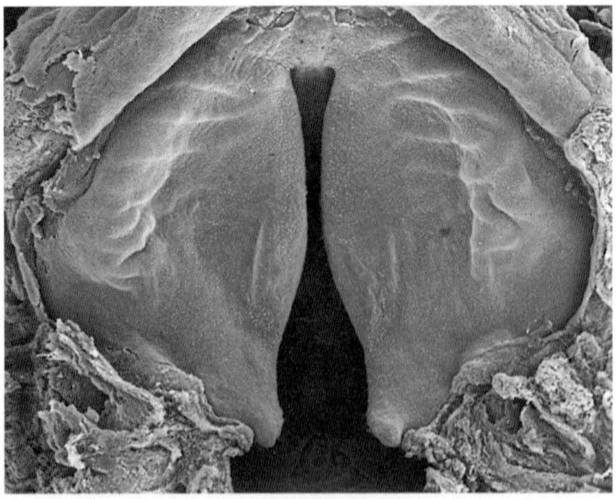

Figure 3-9 Scanning electron microscopy of the secondary palate in a 53-day-old embryo. Fusion with the primary palate has occurred. Courtesy of L. Russell, Chapel Hill, North Carolina.

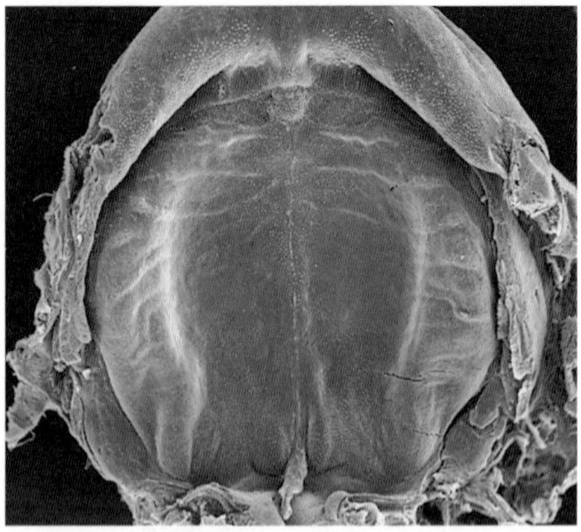

Figure 3-10 Scanning electron microscopy. At 59 days, complete fusion of the hard and soft palate has occurred. Courtesy of L. Russell, Chapel Hill, North Carolina.

The definitive (secondary) palate is formed from two medial shelf-like outgrowths of the maxillary swellings. These are the vertical palatine shelves, which appear during the sixth week and elevate to a horizontal position during the seventh week (Fig. 3-9). The shelves fuse with the triangular portion of the intermaxillary segment and fuse with each other by programmed cell death of the medial edges, allowing mesenchyme to join in the midline. Complete fusion occurs by the tenth week (Fig. 3-10). The soft palate and uvula form by merging (37, 49, 66, 73, 99).

PHARYNGEAL ARCHES

Embryology

The pharyngeal arches appear during the fourth and fifth weeks of development. Deep pharyngeal clefts separate the bars of mesenchyme. Accompanying the arches and externally located clefts, the pharyngeal pouches appear along the lateral walls of the pharyngeal portion of the gut (Fig. 3-11). The maxillary and mandibular swellings develop from the first (mandibular) arch (74, 97). The pharyngeal arches and their nerves, muscles, cartilages, bones, ligaments, and pharyngeal pouches are summarized in Table 3-2.

The ears develop from six mesenchymal proliferations, the auricular hillocks, three on each side of the first pharyngeal cleft. The definitive auricle results from fusion of the hillocks. With development of the mandible, the ears, which initially form in the lower neck region, ascend to the sides of the head at the level of the eyes (98).

The second pharyngeal arch grows down over the third and fourth arches, leaving a space that becomes the cervical sinus. The inferior parathyroid gland and thymus arise from the third pharyngeal pouch. The superior parathyroid gland arises from the fourth pharyngeal pouch. The ultimobranchial body arises from the sixth pouch (74) (Fig. 3-12).

Molecular Correlates of Pharyngeal Arch Development

Cranial neural crest, originating at the dorsal margin of the neural tube, migrates and populates the pharyngeal arches. Other embryonic germ layers (ectoderm, endoderm, and mesoderm) are involved in pharyngeal arch development. Genes that direct such development are summarized in Table 3-3. *Otx2* and *Gbx2* determine the midbrain–hindbrain boundary at the isthmus (Fig. 3-13), where Fgf8 acts to suppress Hox gene expression, and together with other factors acting locally, permit the mandibular arch to develop. However, positive Hox gene expression is essential for the developing hyoid arch.

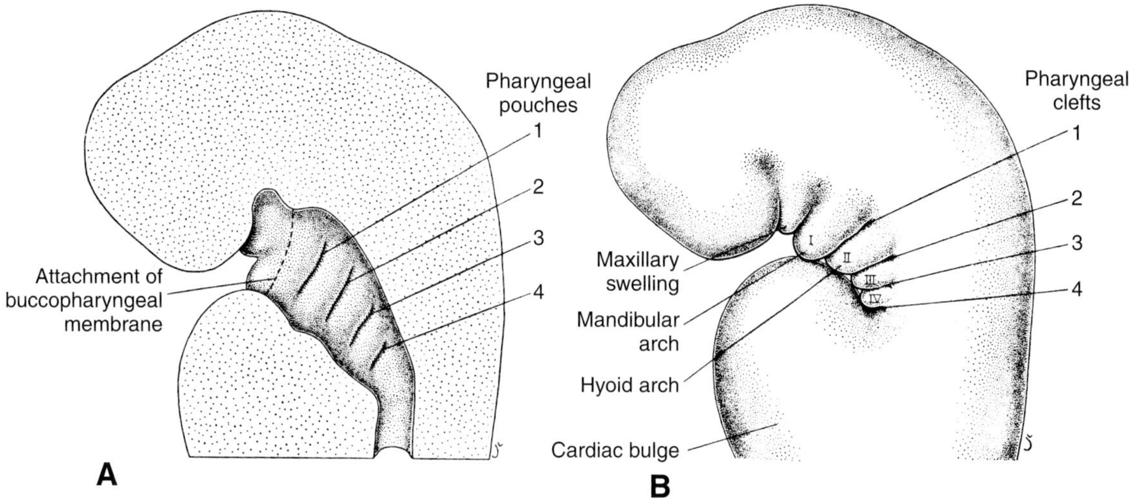

Figure 3-11 A, Sagittal section through the cephalic end of a 5-week embryo (approximately 6 mm), showing the pharyngeal pouches along the right lateral wall of the pharyngeal gut. Broken line represents the approximate site of attachment of the buccopharyngeal membrane. B, The pharyngeal arches and clefts in a 5-week embryo. From Langman, 1963 (73).

Table 3-2 Pharyngeal Arches

Pharyngeal Arches	Nerves	Muscles Derived from Somitomeres	Skeletal, Cartilaginous, and Ligamentous Structures of Neural Crest Origin	Pharyngeal Pouches of Endodermal Origin
First (mandibular)	Trigeminal (V)	Mastication (temporalis, masseter, medial and lateral pterygoids), myelohyoid, anterior belly of digastric, tensor palatini, tensor tympani	Meckel's cartilage, spheno-mandibular ligament, malleus, incus	Eustachian tubes
Second (hyoid)	Facial (VII)	Facial expression (buccinator, auricularis, frontalis, orbicularis oculi, orbicularis oris), posterior belly of digastric, stylohyoid, stapedius	Styloid process, stapes, stylohoid ligament, lesser cornua of hyoid, upper portion of body of hyoid	Palatine tonsils
Third	Glossopharyngeal (IX)	Stylopharyngeus	Greater cornua of hyoid, lower portion of body of hyoid	Thymus, inferior parathyroid glands
Fourth	Superior laryngeal branch and recurrent laryngeal branch of vagus (X)	Levator veli palatini, pharyngeal constrictors	Laryngeal cartilages	Superior parathyroid glands
Sixth				Ultimobranchial body

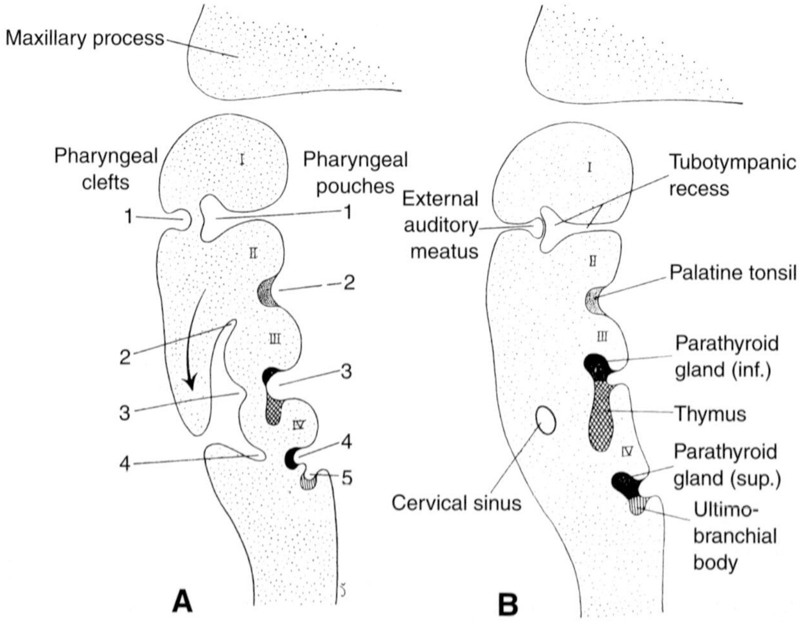

Figure 3-12 A, Schematic representation of development of the pharyngeal clefts and pouches. The second arch grows over the third and fourth arches, thereby burying the second, third, and fourth pharyngeal clefts. B, The remnants of the second, third, and fourth pharyngeal clefts form the cervical sinus. Note also the differentiation of the epithelium in the wall of the pharyngeal pouches. From Langman, 1963 (73).

Bmp4 causes neural crest cell death at two points, segregating three separate streams of neural crest migration into the pharyngeal arches. Despite the many genes that direct neural crest migration into the pharyngeal arches (Table 3-3), the neural crest itself is also an intrinsic source of molecular information that varies between species (51, 52, 59, 91, 93, 101).

Cranial Neural Crest

The cranial neural crest originates at the dorsal margin of the neural tube. *Wnt* is necessary and sufficient for the induction of neural crest. In the absence of *Wnt* signaling, neural crest cells are not generated (42). ErbB4, which is expressed in the neural ectoderm, is necessary for neural crest cell migration (45, 48).

The developing hindbrain is subdivided into cell-lineage restricted units—the rhombomeres (Fig. 3-14). Neural crest cells from rhombomeres 1 and 2 together with caudal midbrain-derived crest cells populate the first (mandibular) arch. Crest cells from rhombomere 4 populate the second (hyoid) arch. Rhombomeres 6 and 7 contribute to the third, fourth, and sixth pharyngeal arches (6, 51). Rhombomeres 3 and 5 are depleted of neural crest production, which allows three distinct streams of crest cells to migrate into the pharyngeal arches (51) (*vide supra*) (Fig. 3-13).

Although rhombomeres 3 and 5 produce some crest cells, the majority die by apoptosis caused by Bmp4. Decreased Msx2 expression is also found in the apoptotic cells, but Bmp4 plays the pivotal role. Territories opposite rhombomeres 3 and 5 express Sema3A, which also inhibits crest cell migration, as does mesenchyme adjacent to these segments. Bmp receptors and their transducers are present in the other rhombomeric crest as well, but they are antagonized by Noggin, thus restricting Bmp4-mediated apoptosis (29, 36, 51, 52, 94) (Table 3-3).

Table 3-3 Genes That Direct Pharyngeal Arch Development

Genes	Locations	Functions
Otx2, *Gbx2*	Isthmus	Positions isthmic organizer at midbrain–hindbrain boundary
Fgf8	Isthmus	Suppresses Hox gene expression in delaminating neural crest from rhombomere 1
Hoxa2	Rhombomere 4	Selector gene that confers identity to rhombomere 4 neural crest migrating into the second (hyoid) pharyngeal arch
Bmp4	Rhombomeres 3 and 5	Neural crest cell death, restricting most rhombomere 3 and 5 crest cells
Msx2	Rhombomeres 3 and 5	Associated with apoptotic rhombomere 3 and 5 crest cells
Sema3A	Opposite rhombomeres 3 and 5	Inhibits crest migration from rhombomeres 3 and 5
Nog	Hindbrain neural crest	Antagonist that restricts Bmp4-mediated apoptosis in hindbrain crest (except from rhombomeres 3 and 5)
Bmp7, *Fgf8*, *Pax1*, *Shh*	Endoderm of pharyngeal pouches	Formation and regionalization of pharyngeal arches
Dlx5, *Dlx6*	First pharyngeal arch	Specification of proximal maxillary division and distal mandibular division

Bmp7, posterior endodermal margin of pharyngeal pouches. *Fgf8*, anterior endoderm of pharyngeal pouches. *Pax1*, proximal endoderm of pharyngeal pouches. *Shh*, early posterior endoderm of second pharyngeal pouch and, later, posterior endoderm of third pharyngeal pouch

Data derived from Simeone, 2000 (93), Smith and Graham, 2001 (94), Depew et al., 2002 (27), Trainor et al., 2002 (101), and Graham, 2003 (51).

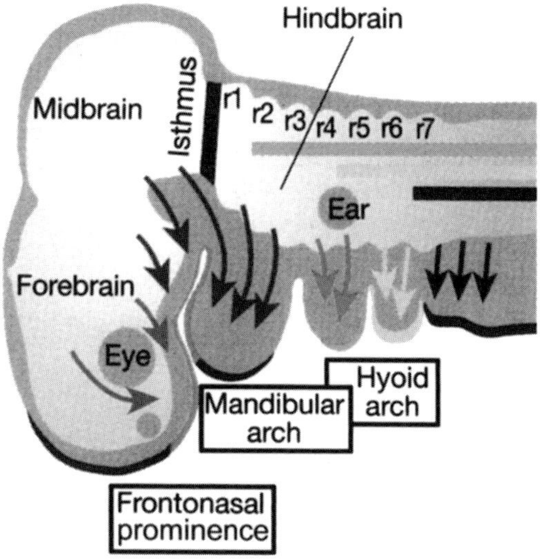

Figure 3-13 Frontonasal process, mandibular arch, hyoid arch, sensory structures (eye, ear), forebrain, midbrain, isthmus, hindbrain, and rhombomeres (r1 to r7) in the developing avian head. Neural crest cells migrate into the pharyngeal arches in three streams (arrows). See text for details. Modified from Helms and Schneider, 2003 (59).

Rhombomeric *Hox* Gene Identity

Each rhombomere has a unique *Hox* gene identity (Fig. 3-14). *Hoxa2* is expressed up to the rhombomere 1–2 boundary and *Hoxb2* is expressed up to the rhombomere 2–3 boundary. However, no *Hox* gene expression is found in neural crest anterior to rhombomere 3. Thus, rhombomere 1 and 2-derived crest, which populates the first pharyngeal (mandibular) arch,

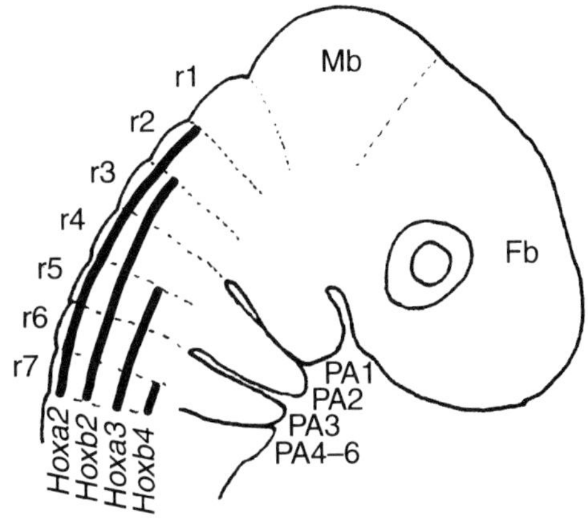

Figure 3-14 Developing avian hindbrain and pharyngeal arches. Each rhombomeric segment has a unique *Hox* gene identity, which is reflected in how the migrating neural crest affects the pharyngeal arches. See text for details. Rhombomeres: r1 to r7. *Hox* genes: *Hoxa2*, *Hoxb2*, *Hoxa3*, *Hoxab4*. Pharyngeal arches: PA1, PA2, PA3, PA4–6. Fb, Forebrain. Mb, Midbrain. Modified from Chambers and McGonnell, 2002 (6).

has no *Hox* gene expression. The midbrain–hindbrain boundary, known as the isthmus, expresses Fgf8, which suppresses Hox expression in delaminating crest from rhombomere 1. Other factors, acting more locally, probably downregulate Hoxa2 expression in deliminating crest from rhombomere 2 (59, 77, 101-103).

Hoxa2 is essential for second pharyngeal (hyoid) arch development. If *Hoxa2* is knocked out, rhombomere 4 crest in the hyoid skeletal identity is lost and replaced by mandibular skeletal patterning. On the other hand, if Hoxa2 is overexpressed in rhombomeres 1 and 2, mandibular skeletal patterning is lost in the mandibular arch and replaced by hyoid skeletal patterning (46, 53, 84, 89).

Endoderm, Ectoderm, and Mesoderm

The order of embryonic germ layers—endoderm, ectoderm, and mesoderm—reflects their relative importance in patterning pharyngeal arch development. In the pharyngeal pouches, the endoderm bulges out and contacts the ectoderm. At these points, ectoderm and endoderm remain in close contact and expand along the proximal–distal axis, separating the pharyngeal arches. A number of genes are expressed in the endoderm: Bmp7 at the posterior endodermal margin of the pharyngeal pouches; Fgf8 in the anterior endoderm of the pharyngeal pouches, Pax1 in the proximal endoderm of the pharyngeal pouches; and Shh in the early posterior endoderm of the second pharyngeal pouch and, later, in the posterior endoderm of the third pharyngeal pouch (51, 104) (Table 3-3).

Jaw patterning signals come from the anterior endoderm, which underlies the neural tube and adjacent paraxial mesoderm between the anterior tip of the embryo and rhombomere 2. If the endoderm under the anterior midbrain is removed, Meckel's cartilage is lost in the lower jaw. Endoderm is not only required for skeletal differentiation, but also for skeletal orientation relative to the body axis (6, 25). Endoderm also induces epibranchial placodes. Furthermore, endoderm is regionalized, forming the thyroid, parathyroids, and thymus (51).

A zone of frontonasal ectoderm regulates patterning and growth of the face. A molecular boundary has been identified by Fgf8- and Shh-juxtaposed domains, which presage the initial site of frontonasal outgrowth (60). Vascular endothelial cores and skeletal muscles in the pharyngeal arches are of mesodermal origin.

Intrinsic Neural Crest Patterning

Thus far, the roles of various factors in directing patterning of cranial neural crest in the pharyngeal arches have been emphasized. However, intrinsic neural crest patterning also occurs, and is species specific. For example, in birds, when neural crest cells that participate in beak morphogenesis are exchanged between two species, crest itself determines beak form; quail neural crest cells produce quail beaks in duck hosts and duck neural crest cells produce duck bills in quail hosts (91). In six species of finches with different types of beaks, Bmp4 expression (Table 3-3) varies with the different types. By altering Bmp4 levels experimentally during development, misshapen beaks can be generated (85a).

SKELETAL DEVELOPMENT

Epithelial–Mesenchymal Interaction, Condensation, and Differentiation

Skeletogenesis and chondrogenesis result from a sequence of events involving epithelial–mesenchymal interaction, condensation, and differentiation (Fig. 3.15, Table 3-4). The connection between epithelial–mesenchymal interaction is a close one in skeletogenesis of the craniofacial region, whereas it is less close in the developing skeleton of the limbs (53a, 55).

Condensation resulting from epithelial–mesenchymal interaction may result from an increased number of cells within the condensation and/or a decrease in the number of cells surrounding it. The following correlates, which are not mutually exclusive, include (a) shorter cell cycle within the condensation and/or longer cell cycle around it, (b) larger fraction of the cell population within the condensation and/or smaller fraction around it, and (c) migration of cells towards a center and/or failure of cells to move away from a center (53b).

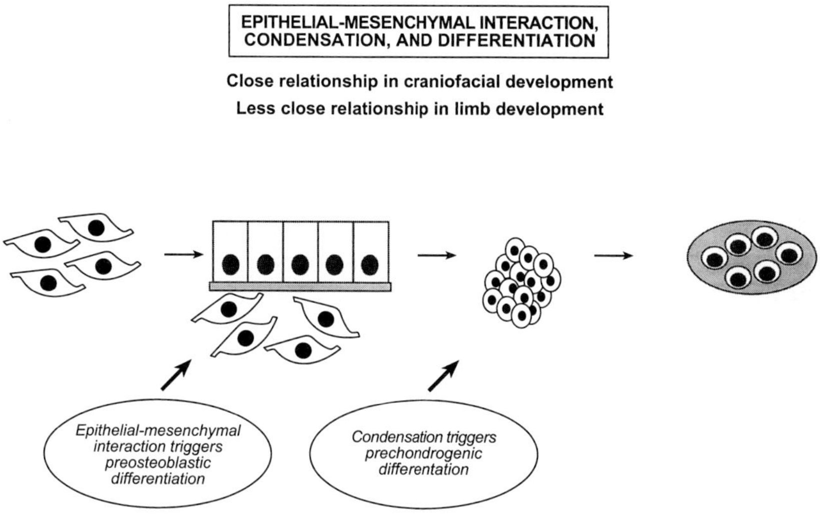

Figure 3-15 Four phases of development of a skeletal element (from left to right): migration of preskeletal cells to site of future skeletogenesis; epithelial-mesenchymal interaction; condensation; differentiation into chondroblasts or osteoblasts. See text. Modified from Hall and Miyake, 2000 (56).

Patterning within condensations may be complex. For example, more than one bone or cartilage can arise from a single condensation, such as two cartilages of the lower jaw, or the malleus and incus of the middle ear (56).

Condensation lead to chondrogenesis and osteogenesis. In chondrogenesis, condensation triggers prechondrogenic differentiation. In contrast, with osteogenesis, epithelial-mesenchymal interaction initiates preosteoblastic differentiation before condensation. Preosteoblasts then condense, become osteoblasts, and deposit bone matrix (28, 54, 55) (Fig. 3-15).

Defective condensations can have important consequences. A small condensation may result in an abnormally small bone. An even smaller condensation may result in an absent bone (53a).

Cell migration is an important first step in the development of bone. Two examples include (a) cranial neural crest cells producing osteoblasts of the membrane bones of the craniofacial skeleton and (b) somatic cells producing chondroblasts of the cartilage models of the vertebral column. In *TCOF1* mutations, which cause Treacher Collins syndrome, neural crest migration into the zygomatic region is defective (50,53a).

Calvarial and Sutural Initiation

The neurocranium is divided into a calvarial component of flat bones, which have intramembranous bone formation, and a cranial base component, which arises from endochondral bone formation. The sides and roof of the skull arise as membranous ossification characterized by needle-like bone spicules that radiate peripherally. At the margins of the frontal, parietal, and occipital bones, presumptive sutures and fontanels appear.

Five major sutures are present in the calvaria. Three—the coronal, lambdoid, and squamosal—are paired, and two—the sagittal and metopic—are single. The frontal bone, facial sutures, metopic suture, and coronal suture are of neural crest origin. The parietal bone, occipital bone, sagittal suture, and lambdoid suture are of mesodermal origin (23, 63a) (Table 3-1).

Sutures develop initially by a wedge-shaped proliferation of cells at the periphery of the extending bone fields—the osteogenic fronts (26), which appear to govern morphogenetic determination of sutural architecture (26, 57, 65). They approximate each other in one of two ways. They may overlap each other, forming a beveled suture, or they may approach each other in the

Table 3-4 Major Classes of Genes and Gene Products Associated with Skeletogenic Condensations, Their Functions, and Their Stages of Action

Gene/Gene Product	Function	Stage
Growth factors		
BMPs	Regulate *Hox* genes (*Hoxa2, Hoxd11, Pax2*) in response to *Shh*	Growth
	Regulates Msx1, Msx2	Transition to differentiation
FGF2	Regulates N-CAM	Initiation, proliferation, growth
TGFβ	Regulates fibronectin	Initiation
Cell surface, cell adhesion, and extracellular matrix molecules		
Fibronectin	Extracellular glycoprotein regulated by TGFβ regulates N-CAM	Initiation, proliferation
N-cadherin	Cell adhesion molecule	Adhesion
N-CAM	Cell adhesion molecule regulated by FN, *Prx1*, *Prx2*, and FGF	Initiation, adhesion
Noggin	Secreted protein that binds to and inactivates *BMP2, BMP4, BMP7*	Slows or stops growth
Syndecan	Receptor that binds to tenascin; binds to fibronectin to inactivate N-CAM	Sets boundary
Tenascin	Extracellular glycoprotein that binds to syndecan	Stops condensation growth, sets boundary
Hox genes		
Hoxa2	Regulated by BMPs, downregulates *Runx2*	Sets boundary, growth, prevents differentiation
Hoxa13	Alters adhesive properties	Adhesion
Hoxd11	Regulated by BMP	Proliferation, growth
Hoxd11–13	Transcriptional activation	Transition to differentiation
Transcription factors		
Runx2	Transcriptional activating protein inhibited by *Hoxa2* Regulated by *BMP7* and vitamin D_3	Differentiation of chondroblasts Switches cells into osteoblastic pathway
CFKH1	Chicken forkhead–helix transcription factor that regulates TGFβ and interacts with *Smad* transcription factors	Initiation, proliferation
MFH1	Mesenchymal transcription factor	Proliferation
Pax1, Pax9	Encode nuclear transcription factors, regulated by *BMP7*	Growth
Prx1, Prx2	Upstream regulation of N-CAM	Initiation
Scleraxis	Basic helix–loop–helix protein	Proliferation
Sox9	Regulates *Col2α1*	Proliferation

Adapted from Hall and Miyake, 2000 (56).

same plane with an intervening zone of immature fibrous connective tissue, which leads to an end-to-end type of suture (39, 65) (Fig. 3-16A–D).

End-to-end sutures, such as the sagittal and metopic, are formed in the midline (70, 71). Midline initiation may be of the end-to-end type because the biomechanical forces on either side of the initiating suture are likely to be equal in magnitude. In contrast, sutures away from the midline, such as the coronal and frontozygomatic, have biomechanical forces of unequal magnitude acting on them and hence are of the overlapping, beveled type (15).

Development of Cranial and Facial Sutures

At birth, areas where cranial sutures will develop permit adjustive overlap of the calvarial bones

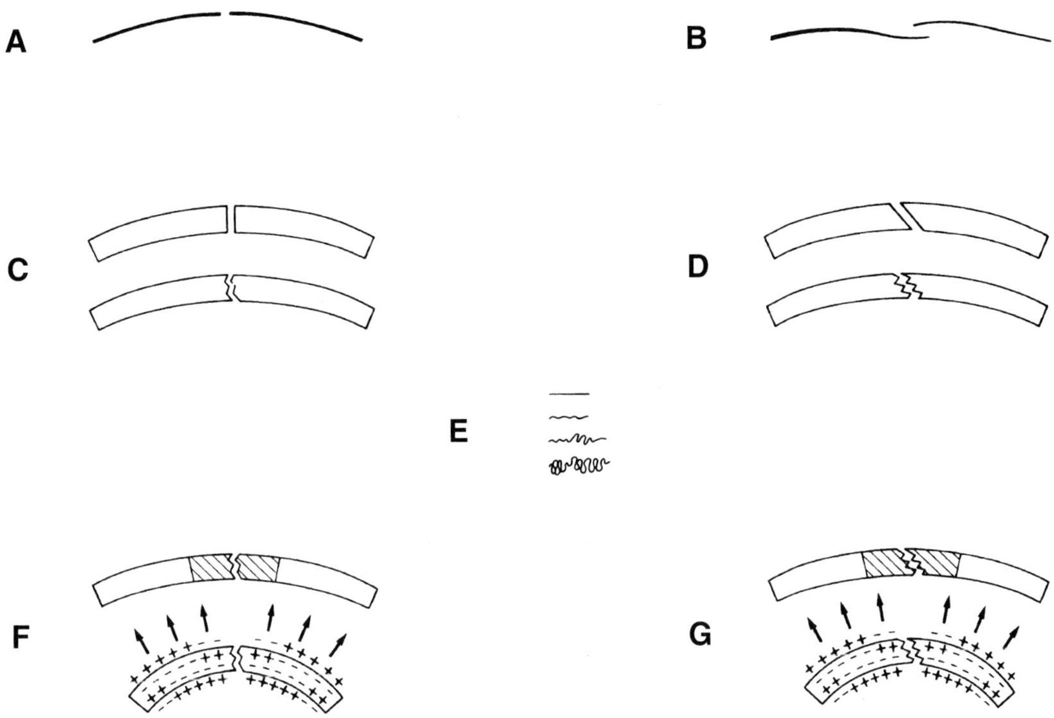

Figure 3-16 Diagram of sutural changes. Left column (A, C, F) represents midline sutures (sagittal, metopic). Right column (B, D, G) represents non-midline sutures (e.g., coronal). A, End-to-end sutural initiation. B, Overlapping sutural initiation. C, Development of interdigitations in midline suture. D, Development of interdigitations in overlapping suture. E, Development of interdigitations, from straight to slightly interdigitated to interdigitated to very interdigitated. F, Changes in shape of calvarial bones with brain expansion. Also note bone deposition (diagonal lines) at midline sutural margins. G, Changes in shape of calvarial bones with brain expansion. Also note bone deposition (diagonal lines) at non-midline sutural margins. From Cohen, 1993 (10).

(Fig. 3-17, see Fig. 1-6) as the human head becomes compressed during passage through the birth canal. The resultant molding normalizes during the first week of life by cranial re-expansion and widening of the sutural areas (Fig. 3-18). Engineering structural analysis has shown that the fetal parietal bones are capable of deforming under load distributions typical of those found during normal labor (78). Thus, molding is a combination of displacement of the cranial bones as well as their intrinsic deformation.

As cranial sutures develop further during infancy, cranial adjustment to the expanding brain takes place by bone deposition at the sutural margins while the sutures proper remain patent. Changes in the regional curvature of the calvarial bones also occur (15) (Figs. 3-16FG, 3-19).

Cranial and facial sutures develop differently (86). Facial bones have fibrous periosteal capsules surrounding them that are fully established by week 17 *in utero*. In contrast, cranial bones develop in a preformed continuous fibrous membrane, the ectomeninx, and do not develop fibrous capsules until after birth. Thus, the more mature fibrous periosteal layers of the fetal facial sutures may be a more effective barrier against osseous fusion than the as yet immature, unlayered cranial sutures (72).

Cranial and facial sutures differ in other respects. Except for the midpalatal suture, facial

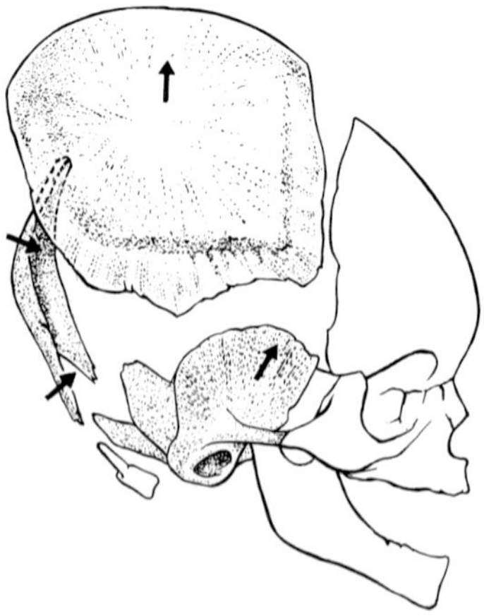

Figure 3-17 Diagram of neonatal skull on first day of life, demonstrating molding of calvarial bones with overlapping edges caused by compression during passage of head through birth canal. Courtesy of Mead Johnson Laboratories.

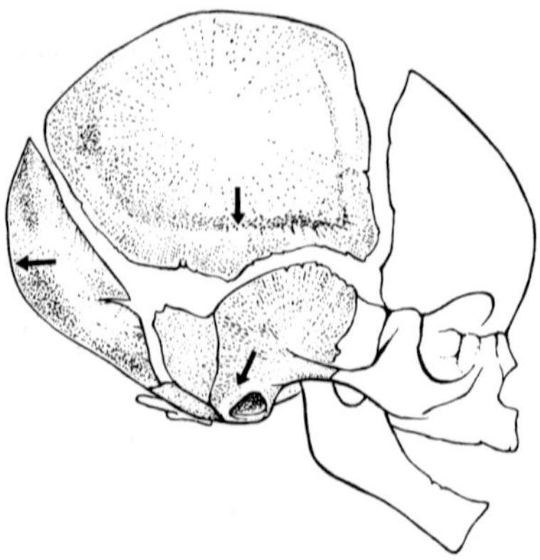

Figure 3-18 Diagram showing re-expansion of the cranium that takes place during the first week of life. The parietal, occipital, and temporal bones have returned to normal positions. Courtesy of Mead Johnson Laboratories.

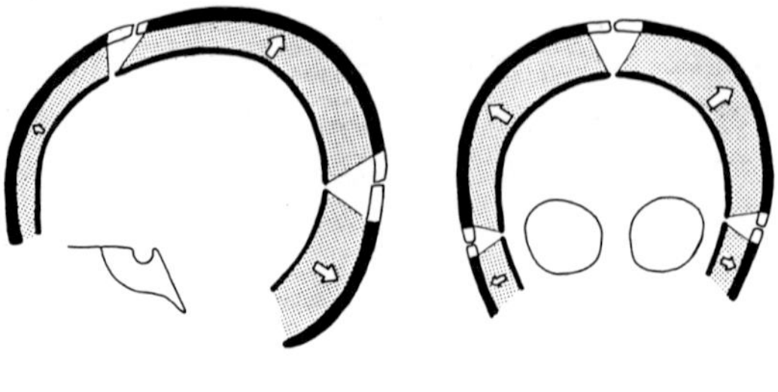

Figure 3-19 Calvarial adjustment to expanding brain by bone deposition at sutural margins. From Friede, 1981 (38).

sutures do not fuse before the seventh to eighth decades (70, 79). In contrast, cranial sutures close relatively early in adult life (Table 3-5). Forces developed by the masticatory muscles are transmitted from the mandible via the facial bones to the cranial vault (Fig. 3-20). Thus,

Table 3-5 Suture Closure in Humans

Suture	Closure begins (years)
Cranial	
Metopic	2[a]
Sagittal	22
Coronal	24
Lambdoid	26
Squamosal	35–39
Sphenofrontal	22
Sphenoparietal	29
Sphenotemporal	28–32
Masto-occipital	26–30
Facial	
Palatal	30–35
Frontomaxillary	68–71
Frontozygomatic	72
Zygomaticotemporal	70–71
Zygomaticomaxillary	70–72
Frontonasal	68
Nasomaxillary	68

[a]Usually obliterated by third year; persists throughout life in 10%.
From Cohen and MacLean, 2000 (23).

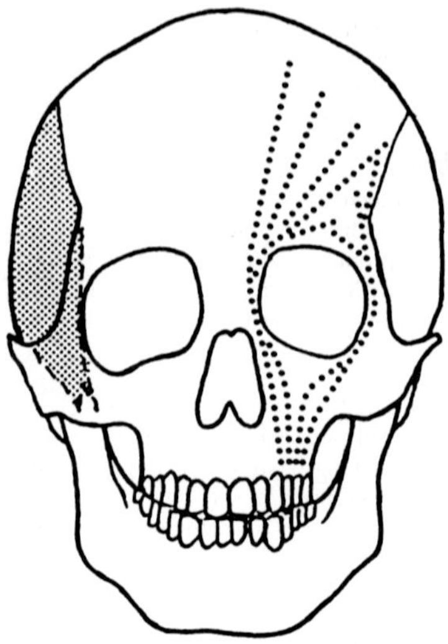

Figure 3-20 Forces transmitted (dotted lines) from mandible via facial bones to cranial vault. Forces are developed by masticatory muscles, in particular, the temporalis (shaded area). Modified after Campbell, 1970 (5).

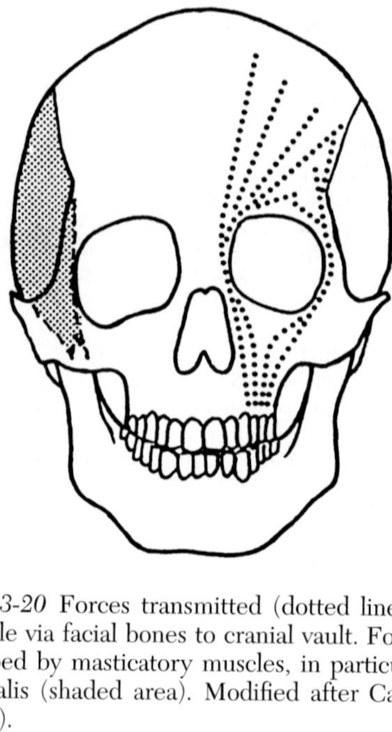

Table 3-6 Genes Expressed in Craniofacial Development

Genes	Expression
Homeobox	
Hoxa2	Identifies rhombomere 4 neural crest migrating to hyoid arch
Hoxb2	Expressed up to rhombomere 2–3 boundary
Gastrulation brain homeobox	
Gbx2	Midbrain–hindbrain boundary (isthmus)
Orthodentical-related homeobox	
Otx1	Hindbrain
Otx2	Forebrain, midbrain, isthmic organizer at the midbrain–hindbrain boundary
Muscle-segment homeobox	
Msx1	Osteogenic fronts of calvarial sutures, mesenchyme of maxillary and mandibular prominences
Msx2	Osteogenic fronts of calvarial sutures, mesenchyme of maxillary and mandibular prominences
Paired (aristaless)-like homeobox	
Alx3	Mesenchymal condensations of mandibular and hyoid arches
Alx4	Mesenchymal condensations of skull, hair, and teeth
Paired box	
Pax1	Pharyngeal pouches, thymus, parathyroids, eustachian tubes, tonsils, ultimobranchial body
Pax2	Midbrain–hindbrain junction, optic stalk
Pax3	Dorsal neural tube, neural crest, thymus, parathyroids, tongue muscles
Pax6	Ocular development
Pax7	Neural crest, skeletal muscles, lateral nasal prominence
Pax8	Thyroid development
Pax9	Neural crest, pharyngeal pouches, thymus, parathyroids, eustachian tubes, tonsils, ultimobranchial body, tooth development
Sonic hedgehog	
Shh	Notochord, floorplate, ventromedial midbrain, ventral forebrain, epithelium of frontal prominence, maxillary prominence, oral epithelium of mandibular prominence, hyoid arch endoderm
Patched	
Ptch1	Epithelial cells adjacent to Shh-expressing cells, mesenchymal cells surrounding epidermal ingrowth, maxillary prominence, median nasal prominence
Ptch2	Co-expressed in epithelium with Shh
Semaphorins	
Sema3A	Inhibits neural crest migration from rhombomeres 3 and 5
Neuregulin receptors	
ErbB4	Neuroectoderm, essential for neural crest cell migration
Gli family	
Gli1	Ventral neural tube, pharyngeal arch mesenchyme
Gli2	Posterior neural tube, pharyngeal arch mesenchyme
Gli3	Posterior neural tube, pharyngeal arch mesenchyme
Wnt family	
Wnt	Neural crest induction
Wnt5a	Maxillary and mandibular prominences
Wnt11	Maxillary prominence
Runx family	
Runx2	Osteogenesis involving membrane and endochondral bone
Fibroblast growth factors	
Fgf2	Osteogenic fronts of calvarial sutures
Fgf4	Osteogenic fronts of calvarial sutures, anterior half of mandibular and hyoid arches
Fgf8	Midbrain–hindbrain boundary (isthmus), suppresses Hox expression in delaminating crest from rhombomere 1, ectoderm of maxillary and mandibular prominences
Fgf10	Maxillary mesenchyme, rostral mandibular mesenchyme, and hyoid arch core mesenchyme
Fibroblast growth factor receptors	
Fgfr1	Osteogenic fronts of calvarial sutures, all facial prominences
Fgfr2	Osteogenic fronts of calvarial sutures, ectoderm of maxillary and mandibular prominences
Fgfr3	Osteogenic fronts of calvarial sutures, brain, frontal prominence, maxillary prominence, lateral nasal prominence
Basic helix–loop–helix	
Twist	Head mesenchyme, including osteogenic fronts of calvarial sutures and all facial prominences, branchial arches
Collagens	
COL1A1	Bone of craniofacial skeleton
COL2A1	Cartilage of cranial base and nasal capsule
Alkaline phosphatase	
ALPL	Calcification of craniofacial skeleton
Osteocalcin	
BGLAP	Craniofacial skeleton
Nel-like	
Nell1	Cranial intramembranous bones
Efrins	
Efnb1	Frontonasal neural crest

(continued)

Table 3-6 (continued)

Genes	Expression
Bone morphogenetic proteins	
Bmp2	Osteogenic fronts of calvarial sutures, mesenchyme of lateral region of mandibular arch, tooth formation
Bmp4	Selective apoptosis of neural crest cells from rhombomeres 3 and 5, osteogenic fronts of calvarial sutures, epithelium of medial ends of maxillary and mandibular prominences, tooth formation
Bmp7	Entire epithelium covering maxillary and mandibular prominences, tooth formation
Bmp antagonist	
Nog	Neural tube, calvarial sutures, cartilage condensations, frontonasal epithelium
Transforming growth factor β	
Tgfb1	Calvarial sutures
Tgfb2	Calvarial sutures, mesenchyme of secondary palate
Tgfb3	Calvarial sutures, medial edge epithelium of palatal shelves
Retinoids	
RALDH2	Frontonasal area
RALDH3	Nasal placode
Goosecoid	
Gsc	Posterior portion of mandibular arch, anterior portion of hyoid arch
Forkhead	
FOXE1	Oronasal membrane, palate, hair shaft, thyroid, epiglottis
Endothelin	
ET1	Epithelium of facial prominences, mesodermal bone of mandibular arch
EDNRA	Hindbrain normal crest, mesenchyme of facial prominences
ECE1	Mesenchyme of facial prominences
Distal-less related	
Dlx1	Mesenchyme of proximal and distal regions of mandibular and hyoid arches
Dlx2	Mandibular arch, distal region of hyoid arch
Dlx3	Distal tips of first and second arch mesenchyme
Dlx5	Distal mesenchyme of first and second arches
Dlx6	Distal mesenchyme of first and second arches

Data from Chen and Behringer, 1995 (7); Wilke et al., 1997 (108); Clouthier et al., 1998 (9); Cohen, 2000 (12), 2000 (13), 2000 (14), 2002 (19), 2003 (20), 2003 (21), 2004 (22); Cohen and MacLean, 2000 (23); Hall and Miyake, 2000 (56); Jernvall and Thesleff, 2000 (63); Bachler and Neubuser, 2001 (2); Boncinelli and Morgan, 2001 (4), Ingham and McMahon, 2001 (62); Chi and Epstein, 2002 (8); Trainor et al., 2002 (101), Lehmann et al., 2003 (75); Richman and Lee, 2003 (88); Warren et al., 2003 (106); OMIM, 2004 (82a).

it would be advantageous for facial sutures to remain patent until old age (15).

MOLECULAR CORRELATES OF CRANIOFACIAL EMBRYOLOGY

Many genes are expressed in craniofacial development including, among others, *Fgf8*, *Fgfr2*, *Shh*, *Ptch*, *Gli1*, *Gli2*, *Gli3*, *Bmp2*, *Bmp4*, *Bmp7*, *Msx1*, *Msx2*, *Twist*, *Dlx1*, *Dlx2*, *Dlx3*, *Dlx5*, and *Dlx6*. These and other genes, together with their expression patterns, are summarized in Table 3-6. Several caveats are important. First, some but not all genes involved in craniofacial development are represented. Second, gene expression is time dependent during development, and this is not reflected in the table. Finally, tissue expression in the table is only representative, but far from complete.

Although craniofacial and limb development are different, the molecular signals regulating both are similar and highly conserved. For example, the genes expressed in craniofacial development listed in the preceding paragraph are also expressed in limb development (83, 92). Table 3-7 summarizes a few gene mutations that result in both craniofacial and limb anomalies.

CRANIOFACIAL GROWTH AND DEVELOPMENT

Facial Development

The changing proportions of head size to body size are illustrated in Figure 3-21 (1, 90). From the two-month embryo to the 22-year-old adult, relative head size decreases significantly, but size is greatest in the embryo (2 months), fetus (3 months), and newborn. In the newborn skull (Fig. 3-22), the cranium and the orbits are relatively large and the facial skeleton is diminutive. In early childhood, the cranium and the orbits remain relatively large, but the eruption of the primary dentition enlarges the facial skeleton. In

Table 3-7 Some Gene Mutations That Result in Both Craniofacial and Limb Anomalies

Mutation-Bearing Genes	Syndrome	Mechanisms	Gene Transcripts in Limbs[a]	Craniofacial Anomalies	Limb Anomalies
FGFR2	Apert syndrome	Gain of function	Apical ectodermal ridge; mesenchyme	Craniosynostosis; facial anomalies	Symmetric syndactyly of hands and feet
TWIST	Saethre-Chotzen syndrome	Haploinsufficiency	Anterior limb bud mesenchyme	Craniosynostosis; facial anomalies	Brachydactyly; mild soft tissue syndactyly
GLI3	Pallister-Hall syndrome	Haploinsufficiency	Anterior limb bud mesenchyme	Hypothalamic hamartoma; facial anomalies	Postaxial polydactyly
	Greig cephalopoly-syndactyly	Haploinsufficiency	Anterior limb bud mesenchyme	Scaphocephaly; frontal bossing; hypertelorism	Broad thumbs/halluces; pre- and postaxial polydactyly of hands and feet

[a]For gene transcripts in the craniofacial region, see Table 3-3.

Data from Biesecker, 1997 (3); Kang et al., 1997 (67), Wild et al., 1997 (107), Cohen, 2000 (17), O'Rourke and Tam, 2002 (83), Wilkie et al., 2002 (109).

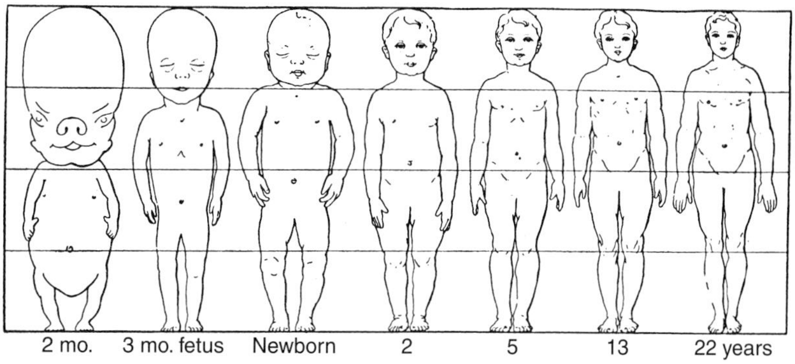

Figure 3-21 Diagram illustrating the changing proportions of head size to body size. From Scammon, 1953 (90).

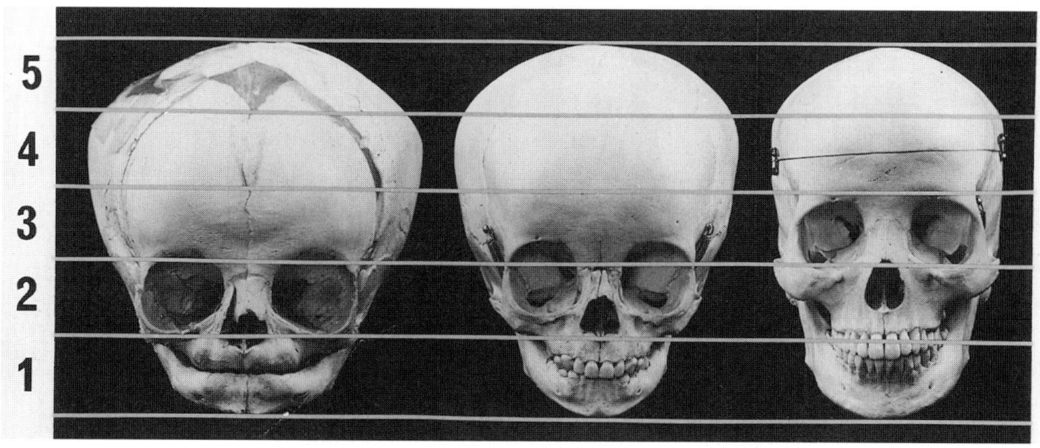

Figure 3-22 Alterations in the skull in the newborn, in childhood, and in adulthood.

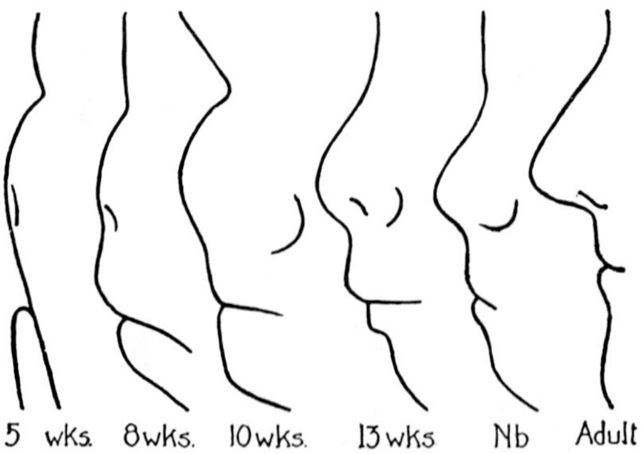

Figure 3-23 Line drawings of the human profile illustrating alterations in form and proportion that occur throughout the life span. From Scammon, 1953 (90).

the adult skull, the facial skeleton is well developed with relatively less prominence of the orbits and the cranium.

Figure 3-23 illustrates line drawings of the human profile showing alterations in form and proportion that occur throughout the life span (1,90). The development of the face occurs chiefly between the fifth and eighth weeks of embryonic life. When first formed, the nose is flat, but during the fetal period and continuing until birth, the nose becomes progressively elevated. By adult life, the nose is well developed with elevation of the nasal bridge. In the facial profile proper (Fig. 3-24), a rounded forehead, small nose, low nasal bridge, and a small recessed mandible characterize infancy. During childhood and adolescence, the nose and mandible become more prominent and the face progressively lengthens. In adulthood, the face is fully developed with elevation of the nasal bridge together with lengthening and prominence of the mandible. The face in the life cycle is illustrated in Figures 1-6 to 1-10.

Figure 3-24 The human face in profile showing postnatal development. From Enlow, 1968 (30).

Craniofacial Bone Remodeling and Displacement

In the craniofacial growth of soft tissues, two interrelated types of growth movement occur in the bony components: (a) remodeling by osteoblastic deposition and osteoclastic resorption, and (b) displacement of bones at their articular junctions. Simple enlargement by periosteal deposition and endosteal resorption is insufficient to account for the complex architectural shape of craniofacial bones, which tend to maintain their own shapes while enlarging at the same time. About half of the external surfaces of most

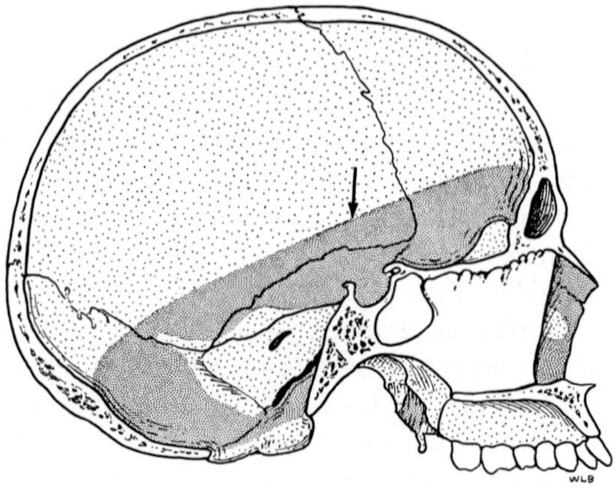

Figure 3-25 A circumcranial reversal line (arrow) encircles the endocranial surface of the calvaria. Below this line, the bone surface is resorptive; above it, the surface is depository. From Enlow, 1968, (30).

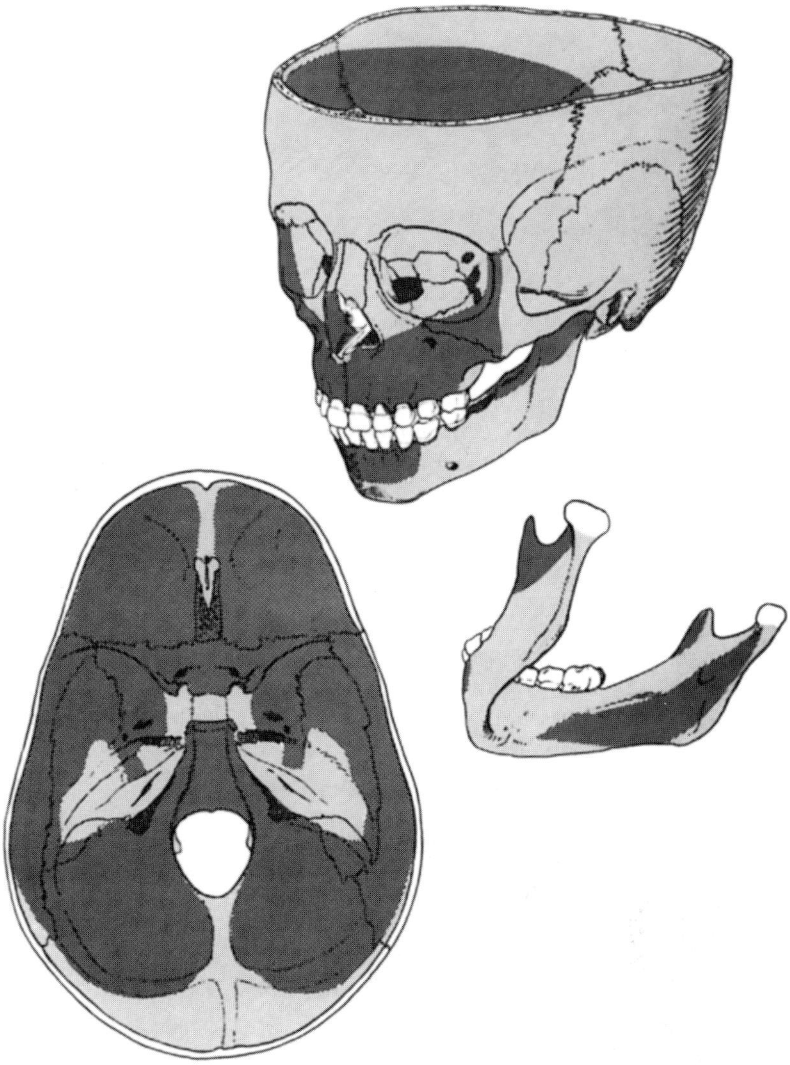

Figure 3-26 Surface remodeling fields that are resorptive (dark shading) and depository (light shading). From Enlow et al., 1971 (35).

craniofacial bones are actually resorptive and, in such locations, deposition occurs on the endosteal surfaces (30–35). Examples include the endocranial surfaces of the calvaria, maxilla, and mandible (Figs. 3-25 to 3-29).

Displacement occurs for all craniofacial bones at the articular contacts of sutures, synchondroses, and condyles. Bones spread apart in conjunction with soft tissue enlargement. The spaces opened up create room for bones to grow (10, 30–32). Examples include the cranial sutures with expansion of the brain (Figs. 3-16FG, 3-19) and the downward and foreward displacement of the maxillary region from the cranium by expansive growth of the midfacial soft tissues (Fig. 3-29).

Bone remodeling occurs throughout life. Table 3-8 summarizes normal growth and modification of the craniofacial skeleton with age. Table 3-9 lists some examples of abnormal alterations in the craniofacial skeleton with age.

The Skull as a Community of Bones vs. the Skull as a Community of Joints

According to Pruzansky (87), the skull is a community of bones separated by joints, but according to Moffett (80), the skull is a community of joints separated by bones. Which view is correct? The answer is that they are both correct. They are simply different contexts in which to view development of the skull.

Pruzansky (87) stated:

> The skull is a community of bones and organ systems of diverse phylogenetic origin and variable patterns of development, altogether relating to several functions vital to the life and well-being of the organism. If in the course of development one member of this community is affected adversely, inevitably other parts will suffer.

Achondroplasia, caused by several *FGFR3* gain-of-function mutations, particularly Gly380 Arg (22), illustrates an abnormal skeletal complex with effects on the skull as a whole. Specifically, endochondral bone formation is affected, and during craniofacial development, the nasal capsule and cranial base, both cartilaginous in origin, are hypoplastic. These have a secondary effect on membrane bones in the skull because all craniofacial bones articulate with one another. Thus, in achondroplasia, the hypoplastic nasal capsule and short anterior cranial fossa result in midface deficiency (Fig. 3-30).[a]

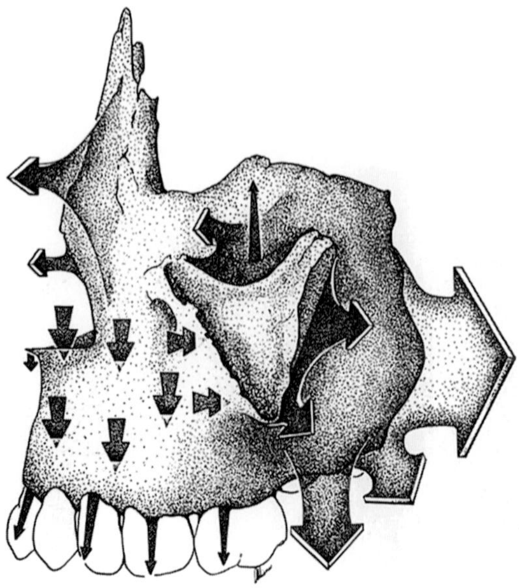

Figure 3-27 Directions of maxillary remodeling. Arrows that enter a regional bone surface represent resorptive fields and arrows that leave a bone surface are depository. From Enlow and Hunter, 1968 (34).

The skull may be considered a community of joints separated by bones; the types of joints, known as craniofacial articulations, include synovial, cartilaginous, fibrous, and dental (80). Table 3-10 summarizes craniofacial articulations together with their physiological and mechanical functions and their remodeling responses.

Asymmetry in Normal Craniofacial Development

Human structural asymmetry is exemplified by many studies of the brain (Table 3-11, Fig. 3-31). For example, Brodmann's area 22 is involved in processing the auditory signals associated with speech between the two hemispheres. Area 22 in the left hemisphere is crucial for word detection and generation, and is preferentially activated. However, area 22 in the right hemisphere aids in discriminating between melody, pitch, and sound intensity, and is activated to a lesser extent. Neuronal clusters in area 22 of the left hemisphere are spaced approximately 20% further apart and are interconnected by longer axons than found with neuronal clusters in the right hemisphere (18, 41).

With respect to the normal face, subtle degrees of asymmetry become particularly evident when properly oriented frontal photographs are divided along the median plane and reprocessed, each side being paired with its mirror image, yielding two different faces (85) (Fig. 3-32).

Some developmental disturbances can augment the normal degree of asymmetry. Left/right tooth crown size asymmetry, evident by measurement but not by visual inspection, is a normal state in the general population (44). In trisomy 21 syndrome, left/right crown size asymmetry is increased, although the asymmetry still cannot be appreciated visually (11, 18, 43).

Preferential laterality for some anomalies is striking, for example, left sidedness for cleft lip. To date, no adequate hypotheses have been put forth to explain such phenomena (11, 18).

[a]The foramen magnum is also reduced in size.

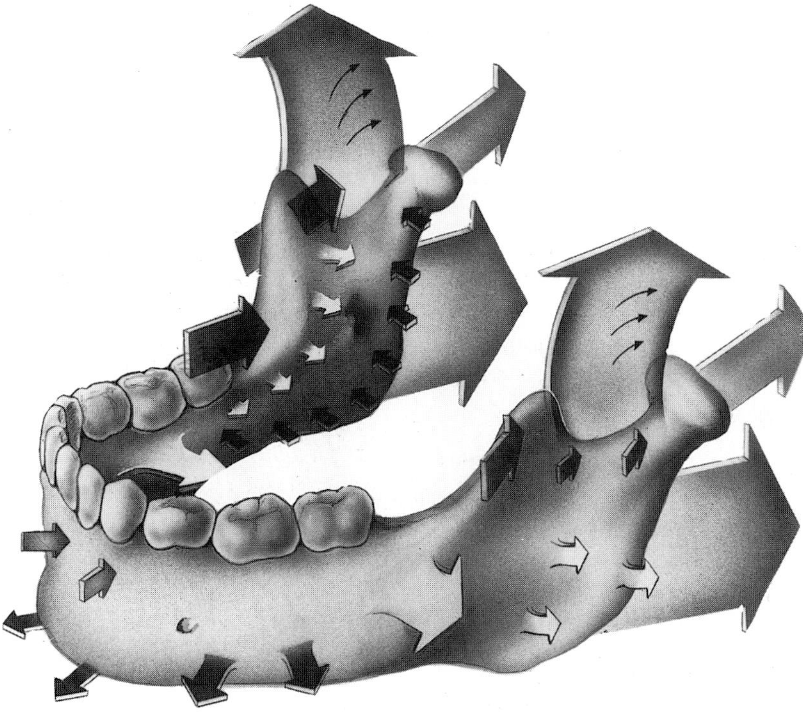

Figure 3-28 Direction of mandibular remodeling. Arrows that enter a regional bone surface represent resorptive fields and arrows that leave a bone surface are depository. From Enlow and Harris, 1964 (33).

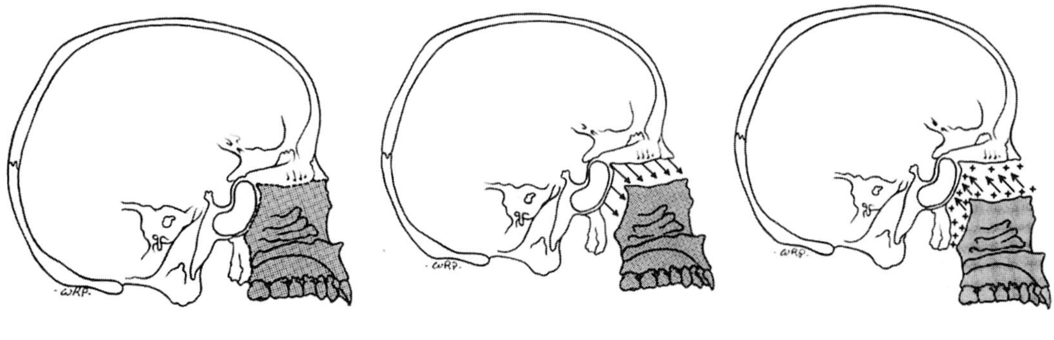

Figure 3-29 Left, Maxillary complex (gray). Middle, Maxillary displacement (arrows); maxillary complex shows downward and forward movement away from the cranium by expansive growth of midfacial soft tissues. Right, Bone deposition (arrows, +) in space created by soft tissue expansion. From Enlow, 1996 (31).

Table 3-8 Normal Growth and Modification of the Craniofacial Skeleton with Age

	Cartilaginous Growth	Sutural Growth	Skeletal Mass	Bone Remodeling
Infancy and Childhood	Active	Passive	Increases	Deposition greater than resorption
Adulthood	Absent	Absent	Remains the same	Deposition equals resorption
Old Age	Absent	Absent	Decreases	Resorption greater than deposition

Table 3-9 Some Examples of Abnormal Alterations in the Craniofacial Skeleton with Age

	Abnormal Skeletal Condition	Pathologic Process
Infancy and Childhood	Achondroplasia	Bony midface deficiency secondary to hypoplasia of cartilaginous nasal capsule and cranial base
	Craniosynostosis	Premature sutural fusion by bone deposition
Adulthood	Acromegaly	Bone deposition exceeds bone resorption
	Hemifacial atrophy	Localized bone resorption exceeds bone deposition
Old Age	Paget's disease of bone	Bone deposition exceeds bone resorption
	Senile osteoporosis	Bone resorption exceeds bone deposition

Alterations of Head Posture

The head participates in controlling visual gaze and in stabilizing/righting reflexes, but inappropriate head-locking posture occurs in torticollis. The interstitial nucleus of Cajal functions as a neural integrator for head posture, suggesting that bilateral imbalance of this structure from damage or inappropriate input might be one mechanism underlying torticollis (69a).

Differences in craniofacial morphology found with the greatest normal head extension and the greatest normal head flexion in relation to the cervical spine are summarized in Table 3-12 and illustrated in Figure 3-33. Table 3-13 defines the lowercase letters used in Figure 3-33 (96). The association of craniofacial morphology with head posture involves several interrelated factors (Fig. 3-34), any of which may initiate the alterations (95).

Table 3-10 Craniofacial Articulations

Type of Articulation	Example	Physiological Function	Mechanical Function	Remodeling Response
Synovial	Temporomandibular joint	Jaw movement	Resists compression and shear to some extent	Limited, avascular
Cartilaginous	Cranial base synchondroses	Active growth	Resist compression	Limited, avascular
Fibrous	Cranial sutures	Allow passage through narrow birth canal; passive growth secondary to brain enlargement	Respond to tension	Great, vascularized
	Facial sutures	Mastication	Sutures remain patent; shock absorbers for forces of mastication	Great, vascularized
	Periodontal fibers	Eruption of teeth; anchoring support of teeth	Respond to tension, compression, and shear	Great, vascularized
Dental	Occlusal and interproximal articulations	Mastication and speech	Subject to compression and shear	None, acellular

Modified from Moffett, 1973 (80).

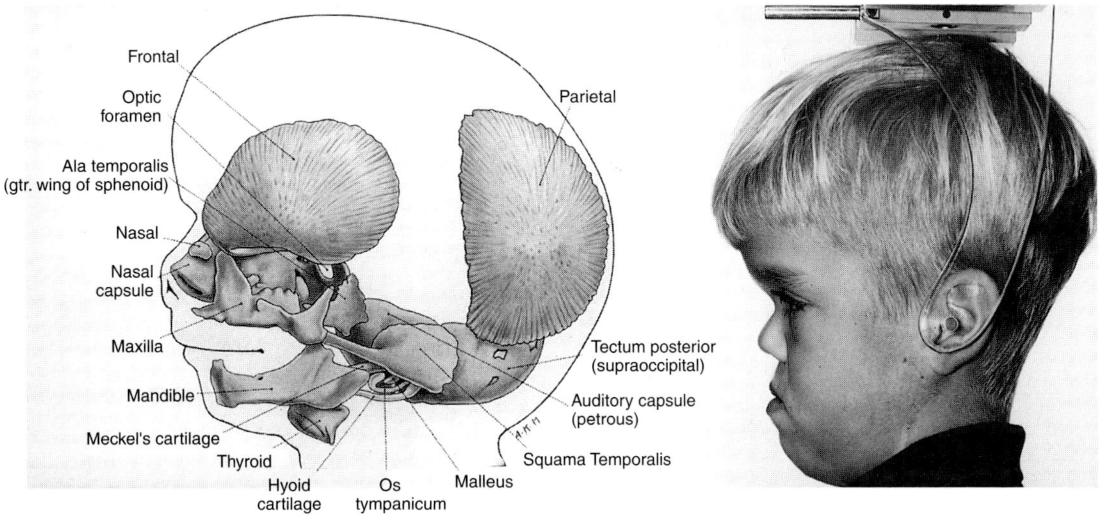

Figure 3-30 Left, Normal fetus of 80 mm. Note continuity of cartilaginous cranial base and nasal capsule together with developing membrane bones (frontal and parietal). From Hamilton et al., 1959 (58). Right, Profile in achondroplasia. Midface deficiency secondary to hypoplastic nasal capsule and cranial base. From Cohen, 2000 (16).

Table 3-11 Left–Right Asymmetry of the Brain

Findings	References
Planum temporale	Geschwind and Levitsky, 1968 (47)
Left side larger (65%)	
Right side larger (11%)	
Equal size (24%)	
Temporoparietal cortex	Galaburda et al., 1978 (40)
Larger volume left (88%)	
Larger volume right (12%)	
Medulla	Kertesz and Geschwind, 1971 (68)
Decussation of left pyramid rostral to that of right pyramid (82%)	
Decussation of right pyramid rostral to that of left pyramid (18%)	
Occipital lobe	LeMay and Clebras, 1972 (76)
Wider on left side (more common)	
Frontal lobe	LeMay and Clebras, 1972 (76)
Wider on right side (more common)	

From Cohen, 1997 (11).

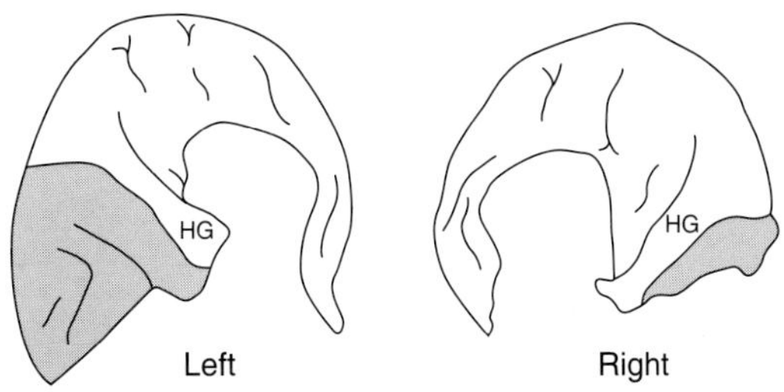

Figure 3-31 Diagram of the superior temporal surface of the brain showing the planum temporale (gray) lying caudal to Heschl's gyrus (HG). From Galaburda et al., 1978 (40).

Table 3-12 lists a small nasopharyngeal airway associated with extension of the head. A compromised airway is found with large adenoids, glossoptosis in Robin sequence, achondroplasia, Apert syndrome, and Hallermann–Streiff syndrome.

Table 3-12 Average Craniofacial Morphology Characterizing Normal but Extreme Postural Extension and Normal but Extreme Postural Flexion of the Head in Relation to the Cervical Spine

Extension of the Head	Flexure of the Head
Large anterior facial height	Small anterior facial height
Small posterior facial height	Large posterior facial height
Large inclination of the mandible	Small inclination of the mandible
Facial retrusion	Facial protrusion
Small nasopharyngeal space	Large nasopharyngeal space

Adapted from Solow and Tallgren, 1976 (96).

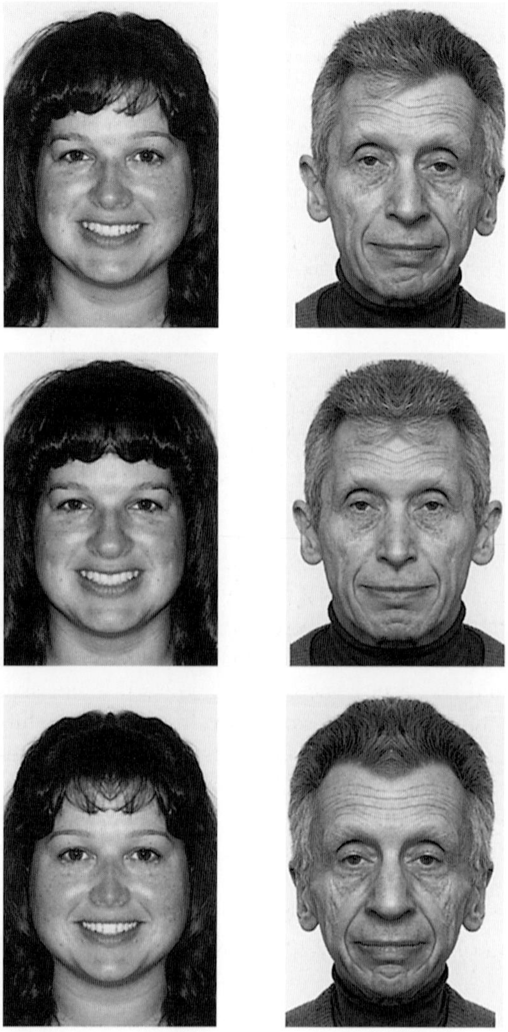

Figure 3-32 Normal facial asymmetry. Frontal photographs are divided along the median plane and reprocessed, each side being paired with its mirror image, yielding two different faces. Top, Normal faces of a woman and a man. Middle, Left side reprocessed. Bottom, Right side reprocessed. Original photographs were scanned into a computer and oriented straight up and down, with no tilting. Vertical lines were placed at the level of both inner canthi. A vertical line was then placed exactly midway between the two inner canthal lines and the photographs were reprocessed from this midline.

Table 3-13 Lowercase Letter Definitions Used in Figure 3-33

Anatomic Measurements from Cephalometrics	Lowercase Letters	Definitions
Anterior facial height	n-gn	Nasion to gnathion
Posterior facial height	s-go	Sella to gonion
Inclination of the mandible	ans-pns/ gn-go	Angle between anterior nasal spine–posterior nasal spine line and gnathion–gonion line
Facial protrusion/ retrusion	s-n-a, s-n-b	Sella–nasion–subspinale angle, Sella–nasion–supramentale angle
Pharyngeal space	/n-s-ba\ ans - pns	Area from nasion to sella to basion to posterior nasal spine to anterior nasal spine to nasion

Adapted from Solow and Tallgren, 1976 (96).

Abnormal head postures are found with various ocular alterations: (a) strabismus, (b) nystagmus, (c) uncorrected refractive errors, and (d) factors such as surgical restrictions, endocrine myopathy, and traumatic orbital edema (69).

The head is hyperextended in achondroplasia for communication with normal-sized people. Increased neck muscle use produces an exaggerated bony occipital convexity (24).

Torticollis is associated with craniofacial asymmetry, and both torticollis and scoliosis are associated with a lateral dental crossbite (61).

FACIAL AGING

Aging does not occur at a uniform rate, but is highly individualized and is subject to genetic, physiological, psychological, pathological, and environmental variation. Facial aging becomes apparent at 20–30 years of age and thereafter. Features include forehead creases, drooping of the upper eyelids, crow's feet, and nasolabial laugh lines. Circumoral striae and hollowing of the cheeks also occur. By 50 years of age and thereafter, these lines progressively deepen, more lines develop, the skin becomes thinner with a tendency to sag, and wrinkles appear on the neck. Age-related sagging and gauntness

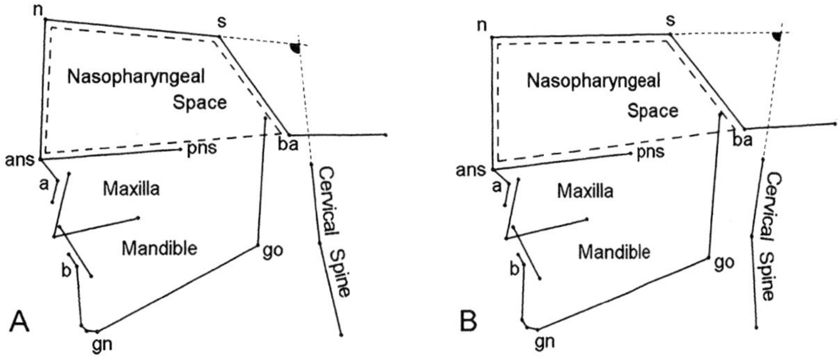

Figure 3-33 Facial diagrams based on cephalometric studies, showing mean craniofacial measurements of 10 normal subjects with the greatest head extension (A) and 10 normal subjects with the greatest head flexion (B) in relation to the cervical spine. From Solow and Tallgren, 1976 (96). See Tables 3-12, 3-13.

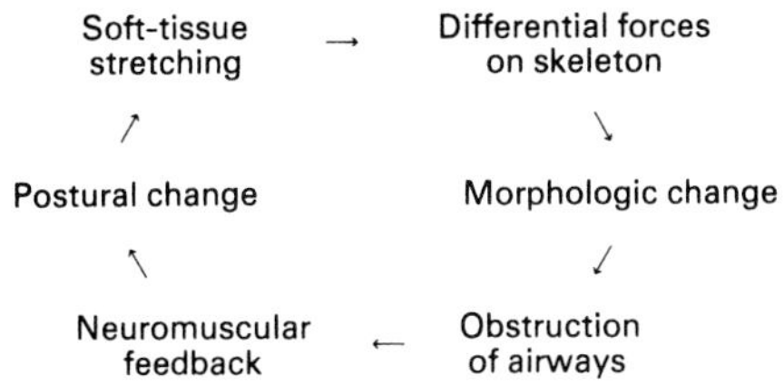

Figure 3-34 The association of craniofacial morphology and head posture involves several interrelated factors, any one of which may initiate the alterations. From Solow and Kreiborg, 1977 (95).

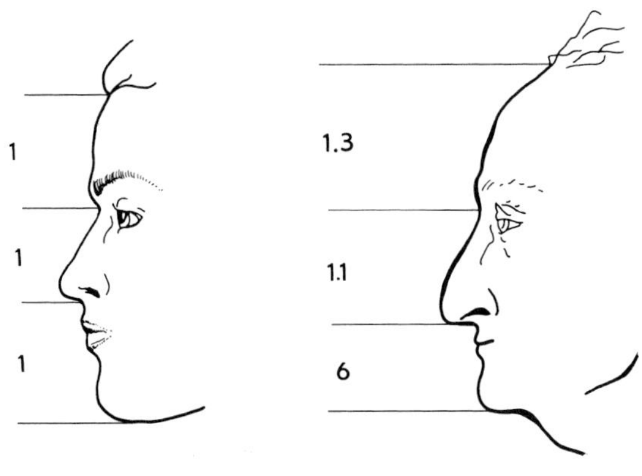

Figure 3-35 Alterations in facial proportions that accompany aging. Note the receded hairline and change in the nasal profile. From Powell and Humphreys, 1984 (85b).

tend to occur in those who are thin, whereas less wrinkling and a double chin tend to occur in those with more fat. Other changes with age include decreased muscle tone and skin elasticity; increased dryness, roughness, and courseness; and the appearance of age spots and other skin alterations. Men experience hair graying, whitening, and often balding. The eyebrows may become bushier and hair growth may occur in the nose and on the ears. Women tend to have thinner hair and may develop hair growth on the upper lip and chin during old age (100). The dentition may be lost with resorption of the alveolar ridges, recessed lower facial contour, and shortening of the lower facial height (85b) (Fig. 3-35).

REFERENCES

1. Arey LB: Developmental Anatomy. Sixth Edition. Philadelphia: WB Saunders, 1954.
2. Bachler M, Neubuser A: Expression of members of the Fgf family and their receptors during midfacial development. Mech Dev 100:313–316, 2001.
3. Biesecker LG: Strike three for *GLI3*. Nat Genet 17:259–260, 1997.
4. Boncinelli E, Morgan R: Downstream of *Otx2*, or how to get a head. Trends Genet 17:633–636, 2001.
5. Campbell B: Human Evolution. Chicago: Aldine Publishing, 1970.
6. Chambers D, McGonnell IM: Neural crest: Facing the facts of head development. Trends Genet 296:381–384, 2002.

7. Chen ZF, Behringer RR: Twist is required in head mesenchyme for cranial neural tube morphogenesis. Genes Dev 9:686–699, 1995.
8. Chi N, Epstein JA: Getting your Pax straight: Pax proteins in development and disease. Trends Genet 18:41–47, 2002.
9. Clouthier DE, Hosoda K, Richardson JA, Williams SC, Yamagisawa H, Kuwaki T, Kumada M, Hammer RE, Yanagisawa M: Cranial and cardiac neural crest defects in endothelin-A receptor-deficient mice. Development 125:813–824, 1998.
10. Cohen MM Jr: Sutural biology and the correlates of craniosynostosis. Am J Med Genet 47: 581–616, 1993.
11. Cohen MM Jr: The Child with Multiple Birth Defects. Second Edition. New York: Oxford University Press, 1997.
12. Cohen MM Jr: Craniofacial disorders caused by mutations in homeobox genes *MSX1* and *MSX2*. J Craniofac Genet Dev Biol 20:19–25, 2000.
13. Cohen MM Jr: Merging the old skeletal biology with the new. II. Molecular aspects of bone formation and bone growth. J Craniofac Genet Dev Biol 20:94–106, 2000.
14. Cohen MM Jr: *RUNX* genes, neoplasia, and cleidocranial dysplasia. Am J Med Genet 104:185–188, 2000.
15. Cohen MM Jr: Sutural biology. In Craniosynostosis, Diagnosis, Evaluation, and Management. Second Edition. M.M. Cohen, Jr. and R.E. MacLean, Editors. New York: Oxford University Press. Chapter 2, pp. 11–23, 2000.
16. Cohen MM Jr: Thanatophoric dysplasia. In Craniosynostosis: Diagnosis, Evaluation, and Management. Second Edition. M.M. Cohen, Jr. and R.E. MacLean, Editors. New York: Oxford University Press. Chapter 27, pp. 366–373, 2000.
17. Cohen MM Jr: *TWIST* and *MSX2* mutations. In Craniosynostosis: Diagnosis, Evaluation, and Management. Second Edition. M.M. Cohen, Jr. and R.E. MacLean, Editors. New York: Oxford University Press. Chapter 8, pp. 95–99, 2000.
18. Cohen MM Jr: Asymmetry: Molecular, biologic, embryopathic, and clinical perspectives. Am J Med Genet 101:292–314, 2001.
19. Cohen MM Jr: Bone morphogenetic proteins with some comments on fibrodysplasia ossificans progressiva and NOGGIN. Am J Med Genet 109:87–92, 2002.
20. Cohen MM Jr: The hedgehog signaling network. Am J Med Genet 123A:5–28, 2003.
21. Cohen MM Jr: TGFb/Smad signaling system and its pathologic correlates. Am J Med Genet 116A:1–10, 2003.
22. Cohen MM Jr: FGFs/FGFRs and associated disorders. In Inborn Errors of Development. C. Epstein, R. Erickson, and B. Wynshaw-Boris, Editors. New York: Oxford University Press. Chapter 33, pp. 380–400, 2004.
23. Cohen MM Jr, MacLean RE: Craniosynostosis: Diagnosis, Evaluation and Management, Second Edition. New York: Oxford University Press, 2000.
24. Cohen MM Jr, Walker GF, Phillips C: A morphometric analysis of the craniofacial configuration in achondroplasia. J Craniofac Genet Dev Biol 1 (Suppl): 139–165, 1985.
25. Couly G, Creuzet S, Bennaceur S, Vincent C, Le Douarin NM: Interactions between Hox-negative cephalic neural crest cells and the foregut endoderm in patterning the facial skeleton in the vertebrate head. Development 129:1061–1073, 2002.
26. Decker JD, Hall SH: Light and electron microscopy of the newborn sagittal suture. Anat Rec 212:81–89, 1985.
27. Depew MJ, Lufkin T, Rubenstein JLR: Specification of jaw subdivisions by *Dlx* genes. Science 298:381–385, 2002.
28. Dunlop L-LT, Hall BK: Relationships between cellular condensation, preosteoblast formation and epithelial–mesenchymal interactions in initiation of osteogenesis. Int J Dev Biol 39:357–371, 1995.
29. Eickholt BJ, Mackenzie SL, Graham A, Walsh FS, Doherty P: Evidence for collapsin-1 functioning in the control of neural crest migration in both trunk and hindbrain regions. Development 126:2181–2189, 1999.
30. Enlow DH: The Human Face. New York: Harper & Row, 1968.
31. Enlow DH: Essentials of Facial Growth. Fourth Edition. Philadelphia: W.B. Saunders, 1996.
32. Enlow DH: Normal craniofacial growth. In Craniosynostosis, Diagnosis, Evaluation, and Management. Second Edition. M.M. Cohen, Jr. and R.E. MacLean, Editors. New York: Oxford University Press. Chapter 4, pp. 35–47, 2000.
33. Enlow DH, Harris D: Postnatal growth of the human mandible. Am J Orthodont 50:25–50, 1964.
34. Enlow DH, Hunter WS: Growth of the face in relation to the cranial base. 44th Congress Report, European Orthodontic Society, 1968.
35. Enlow DH, Kuroda T, Lewis AB: The morphological and morphogenetic basis for cranio-facial form and pattern. Angle Orthodont 41:161–188, 1971.
36. Farlie PG, Kerr R, Thomas P, Symes T, Minichiello J, Hearn CJ, Newgreen D: A paraxial exclusion zone creates patterned cranial neural crest cell outgrowth adjacent to rhombomeres 3 and 5. Dev Biol 213:70–84, 1999.
37. Ferguson MWJ: Palate development. Development 103(Suppl):41–60, 1988.
38. Friede H: Normal development and growth of the human neurocranium and cranial base. Scand J Plast Reconstr Surg Hand Surg 15:163–169, 1981.
39. Furtwangler JA, Hall SH, Koskinen–Moffett

LK: Sutural morphogenesis in the mouse calvaria: The role of apoptosis. Acta Anat 124:74–80, 1985.
40. Galaburda AM, LeMay M, Kemper TL, Geschwind N: Right–left asymmetries in the brain. Science 199:852–856, 1978.
41. Galuske RA, Schlote W, Bratzke H, Singer W: Interhemispheric asymmetries of the modular structure in human temporal cortex. Science 289:1946–1949, 2000.
42. García–Castro MI, Marcelle C, Bronner–Fraser M: Ectodermal Wnt function as a neural crest inducer. Science 297:848–851, 2002.
43. Garn SM, Cohen MM, Gekauskas MA: Increased crown-size asymmetry in trisomy G. J Dent Res 49:465, 1970.
44. Garn SM, Lewis AB, Kerewsky RS: The meaning of bilateral asymmetry in the permanent dentition. Angle Orthodont 36:55–62, 1966.
45. Gassmann M, Casagranda F, Orioli D, Simon H, Lai C, Rüdiger K, Lemke G: Aberrant neural and cardiac development in mice lacking the ErbB4 neureglin receptor. Nature 378:390–394, 1995.
46. Gedron-Maguire M, Mallo M, Zhang M, Gridley T: *Hoxa-2* mutant mice exhibit homeotic transformation of skeletal elements derived from cranial neural creast. Cell 75:1317–1331, 1993.
47. Geschwind N, Levitsky W: Human brain: Left–right asymmetries in temporal speech region. Science 161:186–187, 1968.
48. Golding JP, Trainor P, Krumlauf R, Gassmann M: Defects in pathfinding by cranial neural crest cells in mice lacking the neuregulin receptor ErbB4. Nat Cell Biol 2:103–109, 2000.
49. Gorlin RJ, Cohen MM Jr: The orofacial region. In Textbook of Fetal and Perinatal Pathology. Second Edition. J.S. Wigglesworth and D.B. Singer, Editors. Malden: Blackwell Science. Chapter 22, pp. 732–778, 1988.
50. Gorlin RJ, Cohen MM Jr, Hennekam RCM: Syndromes of the Head and Neck. Fourth Edition. New York: Oxford University Press, 2001.
51. Graham A: Development of the pharyngeal arches. Am J Med Genet 119A:251–256, 2003.
52. Graham A, Francis–West P, Brickell P, Lumsden A: The signalling molecule BMP4 mediates apoptosis in the rhombencephalic neural crest. Nature 372:684–686, 1994.
53. Grammatopoulos GA, Bell E, Toole L, Lumsden A, Tucker AS: Homeotic transformation of branchial arch identity after Hoxa-2 overexpression. Development 127:5355–5365, 2000.
53a. Hall BK: Bone, embryonic development. In Encyclopedia of Human Biology. Second Edition. New York: Academic Press. Volume 2, pp. 105–114, 1997.
53b. Hall BK: Bones and Cartilage: Development and Evolutionary Skeletal Biology. San Diego: Elsevier, 2005.
54. Hall BK, Miyake T: The membranous skeleton: The role of cell condensations in vertebrate skeletogenesis. Anat Embryol 186:107–124, 1992.
55. Hall BK, Miyake T: Divide, accumulate, differentiate: Cell condensation in skeletal development revisited. Int J Dev Biol 39:881–893, 1995.
56. Hall BK, Miyake T: All for one and one for all: Condensations and the initiation of skeletal development. BioEssays 22:138–147, 2000.
57. Hall SH, Decker JD: Calvarial suture morphogenesis: Cellular and molecular aspects. In Scientific Foundations and Surgical Treatment of Craniosynostosis. J.A. Persing, M.T. Edgerton, and J.A. Jane, Editors. Baltimore: Williams & Wilkins. Chapter 4, pp. 27–37, 1989.
58. Hamilton WJ, Boyd JD, Mossman HW: Human Embryology. Cambridge: Heffer & Sons, p. 402, 1959.
59. Helms JA, Schneider RA: Cranial skeletal biology. Nature 423:326–331, 2003.
60. Hu D, Marcucio RS, Helms JA: A zone of frontonasal ectoderm regulates patterning and growth in the face. Development 130:1749–1758, 2003.
61. Huggare J: Postural disorders and dentofacial morphology. Acta Odontol Scand 56:383–386, 1998.
62. Ingham PW, McMahon AP: Hedgehog signaling in animal development: Paradigms and principles. Genes Dev 15:3059–3087, 2001.
63. Jernvall J, Thesleff I: Reiterative signaling and patterning during mammalian tooth morphogenesis. Mech Dev 92:19–29, 2000.
63a. Jiang X, Iseki S, Maxson RE, Sucov HM, Morriss–Kay GM: Tissue origins and interactions in the mammalian skull vault. Dev Biol 241:106–116, 2002.
64. Jirásek JE: Atlas of Human Prenatal Morphogenesis. Boston: Martinus Nijhoff, 1983.
65. Johansen V, Hall SH: Morphogenesis of the mouse coronal suture. Acta Anat 114:58–67, 1982.
66. Johnston MC, Sulik KK: Embryology of the head and neck. In Pediatric Plastic Surgery. D. Serafin and N.G. Georiade, Editors. St. Louis: CV Mosby, pp. 184–215, 1984.
67. Kang S, Graham JM Jr, Olney AH, Biesecker LG: *GLI3* frameshift mutations cause autosomal dominant Pallister–Hall syndrome. Nat Genet 15:266–268, 1997.
68. Kertesz A, Geschwind N: Patterns of pyramidal decussation and their relationahip to handedness. Arch Neurol 24:326–332, 1971.
69. Khawam E, El Baba F, Kaba F: Abnormal ocular head postures: Part I. Ann Ophthalmol 19:347–353, 1987.
69a. Klier EM, Wang H, Constantin AG, Crawford JD: Midbrain control of three-dimensional head orientation. Science 295:1314–1316, 2002.
70. Kokich VG: Age changes in the human fronto-

zygomatic suture. Am J Orthodont 69:411–430, 1976.
71. Koskinen L: Adaptive sutures: Changes after unilateral masticatory muscle resection in rats: A microscopic study. Proc Finn Dent Soc 73(Suppl 7):1–79, 1977.
72. Koskinen–Moffett LK, Moffett BC Jr, Graham JM Jr: Cranial synostosis and intra-uterine compression: A developmental study of human sutures. In Factors and Mechanisms Influencing Bone Growth. A.D. Dixon and B.G. Sarnat, Editors. New York: Alan R. Liss, pp. 365–378, 1982.
73. Langman J: Medical Embryology. Baltimore: Williams & Wilkins, 1963.
74. Larsen WJ: Human Embryology. New York: Churchill Livingstone, 1993.
75. Lehmann OJ, Sowden JC, Carlsson P, Jordan T, Bhattacharya SS: Fox's in development and disease. Trends Genet 19:339–344, 2003.
76. LeMay M, Clebras A: Human brain: Morphologic differences in the hemispheres demonstrable by carotid arteriography. N Engl J Med 287: 168–171, 1972.
77. Mason I, Chambers D, Shamin H, Walshe J, Irving C: Regulation and function of FGF8 in patterning of midbrain and anterior hindbrain. Biochem Cell Biol 78:577–584, 2000.
78. McPherson GK, Kriewall TJ: Fetal head molding: An investigation utilizing a finite element model of the fetal parietal bone. J Biomech 13: 17–26, 1980.
79. Miroue M, Rosenberg L: The human facial sutures: A morphologic and histologic study of age changes from 20 to 95 years. MSD Thesis, University of Washington, 1975.
80. Moffett BC Jr: Syllabus for course in fundamentals of skeletal biology related to clinical dentistry. Unpublished manuscript, University of Washington, 1973.
81. Nakatsu T, Shiota K: Neurulation in the human embryo revisited. Congenital Anomalies 40:93–98, 2000.
82. Nakatsu T, Uwabe C, Shiota K: Neural tube closure in humans initiates at multiple sites: Evidence from human embryos and implications for the pathogenesis of neural tube defects. Anat Embryol 201:455–466, 2000.
82a. OMIM: On-Line Mendelian Inheritance in Man, 2004.
83. O'Rourke MP, Tam PPL: *Twist* functions in mouse development. Int J Dev Biol 46:401–413, 2002.
84. Pasqualetti M, Ori M, Nardi I, Rijli FM: Ectopic Hoxa2 induction after neural crest migration results in homeosis of jaw elements in *Xenopus*. Development 127:5367–5378, 2000.
85. Peck H, Peck S: A concept of facial esthetics. Angle Orthodont 40: 284–318, 1970.
85a. Pennisi E: Bonemaking protein shapes beaks of Darwin's finches. Science 305:1383, 2004.
85b. Powell N, Humphreys B: Proportions of the Aesthetic Face. New York: Thieme-Stratton, 1984.
86. Pritchard JJ, Scott JH, Girgis FG: The structure and development of cranial and facial sutures. J Anat 90:73–86, 1956.
87. Pruzansky S: Clinical investigation of the experiments of nature. ASHA Report 8:62–94, 1973.
88. Richman JM, Lee S-H: About face: Signals and genes controlling jaw patterning and identity in vertebrates. BioEssays 25:544–568, 2003.
89. Rijli FM, Mark M, Lakkaraju S, Dierich A, Dolle P, Chambon P: A homeotic transformation is generated in the rostral branchial region of the head by disruption of Hoxa-2, which acts as a selector gene. Cell 75:1333–1349, 1993.
90. Scammon R: Section 1 in Human Anatomy, by H. Morris. Eleventh Edition. Philadelphia: Blakiston, 1953.
91. Schneider RA, Helms JA: The cellular and molecular origins of beak morphology. Science 299: 565–568, 2003.
92. Schneider RA, Hu D, Helms JA: From head to toe: Conservation of molecular signals regulating limb and craniofacial morphogenesis. Cell Tissue Res 296:103–109, 1999.
93. Simeone A: Positioning the isthmic organizer where *Otx2* and *Gbx2* meet. Trends Genet 16: 237–240, 2000.
94. Smith A, Graham A: Patterning of the pharyngeal arches. BioEssays 23:54–61, 2001.
95. Solow B, Kreiborg S: Soft-tissue stretching: A possible control factor in craniofacial morphogenesis. Scand J Dent Res 85:505–507, 1977.
96. Solow B, Tallgren A: Head posture and craniofacial morphology. Am J Phys Anthrop 44:417–436, 1976.
97. Sulik KK: Craniofacial embryology and dysmorphogenesis. In Holoprosencephaly: An Overview and Atlas of Cases. J.R. Siebert, M.M. Cohen, Jr., K.K. Sulik, C.-M. Shaw, R.J. Lemire, Editors. New York: Wiley-Liss. pp. 59–98, 1990.
98. Sulik KK, Cotanche DA: Embryology of the ear. In Hereditary Hearing Loss and Its Syndromes. R.J. Gorlin, H.V. Toriello, and M.M. Cohen Jr, Editors. New York: Oxford University Press. pp. 22–42, 1995.
99. Sulik KK, Schoenwolf GC: Highlights of craniofacial morphogenesis in mammalian embryos as revealed by scanning electron microscopy. Scanning Electron Microsc 4:1735–1752, 1985.
100. Taister MA, Holliday SD, Borrman HIM: Comments on facial aging in law enforcement investigation. Forensic Sci Comm 2:1–10, 2000.
101. Trainor PA, Ariza–McNaughton L, Krumlauf R: Role of the isthmus and FGFs in resolving the paradox of neural crest plasticity and prepatterning. Science 295:1288–1291, 2002.
102. Trainor PA, Krumlauf R: Hox genes, neural crest

cells and branchial arch patterning. Curr Opin Cell Biol 13:698–705, 2001.

103. Trainor PA, Krumlauf R: Patterning the cranial neural crest: Hindbrain segmentation and Hox gene plasticity. Nat Rev Neurosci 1:116–124, 2000.

104. Veitch E, Begbie J, Schilling TG, Smith MM, Graham A: Pharyngeal arch patterning in the absence of neural crest. Curr Biol 9:1481–1484, 1999.

105. Verwoerd DCA, Van Oostrom CG: Cephalic neural crest and placodes. Adv Anat Embryol Exp Morphol 58:1–75, 1978.

106. Warren SM, Brunet LJ, Harland RM, Economides AN, Longaker MT: The BMP antagonist noggin regulates cranial suture fusion. Nature 422:625–629, 2003.

107. Wild A, Kalft–Suske M, Vortkamp A: Point mutations in human *GLI3* cause Greig syndrome. Hum Mol Genet 6:1979–1984, 1997.

108. Wilke TA, Gubbels S, Schwartz J, Richman JM: Expression of fibroblast growth factor receptors (FGFR1, FGFR2, FGFR3) in the developing head and face. Dev Dyn 210:41–52, 1997.

109. Wilkie AOM, Patey SJ, Kan S-H, van den Ouweland AMW, Hamel BCJ: FGFs, their receptors, and human limb malformations: Clinical and molecular correlations. Am J Med Genet 112: 266–278, 2002.

4

Anatomical and Physiological Perspectives

THE SKULL

The skull (Fig. 4-1) is composed of bones, teeth, and craniofacial articulations[a] (Table 4-1). The 8 cranial[b] and 14 facial bones of the adult skull are flattened or irregular and immovably joined together, except for the mandible. The ossicular bones, 6 in number, are also movable. During late intrauterine life, the cranial sutures are patent and movable, which molds the skull, permitting the head to pass through the birth canal.[c] The articulation of the mandible is unique because of its ginglymo-arthrodial joint capable of both translational and rotational movement (41, 74, 82).

Growth of the skull is complex. Brain growth can be simply assessed by measuring the head circumference. By 2 years of age, 80% of brain growth is completed, and 90% of adult size is attained by 4–5 years of age. The rate of cranial growth exceeds the rest of the skull for the first 6 months, after which growth of the face takes over (69a).

Vectors of maxillary and mandibular growth are shown in Chapter 3 (see Figs. 3-27, 3-28). The ratio of cranial volume to facial volume changes during growth: birth (8:1), age 2 years (5:1), age 6 years (3:1), age 12 years (2.5:1), and age 18 (2:1). The face grows downward and forward (see Fig. 3-29). Because the mandible is farther from its adult dimensions than the maxilla, it grows at a faster rate. Maxillary growth is completed on the average at age 14 in females and at age 15 in males. Mandibular growth is completed on the average at age 18 in females and at age 20 or 21 in males (69a). In male Atlantic salmon, rapid and pronounced growth of the lower jaws occurs *during adult life* when the *starving* salmon migrate upriver for spawning (107).

The facial profile may be orthognathic, retrognathic, or prognathic. With an orthognathic face, the maxilla and mandible produce a straight profile. The retrognathic profile is convex-appearing with a retrusive mandible. The prognathic profile is concave-appearing with a protrusive mandible (33a).

Head shape can be mesocephalic, and at the extremes, brachycephalic or dolichocephalic. With brachycephaly, the cranial base is more upright, resulting in more posterior placement of the maxilla and a more forward placement of the mandible, resulting in a tendency towards prognathism. With dolichocephaly, the maxilla has more protrusive placement because of forward basicranial rotation. The mandible has backward rotation, resulting in a tendency towards retrognathia (33a).

[a]For details about craniofacial articulations, see Chapter 3.
[b]Not included are Wormian bones.
[c]For details about skull molding, see Chapter 3.

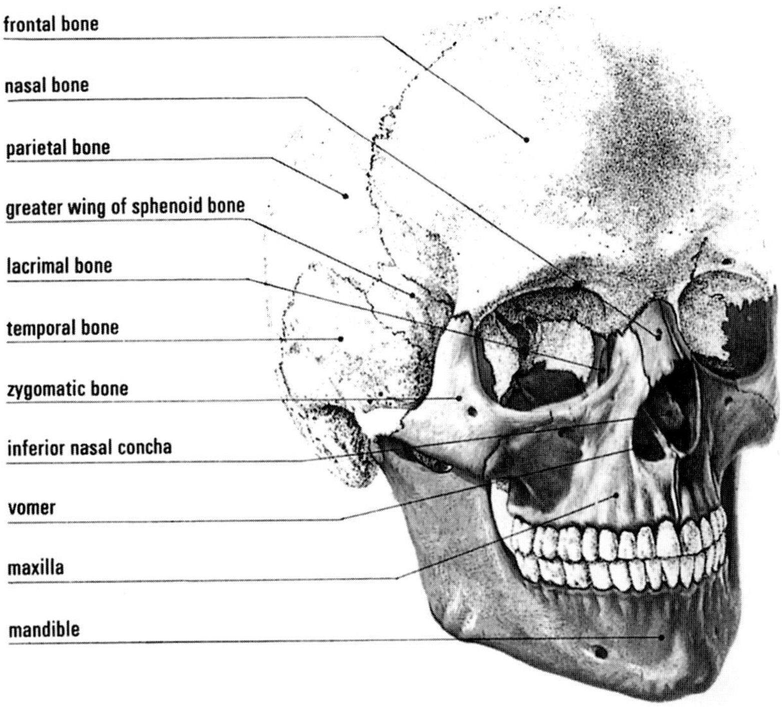

Figure 4-1 The skull. From Vannini and Pogliani, 1980 (101).

THE BRAIN

Cerebral Cortex

Divisions of the central nervous system are shown in Table 4-2. The brain consists of three main regions: the forebrain, midbrain, and hindbrain. Within the forebrain are the diencephalon, which includes the thalamus, hypothalamus, the limbic system, and other associated structures; and the telencephalon, which includes the cerebrum. The diencephalon deals with behavior and emotions, but also with hormonal control and basic drives, such as appetite and sleep. The telencephalon performs sensory functions, motor functions, and functions associated with various mental activities. In the midbrain are the superior colliculi, which receive and integrate visual information, and the inferior colliculi, which receive and integrate auditory information. The hindbrain consists of the cerebellum, which functions in the coordination of movement; the medulla oblongata, which functions in homeostasis; and the pons, which functions in signal conduction (41, 82, 105).

In a lateral view of the cerebral cortex (Fig. 4-2), the frontal, temporal, parietal, and occipital lobes can be identified, in addition to the primary visual, auditory, and olfactory areas, as well as Broca's area, Wernicke's area, and the angular gyrus (38).

Primary Motor and Somatic Sensory Areas

An arch that extends across the roof of the brain is the primary motor cortex, which exercises voluntary control over the muscles. Just behind the primary motor cortex is the primary somatic sensory area, where signals are received from the skin, bones, joints, and muscles. Most of the body can be mapped onto the cerebral cortex, producing two distorted homunculi (Fig. 4-3). The distortions result from the precision required to control various parts of the body. The face and the hands, which require great precision, are the most exaggerated. Only half of each cortical region is shown in Figure 4-3: the left somatic sensory arch, which receives sensations from the right half of the body, and the right motor cortex, which controls the left half of the body (38).

Table 4-1 The Skull

Bones, Teeth, and Craniofacial Articulations	
Bones	28[a]
Cranial	8
Frontal	1
Parietal	2
Temporal	2
Occipital	1
Sphenoid	1
Ethmoid	1
Facial	14
Nasal	2
Maxillary	2
Zygomatic	2
Mandibular	1
Lacrimal	2
Palatine	2
Inferior conchae	2
Vomer	1
Ossicular	6
Mallear	2
Incudal	2
Stapedial	2
Teeth	20, 32
Deciduous	20
Incisors	8
Canines	4
Molars	8
Permanent	32
Incisors	8
Canines	4
Premolars	8
Molars	12
Craniofacial articulations	209[b]
Synovial	8
Temporomandibular	2
Atlanto-occipital	2
Audio-ossicular	4
Cartilaginous	10
Synchondral	7
Condylar	2
Nasal septal	1
Fibrous	109
Craniosutural	18[c]
Faciosutural	35[c]
Audioligamentous	4
Periodontal (deciduous)	20
Periodontal (permanent)	32
Dental	82
Occlusal (deciduous)	14
Occlusal (permanent)	32
Interproximal (deciduous)	8
Interproximal (permanent)	30

[a]The 28 bones in the skull do not include Wormian bones found in some skulls.
[b]For discussion of craniofacial articulations, see Chapter 3, including Table 3-10.
[c]For differences between cranial sutures and facial sutures, see Chapter 3.

Data from Goss, 1961 (41); Moffett, 1973 (74); and Patton and Thibodeau, 2000 (82).

Table 4-2 Divisions of the Central Nervous System

Embryonic Divisions	Adult Derivatives	Cavities
Brain		
Prosencephalon		
Telencephalon	Cerebral hemispheres	Lateral ventricles
Diencephalon	Thalamus, hypothalamus	Third ventricle
Mesencephalon	Midbrain	Cerebral aqueduct
Rhombencephalon		
Metencephalon	Pons, cerebellum	Fourth ventricle
Myelencephalon	Medulla oblongata	
Spinal cord	Spinal cord	Central canal

Modified from Watson, 1995 (105).

Hemispheric Specialization

The left hemisphere generally takes precedence over the right for speech and language function, whereas the right hemisphere primarily processes tonal stimuli and music, as well as visual–spatial integration and other nonverbal functions. The left and right auditory regions are specialized for processing sounds based on acoustic properties. Rapidly changing signals are preferentially processed in the auditory areas of the left hemisphere because of enhanced temporal resolution, whereas tonal stimuli are preferentially processed in the right auditory areas of the right hemisphere because of specialized abilities for spectral resolution. Because of crossed pathways from the ear to the auditory cortex, the reaction time is faster and stimulus identification is more accurate when the right ear is presented with speech-type stimuli or when the left ear is presented with tonal stimuli (56, 93, 108).

Broca's area and Wernicke's area (Fig. 4-2) are involved in the production and comprehension of language; the angular gyrus mediates between the visual and auditory centers of the brain. Linguistic specializations are left-sided and are not found in the corresponding areas of the right hemisphere (38). Infants have right mouth asymmetry when they babble, indicating left hemisphere control. Thus, infant babbling is not random motor activity, but rather prelanguage (47).

When disgust is experienced, right-sided activation of the frontal and temporal regions occurs (27). However, impaired ability to *recognize*

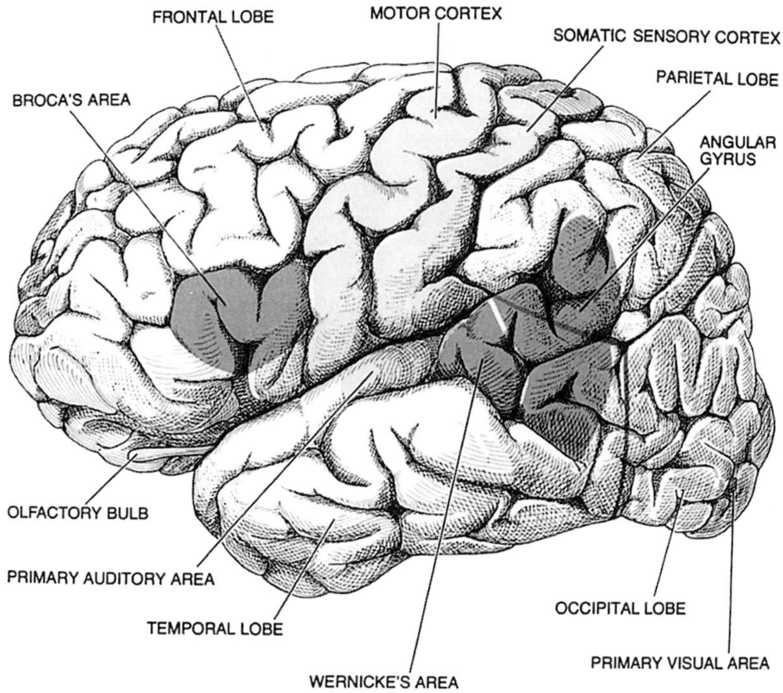

Figure 4-2 Cerebral cortex. From Geschwind, 1979 (38).

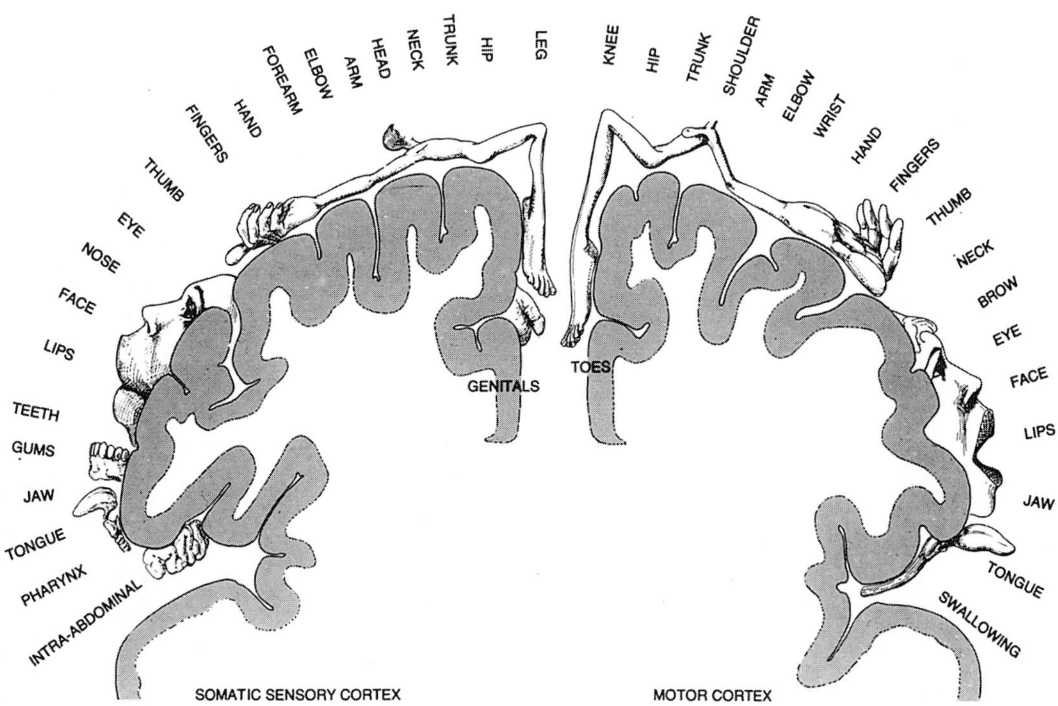

Figure 4-3 Somatic sensory and motor regions of the cerebral cortex. From Geschwind, 1979 (38).

facial disgust occurs with *left-sided* damage to the insula and putamen, but recognition of other kinds of facial expression is unaffected (15). Also, when happiness is experienced, *left-sided* activation of the anterior temporal region occurs (27).

Ventromedial Area

An area that governs facial recognition is found on the ventromedial surface of the temporal and occipital lobes on both sides of the cerebral cortex (see Figure 1-17). Any lesion that destroys this area impairs the ability to identity faces, but has almost no other effects, although dimming of vision may occur. The condition known as prosopagnosia (38) is discussed further in Chapter 1.

LANGUAGE AND SPEECH

Verbal Utterances and Speech

Verbal utterances arise in Wernicke's area and are transferred through the arcuate fasciculus to Broca's area, where it evokes a coordinated program for vocalization. The program is transferred to the adjacent facial area of the motor cortex, activating the appropriate muscles of the mouth, lips, tongue, soft palate, and larynx. Wernicke's area also plays a role in the comprehension of speaking, reading, and writing. Word sounds received in the primary auditory area pass through Wernicke's area to be understood as a verbal message. A written message is received in the primary visual area and transmitted to the angular gyrus, which transforms the auditory form of the word in Wernicke's area. In writing words in response to oral instructions, the information is passed along the same pathways in the opposite direction (38).

A lesion in Broca's area affects speech but has a smaller effect on comprehension. In Broca's aphasia, speech is labored, slow, and articulation-impaired. Difficulty with verbal inflections, pronouns, connective words, and grammatical constructions result in a telegraphic style of speech. For example, if a patient is asked about dental appointments for his family, he might say:

> Yes . . . Monday . . . Dad and Dick . . . Wednesday nine o'clock . . . 10 o'clock . . . doctors . . . and teeth (38)

In Wernicke's aphasia, speech is phonetically and grammatically normal, but semantically deviant and, in some instances, can be grammatically distorted. Words tend to be inappropriate and may include nonsense words or syllables. A patient asked to describe a picture showing two boys stealing cookies behind a woman's back might say:

> Mother is away here working her work to get her better, but when she's looking the two boys looking in the other part. She's working another time (38)

Genetic Factors

The origin of speech and the faculty of language are controversial (6, 44, 46). Some recent genetic trends are noteworthy, although evidence is far from conclusive. Specific language impairment is common, with a prevalence of about 6% to 7% in the general population (65, 100). Concordance is nearly 100% in monozygotic twins and 50% to 70% in dizygotic twins, suggesting a strong genetic basis (5, 25).

A large family with a severe speech and language disorder was found to have a mutation in exon 14 of *FOXP2*, which maps to 7q31. The transcripts of the gene are expressed in the developing brain (35, 60, 61, 92). However, a large linkage study of children with specific language impairment has not reported a mutation in exon 14 (80), but *FOXP2* is a large gene with many exons, making gene sequencing impractical (12); strong association was found to other genes adjacent to *FOXP2*. Some studies have implicated 7q31 in autism (3, 103). *WNT2* has been implicated in a subset of families with autism and more severe language impairment (104).

Tourette syndrome is associated with familial aggregation; parent-offspring inheritance has been reported. Twin studies show 89–94% concordance for the disorder. Several genes have been implicated, such as *IMMP2L*, *OX2R*, and *SLITRK1*. Chromosomal mapping has included 3p14-p21, 4q34-q35, 5q35.2-q35.3, 7q31, 11q23, and 18q22 (1, 7, 81, 83, 99).

Voice and Sounds

Air pressure is raised in the lungs, which is driven through the space between the vocal cords and passages above the glottis, vibrating the vocal cords. Voice adjustments are made by the joints and muscles of the larynx. Motor signals arise in the motosensory cortex and are conducted via the vagus nerve to the larynx. Sensory fibers of the mucosa and muscle spindles of the larynx transmit the position and tension of the vocal cords to the central nervous system. The close connections of the auditory pathways with bulbar and cortical motor speech centers are important for fine voice adjustment (31, 41).

In addition to vocal sounds, there are two types of voiceless sound consonants: fricative and plosive. Fricative sounds, such as s, sh, f, and th, are produced when the vocal tract is partially closed. Plosive sounds such as p, t, and k, are produced when the vocal tract is completely closed (9).

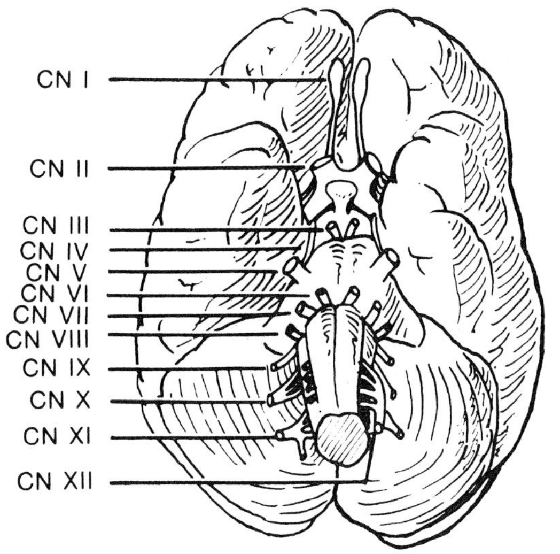

Figure 4-4 Cranial nerves, base of the brain. From Liebgott, 1986 (67).

CRANIAL NERVES

Table 4-3 summarizes the 12 cranial nerves together with their functional components, their innervated structures, their locations at the base of the brain, and their skull foramina exits. The cranial nerves are shown in a diagrammatic inferior view of the brain (Fig. 4-4).

The olfactory tracts arise from the olfactory trigone to become the olfactory bulbs (I). The axons of the ganglion cells of both retinas, which make up the optic nerve (II), meet at the optic chiasm and with partial crossing over enter the optic tracts and run to the lateral geniculate body of the thalamus. The oculomotor nerve (III) arises from the interpeduncular fossa of the midbrain, travels forward, pierces the dura of the triangular fold, enters the lateral wall of the cavernous sinus, and exits through the superior orbital fissure. The trochlear nerve (IV) arises from the midbrain just below the inferior colliculi, encircles the brainstem, passes anteriorly, and then passes through the lateral wall of the cavernous sinus, exiting through the superior orbital fissure (53, 67, 105).

The trigeminal nerve (V) originates from the pons, travels anteriorly over the petrous ridge, and pierces the dura, producing the trigeminal dural cave; it becomes the trigeminal ganglion and gives rise to the ophthalmic, maxillary, and mandibular divisions. The abducens nerve (VI) arises from the pons near the midline just above the junction with the medulla, ascends on the clivus, pierces the dura, runs below it, enters the cavernous sinus, and exits through the superior orbital fissure. The facial nerve (VII) arises from the pons–medulla junction, passes laterally, pierces the dura, passes through the internal acoustic meatus, enters the petrous portion of the temporal bone, traverses the facial canal, and exits through the styloid foramen. The vestibulocochlear nerve (VIII) arises from the pons–medulla junction, runs to the posterior part of the petrous bone to the internal acoustic meatus, and divides into a cochlear portion for hearing and a vestibular portion for the semicircular canals. The glossopharyngeal nerve (IX) arises from the medulla just posterior to the olive and exits through the jugular foramen. The vagus nerve (X) also arises from the medulla just posterior to the olive and exits through the jugular foramen. The accessory nerve (XI) originates from a cranial portion (the medulla just posterior to the olive) and a spinal part (C_1–C_5). The spinal portion forms a single trunk, which enters the foramen magnum and joins the intermediate nerve, exiting through the jugular foramen. The

Table 4-3 Cranial Nerves

Cranial Nerves	Functional Components	Innervated Structures	Locations at Base of Brain	Skull Foramina Exits
Olfactory (I)	Special visceral afferent	Olfactory epithelium	Rhinencephalon	Cribriform plate, ethmoid
Optic (II)	Special somatic afferent	Rods and cones of the retina	Diencephalon, laterally	Optic foramen, sphenoid
Oculomotor (III)	Somatic efferent	Superior, medial, and inferior rectus muscles, inferior oblique muscle, levator palpebrae superioris muscle	Midbrain, periaqueductal, interpeduncular fossa	Superior orbital fissure, sphenoid
	General visceral efferent	Pupillary sphincter and ciliary muscles		
Trochlear (IV)	Somatic efferent	Superior oblique muscle	Midbrain, just below inferior colliculi	Superior orbital fissure, sphenoid
Trigeminal (V)	General somatic afferent	Facial skin, oral and nasal mucosae	Pons, laterally	Ophthalmic division: superior orbital fissure, sphenoid Maxillary division: foramen rotundum, sphenoid Mandibular division: foramen ovale, sphenoid
	Special visceral efferent	Muscles of mastication, tensor tympani, tensor veli palatini, myelohyoid muscle, anterior belly of the digastric muscle		
	General somatic afferent	Proprioception for muscles above		
Abducens (VI)	Somatic efferent	Lateral rectus muscle	Pons, near midline just above junction with medulla	Superior orbital fissure, sphenoid
Facial (VII)	Special visceral efferent	Muscles of facial expression, stapedius muscle, stylohyoid muscle, posterior belly of the digastric muscle	Pons–medulla junction, laterally	Both parts of VII, motor fibers for the muscles of facial expression and taste and secretory fibers of the intermediate nerve, pass through the internal acoustic meatus and enter the facial canal, giving off the greater petrosal nerve, which passes through the petrotympanic fissure, the stapedius nerve for the stapedius muscle in the middle ear, the chorda tympani, which branches off above the stylohyoid foramen, and the facial nerve proper, which leaves the skull through the stylohyoid foramen
	General visceral efferent	Lacrimal gland, submandibular, sublingual, and minor salivary glands, mucosae		
	Special visceral afferent	Taste buds of anterior two-thirds of tongue		
	General visceral afferent	Middle ear		
	General somatic afferent	External ear		
Vestibulocochlear (VIII)	Special somatic afferent	Organ of Corti	Pons–medulla junction, laterally	Internal acoustic meatus (does not leave skull)
	Special somatic afferent	Semicircular canals, utricle, saccule		

(continued)

Table 4-3 (continued)

Cranial Nerves	Functional Components	Innervated Structures	Locations at Base of Brain	Skull Foramina Exits
Glossopharyngeal (IX)	Special visceral afferent	Taste buds of posterior third of tongue	Medulla, laterally, just posterior to olive	Jugular foramen, temporal and occipital
	General visceral efferent	Parotid gland		
	General visceral afferent	Pharynx, carotid sinus, posterior third of tongue, auditory tube, middle ear		
	Special visceral efferent	Stylopharyngeus muscle		
	General somatic afferent	External ear		
Vagus nerve (X)	General visceral efferent	Smooth and cardiac muscles, glands of thoracic and abdominal organs through transverse colon	Medulla, laterally, just posterior to olive	Jugular foramen, temporal and occipital
	General visceral afferent	Carotid and aortic bodies, all structures listed for general visceral efferent above		
	Special visceral efferent	Muscles of soft palate, pharynx, larynx, and esophagus		
	Special visceral afferent	Epiglottis		
	General somatic afferent	External ear		
Accessory (XI)	Special visceral efferent	Muscles of larynx and pharynx with X playing the major role	Cranial part: medulla, laterally, just posterior to olive	Jugular foramen, temporal and occipital
			Spinal part: C_1–C_5, spinal cord, laterally	Enters through foramen magnum
	Special visceral efferent or somatic efferent	Sternocleidomastoid and trapezius muscles		
Hypoglossal (XII)	Somatic efferent	Extrinsic and intrinsic muscles of the tongue	Medulla, near midline between olive and pyramid	Hypoglossal canal, occipital

Modified from Watson, 1995 (105). Data from Kahle, 1986 (53) and Patton and Thibodeau, 2000 (82).

hypoglossal nerve (XII) arises near the midline of the medulla between the olive and the pyramid and exits through the hypoglossal canal (53, 67, 105).

AUTONOMIC NERVOUS SYSTEM

The autonomic nervous system, which is concerned mainly with circulation and internal organs, is composed of sympathetic and parasympathetic divisions. Their main differences are in connections to the central nervous system, location of ganglia, and types of neurotransmitters (31, 41).

The sympathetic system is connected to the central nervous system through the thoracic and upper lumbar segments of the spinal cord; its ganglia tend to be located near the spinal column. The parasympathetic system is connected to the central nervous system through cranial nerves III, VII, IX, and X, and the middle three sacral segments of the spinal cord; its ganglia tend to be placed peripherally near the organs innervated. Although both systems innervate many of the same organs, they are usually antagonistic to each other physiologically (41).

In the sympathetic system, transmission from preganglionic to postganglionic neurons is cholinergic, whereas stimulation of all effector organs except sweat glands is adrenergic. Synaptic transmissions in parasympathetic ganglia and at effector organs are cholinergic (31).

Parasympathetic System

Cranial outflow from the midbrain includes fibers in the oculomotor nerve (III) and from the medulla includes fibers in the facial (VII), glossopharyngeal (IX), and vagus (X) nerves (Fig. 4-5). Preganglionic fibers from III synapse at the ciliary ganglion, where postganglionic fibers activate the pupillary sphincter muscle and contract the ciliary muscle. Preganglionic fibers from VII synapse at the pterygopalatine ganglion, where postganglionic fibers activate the lacrimal glands and mucosal glands of the nasal cavity, oral cavity, and pharynx. Preganglionic fibers from VII also synapse at the submandibular ganglion, where postganglionic fibers activate the submandibular and sublingual salivary glands. Preganglionic fibers from IX synapse at the otic ganglion, where postganglionic fibers activate the parotid gland and oral mucosal glands (41) (Fig. 4-5).

Sympathetic System

In the sympathetic system, preganglionic neurons originate in the hypothalamus with fibers descending through the brainstem and thoracolumbar segments of the spinal cord, exiting through the ventral roots of the spinal cord, and terminating in various ganglia where synapses

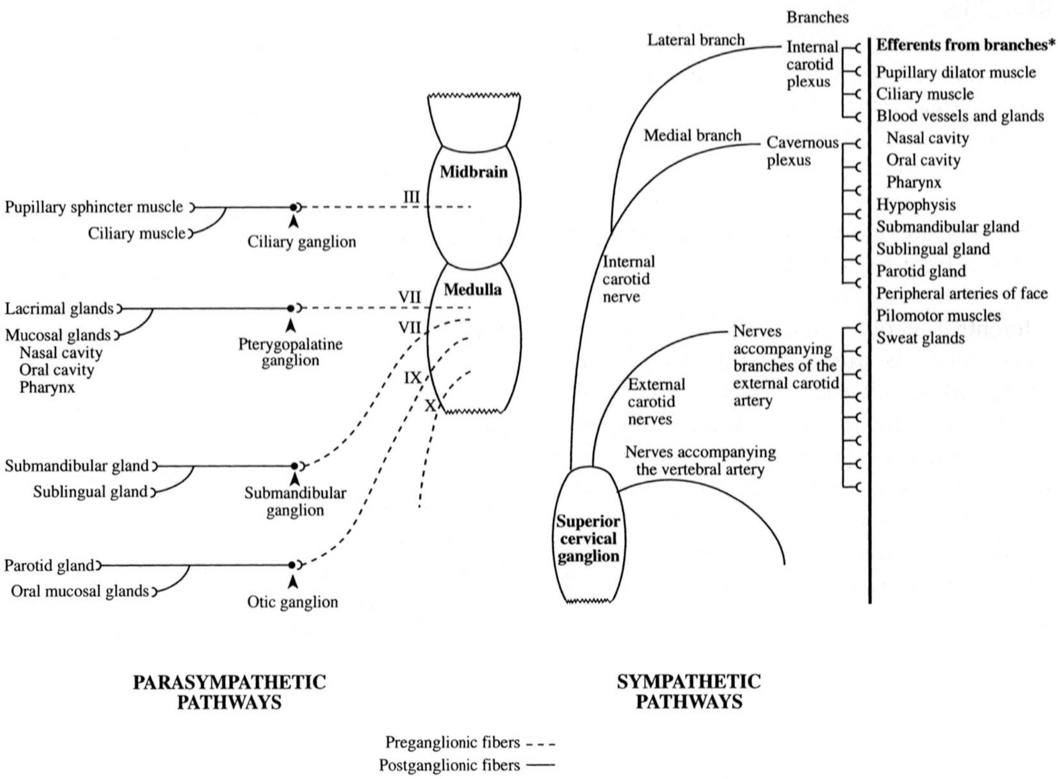

Figure 4-5 Schematic representation of sympathetic and parasympathetic pathways affecting the facial area. Asterisk indicates that efferents from branches are not necessarily in order. See text.

result in postganglionic fibers. The superior cervical ganglion, the cephalic portion of the sympathetic system, gives rise to (a) the internal carotid nerve, which accompanies the internal carotid artery, (b) external carotid nerves, whose branches accompany various branches of the external carotid artery, and (c) nerves that accompany the vertebral artery (Fig. 4-5). The internal carotid nerve has two branches: a lateral branch forming the internal carotid plexus and a medial branch forming the cavernous plexus. These two branches have a number of further communicating branches (41).

Efferents from all the branches of the superior cervical ganglion supply the pupillary dilator muscle; ciliary muscle; blood vessels and glands of the nasal cavity, oral cavity, and pharynx; hypophysis; submandibular, sublingual, and parotid salivary glands; peripheral arteries of the face; pilomotor muscles; and sweat glands (41) (Fig. 4-5).

REFLEXES

Many reflexes involve the head and neck. A reflex, generally referred to as a reflex arc, has five links in its simplest form: (a) a receptor, which reacts to a stimulus; (b) an afferent conductor, which transmits the impulse to a "reflex center"; (c) the "reflex center" itself, where the afferent message from the receptor may converge with afferent impulses from other receptors, or with afferents from other sources, which may modify the effect of the afferent impulses from the receptor; (d) an efferent conductor (nerve fiber) going to the effector organ; and (e) an effector, which produces a reaction (in a muscle, gland, or vessel, or may include several components). Reflexes vary from simple to complex. For example, deglutition is a complex reflex (11). Some craniofacial reflexes are summarized in Table 4-4.

VISION

The senses are summarized in Table 4-5. These include vision, audition, balance, olfaction, gustation, cutaneous–mucosal sensitivity, thermal sensitivity, proprioception, and nociception.

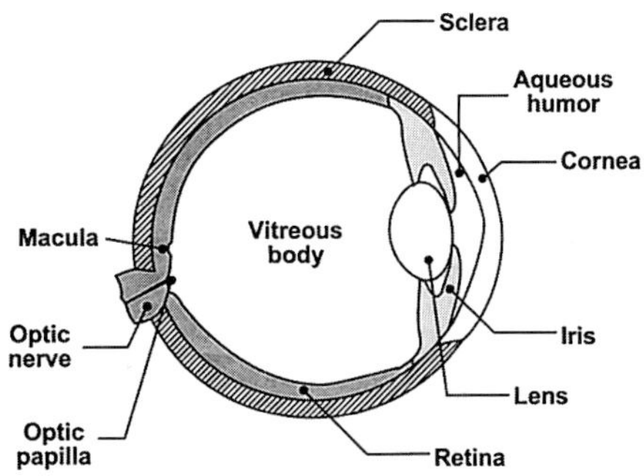

Figure 4-6 Eyeglobe, horizontal plane.

The eyeglobe is shown in cross-section in Figure 4-6. Light must pass through the cornea, aqueous humor, lens, and vitreous body before reaching the retina with its light-sensitive receptors. The extra-ocular and intra-ocular muscles, their actions, and their associated cranial nerve efferents are summarized in Table 4-6. The extra-ocular muscles are shown in Figure 4-7.

Tears

Tears are secreted by the lacrimal glands, which lubricate the cornea continuously by reflex blinking. They drain via the lacrimal ducts, lacrimal sac, and nasolacrimal duct into the nasal sinuses. Tears wash away irritating particles and chemicals, and contain lysozyme and immunoglobulin A, which provide a defense against infections (31).

Iris

The iris contains smooth muscle: circular fibers of the pupillary sphincter and radiating fibers of the pupillary dilator. Cholinergic activation of the pupillary sphincter provides contraction (miosis) and adrenergic activation of the pupillary dilator produces dilatation (mydriasis) (31).

Lens

The lens is held in place by the zonular fibers. When the eyes adjust for far vision, the zonular fibers are stretched by relaxation of the ciliary muscle, flattening the lens. When the eyes are

Table 4-4 Some Craniofacial Reflexes

Reflex	Description	Reflex	Description
Rooting reflex[a]	Initiation of sucking and turning the head toward a stroke to the cheek	*Auditory–palpebral reflex*	Eyelid blink in response to a loud noise
Tongue thrust reflex[a]	Protruding the tongue to eject premature solid food or foreign objects	*Auditory–oculogyric reflex*	Deviation of the eyes in the direction of a sound
Lip reflex[a]	Contraction of the lips following a tap of the upper lip	*Cochleopupillary reflex*	Dilation of the pupils, or constriction followed by dilation, in response to a loud noise
Head retraction reflex[a]	Retraction of the head following a hard tap of the upper lip	*Vestibulo-ocular reflex*	Integrated eye–head–body-orienting response
Glabellar reflex[a]	Eye blinking following a tap to the glabella	*Bregmocardiac reflex*	Pressure on the anterior fontanelle results in bradycardia
Accommodation reflex	Pupillary contraction with ocular convergence	*Lacrimal reflex*	Secretion of tears elicited by touching the conjunctiva over the cornea
Pupillary reflex	Pupillary contraction on sudden exposure to light	*Photic sneeze reflex*	Sneezing on entering bright sunlight
Consensual light reflex	Centrally mediated pupillary contraction in response to retinal illumination of the opposite eye	*Nasal reflex*	Irritation of nasal mucosa provokes sneezing
Cilospinal reflex	Pupillary dilation from sudden pain	*Jaw jerk reflex*	Jaw closure in response to a tap on the chin
Cutaneous–pupillary reflex	Dilation of the pupils on pinching the skin of the cheek or neck	*Salivary–taste reflex*	When a gustatory stimulus is applied to the anterior two-thirds of the tongue, the salivary glands increase their output of saliva
Corneal reflex	Closing the eyelids when a foreign body touches the cornea		
Blink reflex	Blinking and eyelid closure in response to a rapidly approaching object	*Gag reflex*	Touching the posterior wall of the pharynx results in gagging
Oculocardiac reflex	Bradycardia following pressure to the eyeglobes	*Cough reflex*	Coughing in response to irritation of the larynx, trachea, or bronchial tree
General acoustic muscle reflex	Generalized jerking of the body in response to a sudden loud noise	*Carotid sinus reflex*	Bradycardia and momentary hypotension in response to pressure on the carotid sinus in the neck

[a]Transitory infant reflexes normally disappear by 5–7 months. They may appear in adult life on a neuropathological basis.

Data from Brodal, 1981 (11), Despopoulos and Silbernagl, 2003 (31), and Gilman and Newman, 1988 (39).

adjusted for near vision, the zonular fibers are contracted by the ciliary muscles, making the lens more convex (30).

Aqueous Humor

Aqueous humor is secreted by the ciliary processes. It flows through the posterior chamber and the pupil into the anterior chamber. From the angle of the anterior chamber, it drains through the trabecular meshwork into the venous canal of Schemm. Production and drainage of aqueous humor must maintain balance to provide constant ocular pressure (31).

Accommodation

When the eyes are adjusted for vision, parallel light rays from a distant point are imaged at a point located on the retina. For close vision, the curvature of the lens increases and the image at the nearby point moves to the retinal plane. The refractive power around the edge of the optical apparatus is higher than near the optical axis. The spherical alteration can be minimized by narrowing the pupils. The refractive power is the reciprocal of the anterior focal length in meters and is measured in diopters (dpt) (31).

Table 4-5 The Senses

Modalities	Sense Organs	Senses	Stimuli	Sense Qualities	Central Nervous System
Vision	Eyes	Rods and cones of the retina	Photic energy	Rods: black and white vision and night vision Cones: color vision and sharp vision	First synapse occurs between bipolar cells and ganglion cells of the retina. Second order neurons converge on the optic disc, leave the eyeball, and enter the optic nerve with partial crossing at the optic chiasm. Second synapse occurs in the lateral geniculate body of the thalamus. Third order neurons end in the visual cortex of the occipital lobe.
Audition	Ears	Hair cells of the organ of Corti	Acoustic energy	Tones and noises	Bipolar cells of the spiral ganglion converge to form the cochlear part of the vestibulocochlear nerve, which enters the pons–medulla junction where the first synapse occurs. Most second order neurons cross to the contralateral side of the pons in the trapezoid body, enter the lateral lemniscus, and ascend through the brainstem, ending in the inferior colliculus (some are third or fourth order neurons for those fibers that crossed the midline in between the two nuclei of the lateral lemniscus). After entering the brachium of the inferior colliculus, third order neurons ascend to the geniculate body of the thalamus. Fourth order neurons end in the primary auditory cortex and the secondary auditory cortex of the temporal lobe.
Balance	Semicircular canals, utricle, saccule	Cilia of the crista ampullaris	Accelerated rotation, nodding, and sideways tilting of the head	Detects changes in head position	Bipolar neurons of the vestibular ganglion, forming the vestibular part of the vestibulocochlear nerve, synapse with the vestibular nuclei. Important tracts extend to the contralateral side, to the ocular muscle nuclei, cerebellum, motor neurons of the skeletal muscles, and to the postcentral gyrus.
Olfaction	Nose	Olfactory epithelium	Odorant molecules	Odors	Olfactory filia collect into bundles–the olfactory nerve. The olfactory tract divides into medial and lateral olfactory striae. The fibers of the former reach the opposite olfactory bulb through the anterior commissure. The fibers of the latter terminate in the olfactory tubercle, uncus, and parahippocampal gyrus. Axons of third order neurons in the primary olfactory cortex project to the secondary olfactory cortex (entorhinal cortex).

(continued)

Table 4-5 (continued)

Modalities	Sense Organs	Senses	Stimuli	Sense Qualities	Central Nervous System
Gustation	Tongue, palate, epiglottis, pharynx, upper esophagus	Taste buds	Soluble substances	Sweet, salty, sour, bitter, umami	Sensory stimuli for the taste buds are conducted to the endings of the facial nerve, glossopharyngeal nerve, and vagus nerve, and are relayed by the nucleus of the tractus solitarius, ending in the postcentral gyrus via the thalamus, hypothalamus, limbic system via the pons, and in the insular cortex.
Cutaneous–mucosal sensitivity	Craniofacial skin, tongue, oral and nasopharyngeal mucosae	Mechanosensors	Mechanical energy	Touch, pressure, vibration	The posterior funiculus–lemniscus system is the principal route by which the postcentral gyrus of the parietal lobe receives input from mechanosensors. Secondary afferent trigeminal fibers end in the posteromedial ventral nucleus of the thalamus. Tertiary afferent fibers end in the somatosensory cortex.
Thermal sensitivity	Craniofacial skin, lips, oral and nasopharyngeal mucosae	Thermosensors	Thermal energy	Temperature changes	The posterior funiculus–lemniscus system is the principal route by which the postcentral gyrus of the parietal lobe receives input from thermosensors. Secondary afferent trigeminal fibers end in the posteromedial ventral nucleus of the thalamus. Tertiary afferent fibers end in the somatosensory cortex.
Proprioception	Masseter, temporalis, and medial pterygoid muscles and their musculotendinous junctions, temporomandibular joint	Propriosensors	Jaw elevation; rotation and translation of the temporomandibular joint	Regulation of muscle length and tension	The posterior funiculus–lemniscus system is the principal route by which the postcentral gyrus of the parietal lobe receives input from proprioceptors. Secondary afferent trigeminal fibers end in the posteromedial ventral nucleus of the thalamus. Tertiary afferent fibers end in the somatosensory cortex.
Nociception	Craniofacial skin, muscles, bones, oral and nasopharyngeal mucosae, teeth, periodontium	Nocisensors	Noxious stimuli of various types	Pain	The posterior funiculus–lemniscus system is the principal route by which the postcentral gyrus of the parietal lobe receives input from nocisensors. Secondary afferent trigeminal fibers end in the posteromedial ventral nucleus of the thalamus. Tertiary afferent fibers end in the somatosensory cortex.

Data from Foster and Danker, 1980 (37), Kahle, 1986 (53); Bradley, 1995 (9); Watson, 1995 (105), Patton and Thibodeau, 2000 (82); and Despopoulos and Silbernagl, 2003 (31).

Table 4-6 Ocular Muscles

Muscles	Cranial Nerves	Actions
Extra-ocular		
Levator palpebrae superioris	III (Oculomotor)	Raises upper eyelid
Superior rectus	III(Oculomotor)	Turns corneal surface upward with medial deviation and slight rotation
Superior oblique	IV (Trochlear)	Corrects medial deviation and slight rotation of the superior rectus
Medial rectus	III (Oculomotor)	Turns corneal surface medially
Lateral rectus	VI (Abducens)	Turns corneal surface laterally
Inferior rectus	III (Oculomotor)	Turns corneal surface downward with medial deviation and slight rotation
Inferior oblique	III (Oculomotor)	Corrects medial deviation and slight rotation of the inferior rectus
Intra-ocular		
Ciliary	III (oculomotor; ciliary ganglion)	Chief muscle of accommodation; allows lens to become more convex on contraction
Pupillary sphincter	III (oculomotor; ciliary ganglion)	Contracts pupil
Pupillary dilator	Long ciliary nerve (sympathetic)	Dilates pupil

Data from Goss, 1961 (41).

Reflexes

Besides the accommodation reflex, in which pupillary contraction occurs with ocular convergence, many other ocular reflexes are known, including the pupillary reflex, consensual light reflex, cilospinal reflex, cutaneous–pupillary reflex, corneal reflex, blink reflex, oculocardiac reflex, auditory–palpebral reflex, auditory–oculogyric reflex, cochleopupillary reflex, vestibulo–ocular reflex, and lacrimal reflex. These are summarized in Table 4–4.

Eye Movements

When the external ocular muscles (Table 4-6) move the eyes in the same direction, the movements are conjugated, whereas opposing eye movements may be either divergent or convergent. The eye axes are parallel when gazing into the distance, but fixation on a nearby object results in convergence of the visual axes, accompanied by pupillary contraction and accommodation of the lens. Most eye movements are quick and jerky, and are known as saccades that

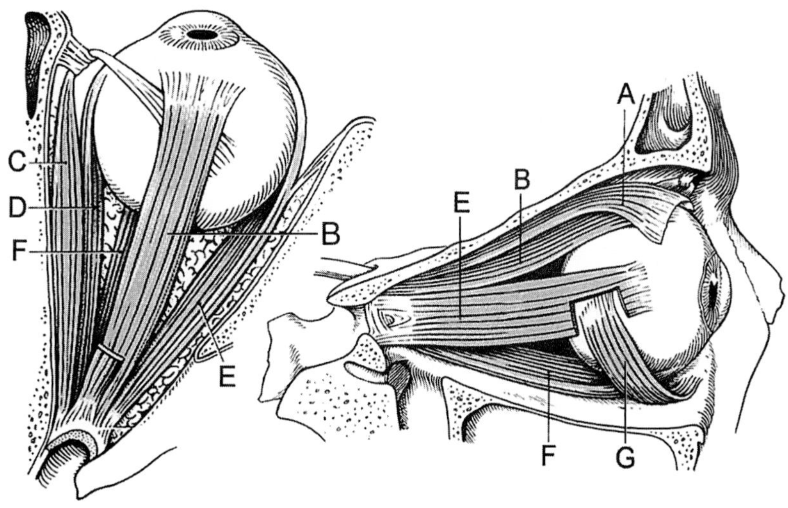

Figure 4-7 Extra-ocular muscles. Left, Eye muscles from above. Right, Lateral view of muscles. A, Levator palpebrae superioris. B, Superior rectus. C, Superior oblique. D, Medial rectus. E, Lateral rectus. F, Inferior rectus. G, Inferior oblique. Modified from Kahle, 1986 (53).

last 10–80 milliseconds to keep an object focused on the macula (30, 31).

Programming of eye movements occurs primarily in the brainstem. Saccades are programmed in the pons, vertical and torsional movements in the mesencephalon, and fine-tuning in the cerebellum. Neurons in the midline of the third nerve (III) nucleus are responsible for vergence movements (31).

Stereoscopic Vision and Depth Perception

The blind spot at the optic disk in one eye is compensated for by the other eye. In near vision, depth vision and three-dimensional vision are achieved primarily through the coordination of both eyes and are limited to the binocular field of vision. When viewing objects from a great distance, size differences, shadows, haze, and overlap are cues for depth perception (31).

Retina

The retina lies on the inner surface of the eyeball except at the optic nerve outlet. The macula is a slight depression across from the pupillary aperture (Fig. 4-6). The layers of the retina from inside out are (a) pigmented epithelium, (b) rods and cones (photoreceptor cells), (c) horizontal cells, (d) bipolar cells, (e) amacrine cells, and (f) ganglion cells,[d] whose processes exit the eyeball as the optic nerve (optic disk) (31) (Fig. 4-8).

Rods and cones have a light-sensitive outer segment connected to their cell bodies by connecting stalks. They synapse with bipolar cells, which, in turn, synapse with ganglion cells. Horizontal and amacrine cells provide additional connections among receptor cells, bipolar cells, and ganglion cells (31) (Fig. 4-8).

Horizontal and amacrine cells, as lateral interacting elements, tend to be inhibitory in their neural interactions. However, rod amacrine cells transmit rod signals to bipolar and ganglion cells. There are at least two major classes of horizontal cells. One class receives most of its input from long wave and medium wave cones. A second class receives its input from all three types of cones, but much more from short wave cones. The inhibitory effects that horizontal cells exert on cones may be to control their overstimulation by large energy gradients (42).

Rods are for black and white vision and night vision, whereas cones are for color vision and sharp vision. Rods and cones are distributed at variable densities throughout the retina. The macula (Fig. 4-6) is exclusively filled with cones and their density decreases rapidly towards the periphery. Rods predominate 20–30° away from the macula. No photoreceptor cells are present at the optic disk, the blind spot in the visual field. Some ganglion cells also contain a light-sensitive pigment (31).

The outer segment of each rod and cone is composed of membranous infoldings, which form a stack of disks. Unlike rod disks, cone disks become progressively smaller towards the tip of the outer segment. For each rod and cone, the outer segment is completely regenerated: old membranous disks are shed at the tip and replaced by new disks from the inner segment. Phagocytic cells of the pigmented epithelium

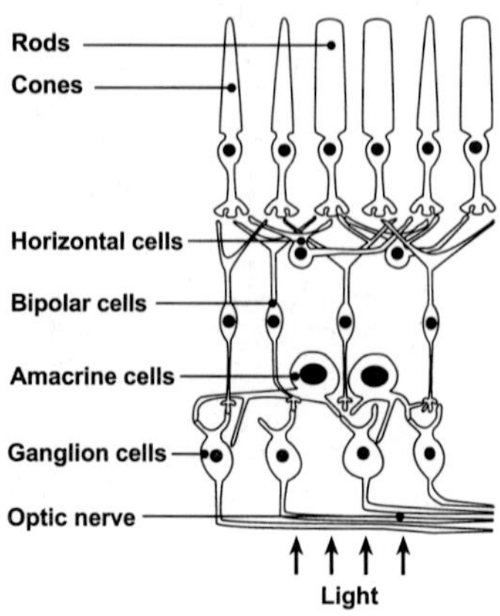

Figure 4-8 Schematic representation of the retina. See text and footnote d.

[d]Sometimes ganglion cells are called *bipolar ganglion cells*, but they should not be confused with *bipolar cells per se* (see Figure 4-8).

engulf disks shed by rods in the morning and disks shed by cones in the evening (31).

Rods and Rhodopsin

The membranous disks of the retinal rods contain rhodopsin, also known as visual purple and opsin 2; it maps to 3q21-q24. The molecular biology is diagrammed in Figures 4-9 to 4-11. Rhodopsin (R) is made up of opsin and 11-cis-retinal. In a photochemical reaction, 11-cis-retinal is converted to all-trans-retinal, and the reaction yields bathorhodopsin (BR), lumirhodopsin (LR), metarhodopsin I (MRI), and metarhodopsin II (MRII) (31, 68) (Fig. 4-9).

Transducin is a G protein that reacts with metarhodopsin II. When guanosine diphosphate (GDP) is replaced by guanosine triphosphate (GTP), transducin breaks down into an activated α-subunit and βγ-subunits (Fig. 4-9). The α-subunit then binds to inhibitory phosphodiesterase (I_{PDE}), which activates cyclic guanosine monophosphate (cGMP) from 5'-guanosine triphosphate (5'-GTP) and opens the sodium

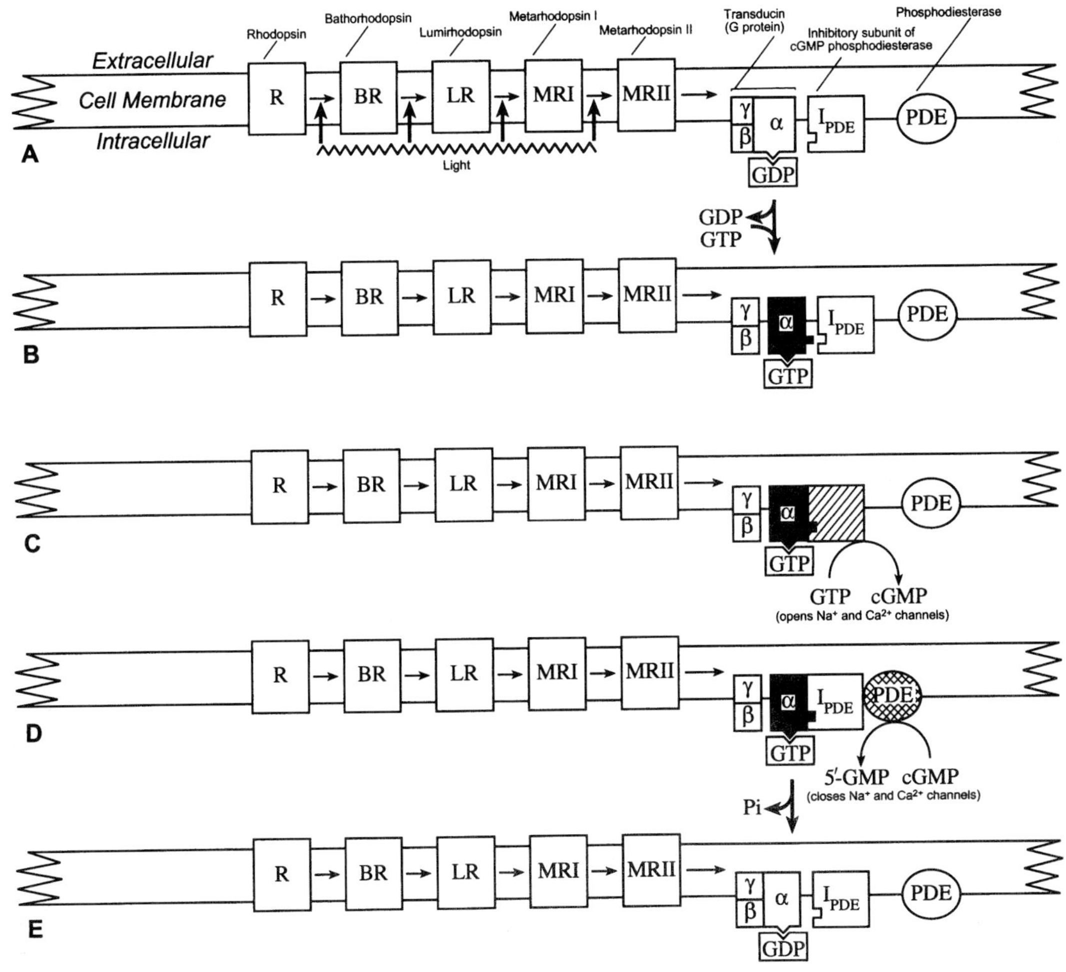

Figure 4-9 Activation of photosensors. A, No activation. B, When GDP is replaced by GTP, transducin splits into an activated α-subunit and a βγ-subunit. C, The α-subunit binds to I_{PDE}, which activates cGMP from 5'-GTP and opens Na^+ and Ca^{2+} channels. D, I_{PDE} binds to PDE and closes Na^+ and Ca^{2+} channels. E, GTP is hydrolyzed to GDP, resulting in the re-association of the α-subunit with the βγ-subunit, resulting in no activation.

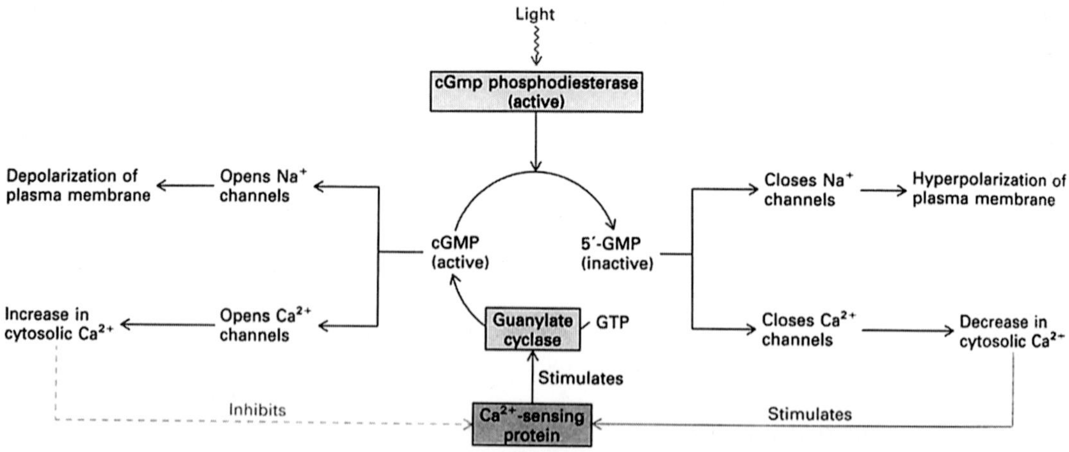

Figure 4-10 Role of Ca^{2+}-sensing protein, guanylate cyclase (GTP), and Ca^{2+} in the adaptation of rods to changes in light levels. Note opening and closing of Na^+ and Ca^{2+} channels. From Lodish et al., 1995 (68).

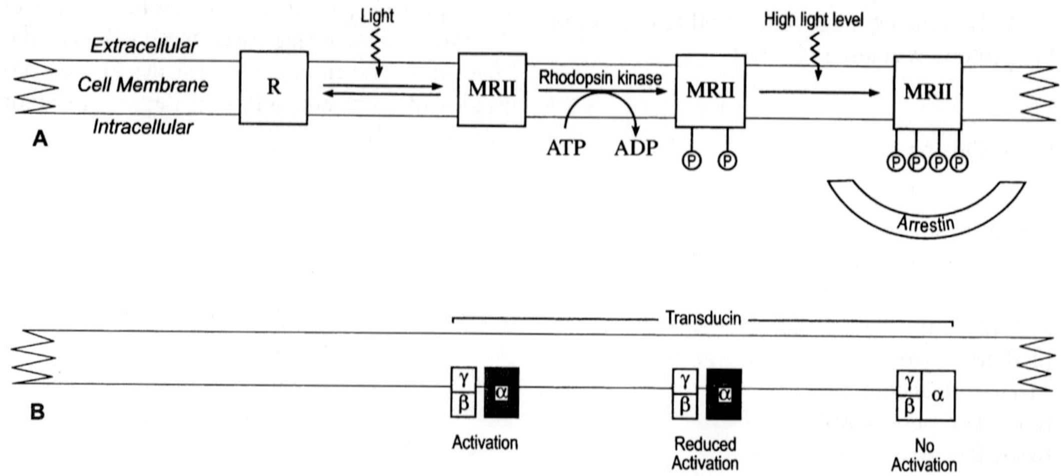

Figure 4-11 Role of phosphorylation in the adaptation of rods to changes in light levels. Circled P, Phosphorylation site. R, Rhodopsin. MRII, Metarhodopsin II. Note changes in light levels in A, resulting in alterations in the activation of transducin in B. Compare with Figure 4-9.

(Na^+) and calcium (Ca^{2+}) channels (Figs. 4-9, 4-10). In another reaction, rhodopsin kinase converts adenosine triphosphate (ATP) to adenosine diphosphate (ADP), phosphorylating MRII and allowing phosphodiesterase (PDE) to lower the cytosolic concentration of cGMP and increase 5'-GMP, which reduces activation of the a-subunit and closes the Na^+ and Ca^{2+} channels (Figs. 4-9–4-11). Arrestin further phosphorylates MRII, which inactivates the a-subunit (Fig. 4-12). Hydrolysis of GTP to GDP results in reassociation of the α-subunit with the βγ-subunits (31, 68) (Fig. 4-9).

In the dark, Ca^{2+} is high; calmodulin-bound Ca^{2+} stimulates phosphorylation of phosducin, aided by cAMP and phosphokinase A. In the light, Ca^{2+} is low; phosducin is dephosphorylated and rapid regeneration of transducin is not

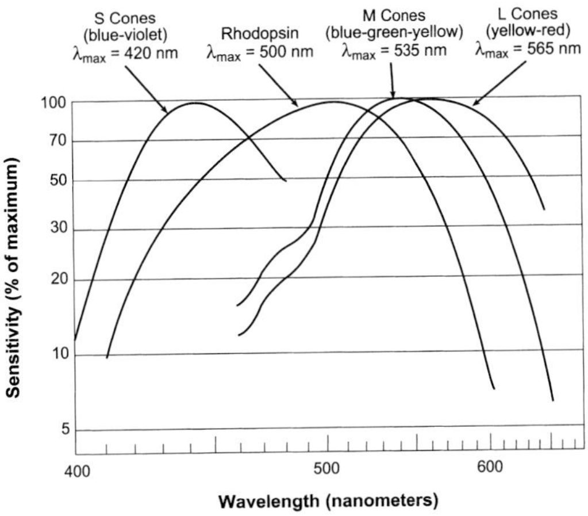

Figure 4-12 Wavelength and sensitivity of S cones, Rhodopsin, M cones, and L cones.

possible. Ca^{2+} also accelerates phosphorylation of MRII in the light, aided by another Ca^{2+} binding protein, recoverin (31, 68).

Cone Opsins

The three color pigments, all known as opsin 1, includes blue-cone pigment, which maps to 7q13.3-q32; green-cone pigment, which maps to Xq28; and red-cone pigment, which also maps to Xq28. Like rhodopsin, color opsins are coupled to 11-cis-retinal for activation. However, 11-cis-retinal combines with different opsin fractions. The photosensitivity of cones is about 100 times less than that of rods, but like rods, the molecular biology is similar, involving transducin, phophodiesterase, cGMP, 5'-GTP, and the opening and closing of Na^{+} and Ca^{2+} channels (31, 68, 81).

Melanopsin

A small population of photosensitive ganglion cells in the retina can regulate nonvisual photic responses, including behavioral response to light, pineal melatonin synthesis, pupillary light reflex, and sleep latency. The involved retinal ganglion cells express the photopigment melanopsin (Opn4), which mediates these nonvisual photic responses (81a).

Light Absorption

The range of detectable light varies from about 400 nanometers (violet) to 700 nanometers (red) (Fig. 4-12). S cones, which absorb short wave (S) blue-violet light, have a maximum wavelength of 420 nanometers (λ_{max} = 420 nm). M cones, which absorb medium wave (M) blue–green light, have a maximum wavelength of 535 nanometers (λ_{max} = 535 nm). L cones, which absorb long wave (L) yellow-red light, have a maximum wavelength of 565 nanometers (λ_{max} = 565 nm). Rhodopsin, which is achromic, absorbs light at about 400 to 600 nanometers, with a maximum wavelength of 500 nanometers (λ_{max} = 500 nm) (42).

Additive Color Mixing

All colors, including white, can be produced by varying the proportions of red, green, and blue; every possible pair of complementary colors can be obtained by mixing these three colors of the spectrum. The human eye can distinguish 200 shades of color and different degrees of color saturation (31).

Hue, Saturation, and Brightness

Color can be described in terms of hue, saturation, and brightness. Hue refers to wavelength contrast of colors, such as yellow or blue and red or green. Saturation defines the mixing of hue with white, gray, or black. A saturated color has a strong hue with little or no white, as, for example, with blood red. An unsaturated color refers to a hue mixed with white; pink is an example. For brightness, yellow and white tend to be bright, whereas blue and black tend to be dark. Texture and gloss may also contribute to color, as with gold or silver (42).

Retinal Processing of Visual Stimuli

Two parallel neural circuits are used in the human trivariant color system. One depends on long wave (L) cones that detect differences in light energy and provide achromic contrast by distinguishing light from dark. The other depends on both long wave (L) cones and medium wave (M) cones, which are responsible for chromic contrast (42).

Four longer wave channels transmit tonic signals to the visual cortex of the brain via two red long wave (L) cone ON/OFF signals and two green medium wave (M) cone ON/OFF signals in small bipolar and ganglion cells (Fig. 4-13). For achromic contrast, signals of neighboring tonic long wave (L) cones channels are compared, whereas for chromic contrast, signals from groups of long wave (L) cone channels are compared with those of medium wave (M) cone channels (42).

In a separate phasic system, red and green cones are mixed in ON/OFF large bipolar and ganglion cells (Fig. 4-13). Although the system plays no role in color vision, it is highly sensitive to achromic contrast and to slow movements; it also contributes to luminance (42).

The region in which photic stimulatory and inhibitory effects on action potential frequency originate is known as the receptor field. Ganglion cells have concentric receptor fields composed of a central zone and a ring-like peripheral zone. Photic stimulation of the central zone increases action potential frequency of ON ganglion cells, whereas stimulation of the peripheral zones decreases action potential frequency. The receptor fields of OFF ganglion cells have the reverse response. For example, red and green light (red ON/green OFF and red OFF/green ON) or blue and yellow light (blue ON/yellow OFF and blue OFF/yellow ON) have opposite effects in the receptor fields of ganglion cells and more centrally located cells of the optic tract (31, 42).

Horizontal cells are responsible for the functional organization of receptor fields. They invert impulses from photosensors in the peripheral zone to sensors in the central zone (31).

In dark adaptation, the center of the receptor field increases in size at the expense of the periphery, which ultimately disappears (31).

Visual Pathways

The first synapse occurs between the axons of bipolar cells and the dendrites of ganglion cells. The axons of ganglion cells converge on the optic disk and leave the eyeball to enter the optic nerve, which exits through the optic canal of the sphenoid bone. The two optic nerves meet at the optic chiasm. Here, fibers whose ganglion cell bodies are located in the nasal half of each retina cross to the contralateral side, whereas fibers from the temporal half of each retina remain uncrossed. Fibers then enter the optic tracts and run to the lateral geniculate body of the thalamus where the second synapse occurs (Fig. 4-14). Axons of third order neurons enter the optic radiation (geniculocalcarine tract), pass through the sublentiform and retrolentiform parts of the

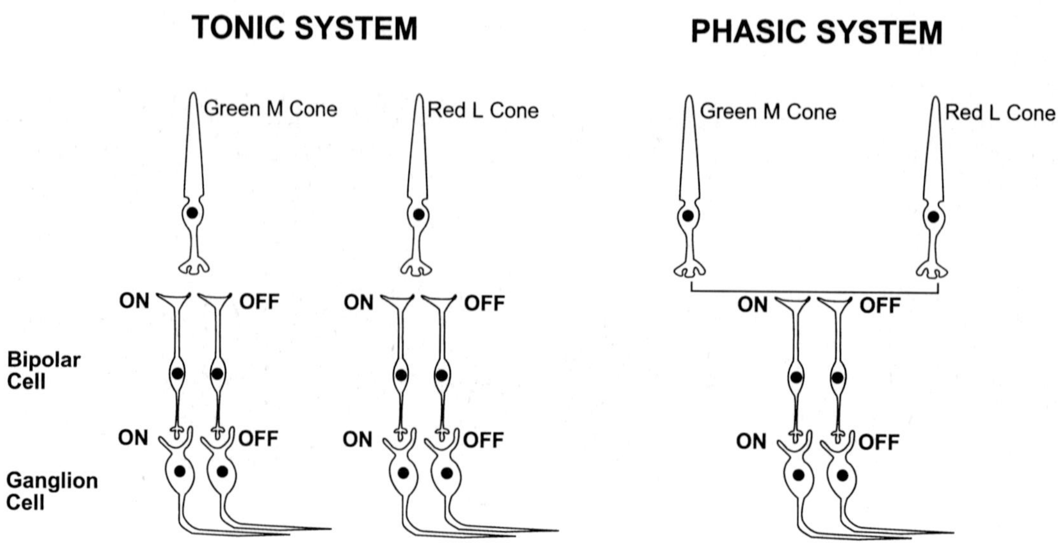

Figure 4-13 ON/OFF signals in the tonic and phasic systems. See text.

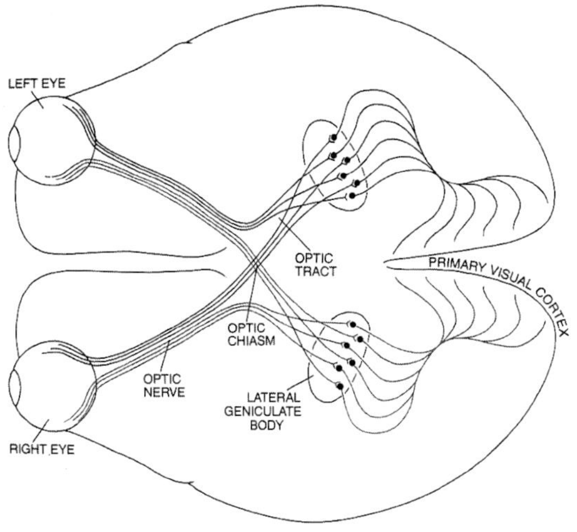

Figure 4-14 Visual pathways. See text. Modified from Hubel and Wiesel, 1979 (48).

internal capsule, and end in the primary visual cortex (Fig. 4-14) on the superior and inferior lips of the calcarine sulcus (area 17) of the occipital lobe. Areas 18 and 19 make up the secondary visual cortex on the remaining medial and lateral surfaces of the occipital lobe (105).

Central Processing of Visual Stimuli

There are three kinds of ganglion cells in the retina: M (or α) cells (10%) of the magnocellular system, which are sensitive to movement; P (or β) cells (80%) of the parvocellular system, which persistently react to constant light, permitting pattern and color analysis; and W (or γ) cells (10%) of the coniocellular system, which regulate pupil diameter and reflex saccades (31).

Fibers of the optic tract extending into the lateral geniculate body are organized into six retinotopic layers (Fig. 4-14). Axons of one eye terminate on layers 2, 3, and 5, while those of the contralateral eye terminate on layers 1, 4, and 6. M cell axons communicate with cells of magnocellular layers 1 and 2, which serve as a relay station for motion-related stimuli that are rapidly conducted to the motor cortex. P cell axons project to parvocellular layers 3 and 6, which process colors and shapes. The neurons of all six layers project through the optic radiation, which is also arranged retinotopically, to the primary visual cortex and pathways to the parietal and temporal lobes. Magnocellular input also reaches the parietal lobes via the superior colliculi and pulvinar (31).

The primary visual cortex is also divided into six retinotopic layers (Fig. 4-14). Color, high-resolution stationary shapes, and stereoscopic depth are processed in some subcortical visual pathways and in the primary visual cortex as separate information channels. Cortical cells are arranged as three-dimensional hypercolumns for the analysis of sensory information from corresponding areas in both retinas. Ocular dominance columns receive alternating input from right and left eyes; orientation columns process the direction of stimulus movement; and cylindrical columns receive information about colors (31).

Ocular Pathology: The Eyes Have It

Pathology affecting the eyes is summarized in Table 4-7 under the following headings: ocular abnormalities, lesions interrupting the visual pathways, color vision deficiencies, albinism, and some syndromes with ocular abnormalities.

AUDITION

Physical Principles of Sound Perception

Air is the main carrier of sound waves, although conduction also occurs through the solid medium of the auditory ossicles, the liquid medium of the cochlea, and the electrical medium of the auditory nerve. Air pressure rises and falls rhythmically at the sound source and travels at a characteristic velocity. Sound waves are best visualized as sinusoidal oscillations (Fig. 4-15). The distance between the tip of one wave and the next is a *wavelength*. A decrease in wavelength raises the tone, whereas an increase lowers the tone. The maximum air pressure deviation from the baseline is the *amplitude*. A decrease in amplitude produces a quieter tone, whereas an increase produces a louder tone (30, 31).

The pitch of a tone is defined by the number of oscillations per unit time, and this is measured in *Hertz* (*Hz*). The young can hear sounds ranging from 16Hz to 20,000 Hz, whereas the upper limit may drop to 5,000 Hz with aging (presby-

Table 4-7 Pathology Affecting the Eyes

OCULAR ABNORMALITIES	
Defects	Definitions
Amblyopia	Dimness of vision without a lesion
Aniridia	Absence of the iris
Anisocoria	Inequality of the pupils in diameter
Anophthalmia	Clinical absence of the eye, but with histological remnants present
Astigmatism	Defective curvature of the refractive surfaces of the eye
Cataract	Opacity of the lens
Coloboma	Fissured defect in any part of the eye Coloboma of the iris, particularly due to persistence of the choroid fissure
Glaucoma	Obstruction of aqueous humor drainage
Hyperopia	Far sightedness
Microphthalmia	Abnormally small eyes
Myopia	Near sightedness
Nystagmus	Involuntary rhythmic, jerking movement of the eyes
Presbyopia	Loss of lens elasticity due to aging
Strabismus	Deviation of the eye; visual axes different than that required by physiological conditions

LESIONS INTERRUPTING THE VISUAL PATHWAY	
Visual Field Defects	Lesions
Blindness of right eye	Complete destruction of the right optic nerve
Bilateral hemianopia	Lesions of the medial optic chiasm (e.g., pituitary tumor, craniopharyngioma)
Left homonymous hemianopia	Destruction of the right optic tract
Left homonymous hemianopia (with some preservation of central vision)	Destruction of the right occipital lobe or right optic radiation (with small lesions in the occipital lobe)
Left homonymous inferior quadrantanopia	Lesions on the upper bank of the right calcarine cortex
Left homonymous superior quadrantanopia	Lesions of the right optic radiation in Meyer's loop in the temporal lobe

COLOR VISION DEFICIENCIES	
Disorders	Comments (Chromosome Map Locations)
Protanopia	Lack of long wavelength red cones (Xq28)
Deuteropia	Lack of middle wavelength green cones (Xq38)
Tritanopia	Lack of short wavelength blue cones (7q31.3-q32)
Protanomaly	Long wavelength red cones present, but sensitivity is shifted spectrally to shorter wavelengths
Deuteranomaly	Middle wavelength green cones present, but sensitivity is shifted spectrally to longer wavelengths
Tritanomaly	Short wavelength blue cones present, but sensitivity is shifted spectrally to longer wavelengths
Cone monochromatism	Red cones only, green cones only, or blue cones only
Rod monochromatism	Autosomal recessive ocular disorders characterized by total color blindness, low visual acuity, photophobia, and nystagmus. Type 1 results from uniparental disomy (chromosome 14). Type 2 is caused by *ACHM2* mutations (2q11). Type 3 is caused by *CNGB3* mutations (8q21-q25). Type 4 is caused by *GNAT2* mutations (1p13).
Retinal cone dystrophy	Progressive loss of visual acuity, photophobia, and defective color vision (6q25-q26)
Cone–rod dystrophy (CORD)	Loss of color vision and visual acuity followed by nyctalopia and loss of peripheral vision. Etiologically heterogeneous. CORD1 (18q21.1-q21.3); CORD2, *CRX* mutation (19q13.3); CORD3, *RPGR* mutation (Xp11.4-q13.1); CORD5 (17p12-p13); CORD6, *GUCY2D* mutation (17p13.1); CORD8 (1q12-q24)
Cone–rod dystrophy with spondylometaphyseal dysplasia	Autosomal recessive
Nyctalopia	Night blindness. Progressive loss of retinal function. Autosomal dominant and X-linked forms. Myopia accompanies the X-linked form and *NYX* mutations (Xp11.4) have been identified.
Retinitis pigmentosa	Progressive loss of retinal function. Etiologically heterogeneous. Most common form involves *RHO* (rhodopsin) mutation (3q21-q24).
Bardet–Biedl syndrome	Pigmentary retinopathy, mental retardation, obesity, and hypogenitalism in males. Etiologically heterogeneous. Linkage to eight different loci (20p12, 16q21, 15q22.3-q33, 14q32.1, 11q13, 4q27, 3p12-p13, 2q31).

(continued)

ALBINISM

Types	Comments (Chromosome Map Locations)
Oculocutaneous albinism, type I	Autosomal recessive disorder characterized by absence of pigment in the eyes, hair, and skin, together with a defect in decussation of the optic tracts. Type IA has complete lack of tyrosinase activity, whereas type IB has reduced tyrosinase activity. Mutations in *TYR* (11q14-q21).
Oculocutaneous albinism, type II	*P* mutations (15q11.2-q12)
Oculocutaneous albinism, type III	*TYRP1* mutations (9p23)
Ocular albinism, type I	Depigmented fundus, *OA1* mutations (Xp22.3)
Ocular albinism, type II	X-linked dominant trait. *CACNA1F* (Xp11.23-p11.4)
Ocular albinism, type III	*P* mutations (6q13-q15). Location of *P* gene for this form of albinism on chromosome 6 seems less likely but not impossible (see oculocutaneous albinism, type II above).
Hermansky–Pudlack syndrome	Autosomal recessive etiologically heterogeneous disorder characterized by oculocutaneous albinism, hemorrhagic diathesis, and pigmented reticuloendothelial cells. Mutations in seven different genes: *HPS1* (10q23.1), HPS2 (*AP3B1*)(Chr.5), *HPS3* (3q24), *HPS4* (22q11.2-q12.2), *HPS5* (11p13-p15), *HPS6* (10q24.32), HPS7 (*DTNBP1*) (6p22.3)

SOME SYNDROMES WITH OCULAR ABNORMALITIES

Syndromes	Comments (Chromosome Map Locations)
Aicardi syndrome	Lacunar retinopathy. Probably X-linked dominant, lethal in male hemizygotes (Xp22)
Alagille syndrome	Posterior embryotoxon, pigmentary retinopathy. Autosomal dominant. *JAG1* mutations (20p12)
Anophthalmia–syndactyly syndrome	Autosomal recessive
Apert syndrome	Ocular proptosis. See Chapter 5 for etiology
Baraitser–Winter syndrome	Iris coloboma, ptosis, hypertelorism. Autosomal recessive?
Biemond syndrome II	Iris coloboma. Autosomal dominant
Blepharophimosis–ptosis–epicanthus inversus	Autosomal dominant. *FOXL2* mutations (3q23)
CHARGE association and Hall–Hittner syndrome	Colobomas, sporadic, some familial cases
CHIME syndrome	Colobomas. Autosomal recessive
Chondrodysplasia punctata	Cataracts. Genetically heterogeneous. Conradi–Hünermann type, X-linked dominant, *EPB* mutations (Xp11.22-p11.23). Rhizomelic type, autosomal recessive, *PEX7* mutations (6q22-q24)
Cohen syndrome	Retinal mottling, constricted visual fields, other ocular abnormalities. Autosomal recessive, *COH1* mutations (8q22-q23)
Crouzon syndrome	Ocular proptosis. See Chapter 5 for etiology
Cryptophthalmos syndrome	Autosomal recessive. *FRAS1* mutations (4q21)
Delleman syndrome	Orbital cysts, microphthalmia, colobomas. All cases sporadic, to date
Ehlers–Danlos syndrome, type VI	Keratoglobus, other ocular abnormalities. Autosomal recessive
Fabry syndrome	Aneurysmal dilatation and tortuosity of conjunctival and retinal vessels, characteristic corneal and lenticular alterations. X-linked recessive. *GLA* mutations (Xq22)
Familial dysautonomia	Decreased tears, absent corneal reflex, other ocular abnormalities. Autosomal recessive. *IKBKAP* mutations (9q31)
Fetal alcohol syndrome	Narrow palpebral fissures, small eyeglobes
Focal dermal hypoplasia	Colobomas. X-linked dominant, lethal in male hemizygotes
Goldenhar syndrome	Epibulbar dermoids, other ocular abnormalities. Most cases are sporadic. Some have autosomal dominant inheritance
Hallermann–Streiff syndrome	Congenital cataracts, microphthalmia, other ocular abnormalities. All cases are sporadic
Holoprosencephaly	Ocular hypotelorism, single eyeglobe in cyclopia, other ocular abnormalities. Etiologically heterogeneous. See Chapter 5 for etiology
Homocystinuria	Ectopia lentis (inferior dislocation), other ocular abnormalities. Autosomal recessive. *CBS* mutations (21q22.3)
Kaufman syndrome	Upslanting palpebral fissures, sparse eyebrows, blepharophimosis, myopia, microcornea, pale optic disks
Lenz microphthalmia syndrome	Microphthalmia, anophthalmia. X-linked (Xq27-q28)
Marden–Walker syndrome	Blepharophimosis, other ocular abnormalities. Autosomal recessive
Marfan syndrome	Ectopia lentis (superior dislocation), other ocular abnormalities. Autosomal dominant. *FBN1* mutations (15q21.1)

(*continued*)

Table 4-7 (continued)

SOME SYNDROMES WITH OCULAR ABNORMALITIES	
Syndromes	Comments (Chromosome Map Locations)
Marshall syndrome	Myopia, cataracts. Dominant inheritance, autosomal dominant vs. X-linked dominant. Basic defect is controversial. It has been suggested that *COL11A1* (1p21) may be a candidate gene, but this has been debated.
Megalocornea and mental retardation	There are four types: Neuhäuser syndrome, Frank–ter Haar syndrome, Verloes syndrome, and Frydman syndrome
Michels syndrome	Blepharophimosis, ptosis, epicanthus. Autosomal recessive
Moebius syndrome	VIth nerve palsy. If the syndrome is defined as involvement of both VIth and VIIth nerve palsies, almost all cases are sporadic.
Mucopolysaccharidoses	Corneal opacities in types I-H (Hurler), I-H/S (Hurler/Scheie), I-S (Scheie), IV-A (Morquio A), IV-B (Morquio B), VI-A (Maroteaux-Lamy A), and VI-B (Maroteaux-Lamy B)
Nance–Horan syndrome	Congenital posterior cataracts. X-linked (Xp22.13-p22.31)
Neurofibromatosis	Type I Lisch nodules, occasionally pulsating exophthalmus due to bony defect of posterosuperior orbital wall, uncommonly optic glioma. Type II senile lens opacities or subcapsular cataracts. Type I and Type II are autosomal dominant. Type I is caused by *NF1* mutations (17q11.2). Type II is caused by *NF2* mutations (22q12.2)
Osteogenesis imperfecta	Blue sclerae in type II and type III. *COL1A1* mutations (17q21.31-q22) or *COL1A2* mutations (7q22.1) in both types II and type III
Peters plus syndrome	Peters anomaly consists of central corneal opacity, thinning of the posterior cornea, and iridocorneal adhesions. Also found are anterior chamber defects and other ocular anomalies. Autosomal recessive. A mutation in*PAX6* (11p13) was found in one case and a *RIEG1* mutation in another
Pfeiffer syndrome	Ocular proptosis. See Chapter 5 for etiology
Rieger syndrome	Abnormal development of anterior chamber of the eye. Autosomal dominant. *RIEG1* mutations (14q25-q26)
Rubella syndrome	Cataracts
Stickler syndrome	Type I (high myopia with retinal detachment, cataracts). Autosomal dominant. *COL2A1* mutations (12q13.11-q13.2), type II (severe early onset and retinal degeneration). Autosomal dominant. *COL11A1* mutations (1p21)
Trisomy 21 syndrome	Upslanting palpebral fissures, epicanthic folds, Brushfield spots, fine lens opacities, convergent strabismus, nystagmus, keratoconus, cataract, rare retinoblastoma
Trisomy 13 syndrome	Microphthalmia, iris coloboma, retinal dysplasia
Trisomy 18 syndrome	Various abnormalities of the eyeglobe, adnexa, and neuro-ophthalmologic system
Tuberous sclerosis	Retinal hamartomas. Autosomal dominant. Mutations in either *TSC1* (9q34) or *TSC2* (16p13.3)
Velocardiofacial syndrome	Narrow palpebral fissures, blue suborbital coloring, tortuous retinal vessels, posterior embryotoxon, small optic disks, cataracts, colobomas. Autosomal dominant (22q11.2)
Waardenburg syndrome	Type I: heterochromia irides, dystopia canthorum. Types IIA and IIB: Heterochromia irides without dystopia canthorum. Type III: Klein-Wardenberg syndrome. Heterochromia irides, dystopia canthorum, limb defects. Type IV: Waardenberg–Shah syndrome. Heterochromia irides without dystopia canthorum, associated Hirschsprung disease. Type I, autosomal dominant; *PAX3*mutations (2q35). Type IIA, autosomal dominant; *MITF* mutations (3p12.3-p14.1). Type IIB, autosomal dominant (1p13.3-p21). Type III, autosomal dominant; *PAX3* mutations (2q35). Type IV, etiologically heterogeneous. Mutations in *EDHRB* (13q22) or in *EDN3* (20q13.2-q13.3), both of which are autosomal recessive, or in *SOX10* (22q13), which is autosomal dominant
Wagner syndrome	Hyaloideoretinal degeneration, chorioretinal atrophy, cataract, peripheral retinal detachment, glaucoma. Autosomal dominant (5q13-q14)
WAGR	*W*ilms tumor, *a*niridia, gonadoblastoma, mental *r*etardation. del(11)(p13)
Walker–Warburg syndrome	Microphthalmia and many other ocular abnormalities. Autosomal recessive. *POMT1* mutations (9q34.1) and *FCMD* mutations (9q31)
Weill–Marchesani syndrome	Spherophakia. Etiologically heterogeneous. Autosomal dominant form with *FBN1* mutations (15q21.1) and autosomal recessive form (19p13.2-p13.3)
Werner syndrome	Senile cataracts. Autosomal recessive. *RECQL2* (8p11.2-p12)
Williams syndrome	Stellate iris, strabismus. Contiguous gene syndrome involving *ELN* (7q11.2)
Wilson disease	Kayser–Fleischer rings. Autosomal recessive. *ATP7B* mutations (13q14.3-q21.1)

Data from Despopoulos and Silbernagl, 2003 (31), Gilman and Newman, 1988 (39) OMIM, 2004 (81), Watson, 1995 (105), Wilson and Wensel, 2003 (106), Sturm and Frudakis, 2004 (98), and Walters et al., 2004 (102).

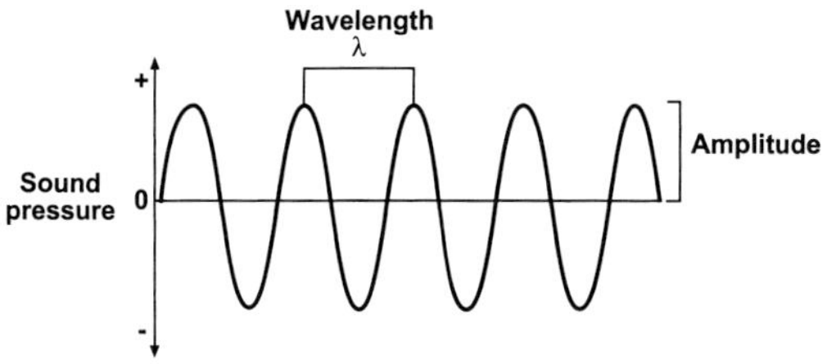

Figure 4-15 Sound waves. λ = wavelength.

cusis). At 1,000 Hz, the absolute threshold, or lowest sound pressure perceived, is 3×10^{-5} *Pa* (*pascals*, or force per unit area). The threshold for hearing someone's voice during a conversation rises dramatically when other tones, such as a boisterous party or loud music, mask ordinary conversation. Sound pressures above 60 Pa are painful (31).

For practical reasons, a logarithmic measure of sound pressure levels is given in *decibels* (*dB*). Using an arbitrary reference sound pressure of $p_o = 2 \times 10^{-5}$ Pa, the *sound pressure level* (*SPL*) can be calculated as

$$dB = 20\log(p_x/p_o)$$

where p_x is the actual sound pressure. A 10-fold increase in sound pressure would mean that the SPL rises by 20 dB (31).

Major Divisions of the Ear

The three major divisions are the external ear, composed of the pinna and the external auditory meatus; the middle ear, consisting of the auditory ossicles and the tympanic membrane; and the inner ear, made up of the cochlea and the semicircular canals (Fig. 4-16).

External Ear

The shape of the pinna helps localize the source of various sounds. Its complex shape probably plays a role in directing sound waves toward the auditory canal, whose surface is somewhat elastic (8, 41, 82).

Middle Ear

Sound traveling down the auditory canal impinges on the tympanic membrane, which causes

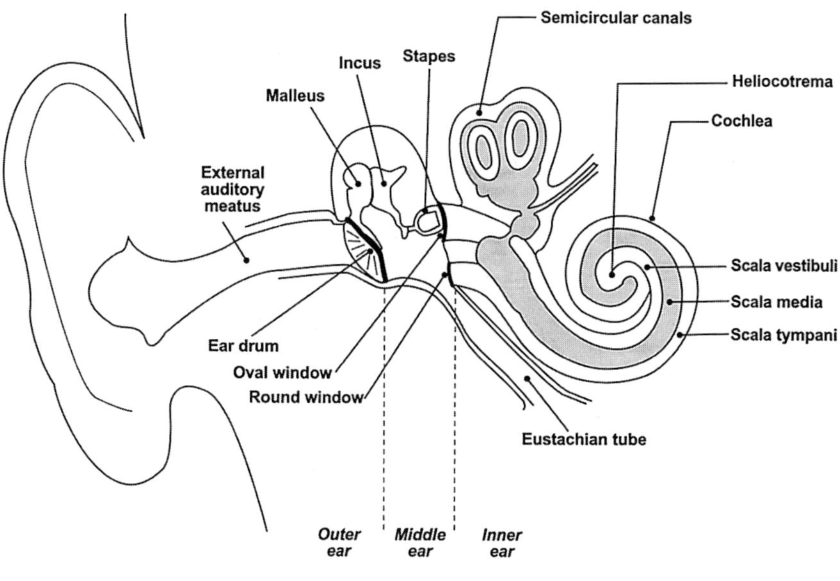

Figure 4-16 Major divisions of the ear: External ear, composed of the pinna and external auditory meatus; the middle ear, consisting of the auditory ossicles and the tympanic membrane; and the inner ear, made up of the cochlea and semicircular canals. The middle ear and the inner ear are enlarged in relation to the external ear to better demonstrate the former.

it to vibrate at an amplitude and frequency corresponding to the sound itself. These are transmitted to the auditory ossicles–the malleus, incus, and stapes–which are firmly attached to one another and function as a unit. The base of the malleus is attached to the tympanic membrane and the stapedial footplate is attached to the membrane covering the oval window (Fig. 4-16). The middle ear contains two small muscles: the tensor tympani, which inserts into the manubrium of the malleus, and the stapedius, which inserts into the stapes (8, 41, 82).

The auditory ossicles conduct sound from the low wave resistance/impedance in air to high resistance in the fluid of the cochlea, with as little loss of energy as possible. The impedance transformation occurs at more than 2,400 Hz and is based on a 22-fold pressure amplification. The ratio of the tympanic membrane area to the oval window area is 17:1, and lever arm action of the auditory ossicles amplifies force by a factor of 1:3 (31).

Hear, Hear, for the Inner Ear

The inner ear is derived from the otic placode. The underlying mesoderm produces FGF19, which induces WNT8C from the overlying neural plate. Subsequently, FGF19 and WNT8C from the hindbrain act in concert on the ectoderm, initiating formation of the otic placode (43, 59).

The cochlea is a tightly coiled, spiraling bony tube with slightly more than 2 1/2 turns. It contains the *scala media*, an endolymph-filled duct accompanied on either side by the *scala vestibuli* and the *scala tympani*, two perilymph-filled cavities that merge at the apex of the cochlea to form the *helicotrema*. The scala vestibuli arises from the oval window and the scala tympani terminates on the membrane of the round window. Perilymph circulates in Corti's tunnel and Nuel's spaces (31).

The organ of Corti consists of outer and inner hair cells. Each outer hair cell contains cilia that touch the tectorial membrane. The base of each hair cell is firmly attached to the basilar membrane by a supporting cell and its cell body that floats in the perilymph of Nuel's spaces. The outer hair cells are, for the most part, innervated by efferent cholinergic neurons from the spinal ganglion. Each inner cell is pear-shaped and surrounded by supporting cells. Cilia project freely into the endolymph. Inner hair cells are arranged in a single row and synapse with more than 90% of the afferent fibers of the spinal ganglion. Efferent axons from the lateral part of the superior olivary nucleus synapse with the afferent endings (31).

The membrane-covered oval and round windows permit displacement of cochlear fluid in response to motion of the stapedial footplate, resulting in a shearing force that causes bending of the cilia of the outer hair cells. Because the basilar membrane of the cochlea gradually changes in width, tension, and viscoelastic properties, the inner ear is tonotopically arranged. Thus, each portion of the basilar membrane vibrates with maximum amplitude to only one specific resonance frequency. Higher frequencies are nearer the oval window and lower frequencies are nearer the helicotrema (8, 31).

One hypothesis about amplification of the sound stimulus within the cochlea stems from somatic electromotility of the outer hair cells attributable to prestin, a motor protein. Accordingly, the outer hair cells contract and elongate synchronously with the sound-evoked receptor potential. However, evidence is accumulating to suggest an alternative hypothesis. Force generation by hair bundles may play a role in cochlear amplification (53a).

Auditory Pathway

Auditory nerve fibers in the spiral ganglion project from the cochlea to the anterolateral, posteroventral, and dorsal cochlear nuclei, whose neurons are tonotopically organized. In these areas, lateral inhibition enhances contrast, suppressing noise. Binaural comparison of intensity and direction of sound takes place at the level of the superior olive and accessory nucleus. Decussations occur at the levels of the nucleus of the lateral lemniscus and the inferior quadrigeminal bodies. Synapses occur with numerous afferent fibers serving as a reflex station, for example, for the muscles of the middle ear. At this level too, sensory information from the cochlear nuclei is compared with spatial information from the superior olive. With connections to the superior quadrigeminal bodies, auditory and visual space are coordinated. At the level of the medial geniculate body, afferent fibers project to the primary auditory cortex and secondary auditory areas (Brodmann's areas 41 and 42) (31).

Ear Abnormalities and Hearing Loss

Abnormalities may affect the pinna, auditory ossicles, or inner ear; they may occur as isolated defects or as components of various syndromes. Hearing loss may be conductive, sensorineural, or mixed, and the degree of loss may be mild, moderate, severe, profound, or total (anacusic). Many conditions with hearing loss are known, both syndromic and non-syndromic; the identification of mutation-bearing genes is progressing at a rapid rate and 75 loci for isolated deafness have been identified to date (84, 95). Some of these are listed in Table 4-8.

Superficial siderosis, in which hemosiderin, caused by repeated episodes of subarachnoid bleeding, is preferentially deposited on the surface of the eighth nerve, cerebellum, and brainstem. This eventually results in hearing loss, ataxia, and myelopathy. Superficial siderosis may contribute to several forms of genetic deafness and possibly to late-onset deafness (32).

BALANCE

The vestibular system of the inner ear is adjacent to and continuous with the cochlea (Fig. 4-16). The perilymph between the bony and membranous labyrinths and the endolymph that fills the membranous labyrinth are continuous with the same fluids in the cochlea. The vestibular apparatus has five divisions: a bulbous utricle, a bulbous saccule, and three semicircular canals, each of which is at right angles to the others. The crista ampullaris in each canal contains hair cells with cilia that project into the cupula, a gelatinous membrane. Each hair cell has a long kinocilium and many stereocilia of variable lengths, whose tips are connected to longer adjacent cilia via "tip links" (31).

The semicircular canals can detect head rotation, nodding, and tilting. When cilia are in the resting state, hair cells release glutamate that fires the action potentials in the nerve fibers of the vestibular ganglion. When the head turns, the semicircular canals also turn, but the endolymph moves more slowly, causing the stereocilia to bend towards the kinocilium and allowing an influx of potassium and sodium ions (K^+, Na^+) between the endolymph and the hair cell interior. This results in open calcium (Ca^{2+}) channels, releasing more glutamate and increasing the action potentials in the nerve fibers of the vestibular ganglion (31).

Both the utricle and the saccule contain maculae with cilia projecting into a gelatinous membrane containing statoconia composed of calcium carbonate. These crystals displace the membrane when the head deviates from the perpendicular axis (31, 39).

Bipolar neurons of the vestibular ganglion, forming the vestibular part of the vestibulocochlear nerve, synapse with the vestibular nuclei. Important tracts extend to the contralateral side, to the ocular muscle nuclei, cerebellum, motor neurons of the skeletal muscles, and to the postcentral gyrus (31, 39, 105) (Table 4-5).

The vestibular reflexes vertically position the eyes when the head tilts or otherwise changes position. There are also vestibular effects on postural motor control to maintain body balance. The vestibular nuclei receive and process visual information as well as that from proprioceptors in the neck muscles. Efferent fibers, projecting to the eye muscles and the vestibulo-ocular reflex, maintain spatial orientation (31).

OLFACTION

Anatomy and Physiology

Within the neuroepithelium are olfactory sensory neurons; basal cells, which are their progenitors, replacing them every 30–60 days; and supporting cells (Fig. 4-17). The dendrites of the sensory neurons branch to form cilia, which project into a layer of mucus secreted by Bowman's glands in the supporting cells. Odorant molecules dissolve in the mucous layer and combine with receptor proteins that are coded by olfactory genes (*vide infra*) (31, 64).

Axons of the same sensory neuron type synapse to dendrites of their respective mitral cells and bristle cells within the glomeruli of the olfactory bulb (Fig. 4-18). The glomeruli

Table 4-8 Some Conditions with Ear Anomalies and Hearing Loss[a]

SYNDROMES WITH EAR ANOMALIES

Beckwith–Wiedemann syndrome
Branchio-oto-renal syndrome
CHARGE association and Hall–Hittner syndrome
Diastrophic dysplasia
Langer–Giedion syndrome
Noonan syndrome
Oculo-auriculo-vertebral spectrum
Potter sequence
Retinoic acid embryopathy
Robinow syndrome
Saethre–Chotzen syndrome
Townes–Brocks syndrome
Treacher Collins syndrome

TERATOGENIC SYNDROMES WITH HEARING LOSS

Rubella syndrome
Fetal alcohol syndrome

NON-SYNDROMIC HEARING LOSS

Locus	Mutation-Bearing Genes
DFNA1	*DIAPH1*
DFNA2	*KCNQ4*
DFNA2	*GJB3*
DFNA3	*GJB6*
DFNA3, DFNB1	*GJB2*
DFNA4	*MYH14*
DFNA5	*DFNA5*
DFNA6/14	*WFS1*
DFNA8, DFNA12, DFNB21	*TECTA*
DFNA9	*COCH*
DFNA11, DFNB2	*MY07A*
DFNA13	*COL11A2*
DFNA15	*POU4F3*
DFNA17	*MYH9*
DFNA48	*MY01A*
DFNB3	*MY015*
DFNB4	*SLC26A4*
DFNB9	*OTOF*
DFNB10	*TMPRSS3*
DFNB12	*CDH23*
DFNB23	*PCDH15*
DFNB29	*CLDN14*
DFNB31	*WI*
DFNB37	*MY06*
DFN3 (X-linked)	*POU3F4*

SYNDROMES ASSOCIATED WITH HEARING LOSS

Syndromes	Mutation-Bearing Genes
Usher syndrome, type 1B	*MY07A*
Usher syndrome, type 1C	*USH1C*
Usher syndrome, type 1D	*CDH23*
Usher syndrome, type 2A	*USH2A*
Usher syndrome, type 1F	*PCDH15*
Waardenburg syndrome, type 1	*PAX3*
Waardenburg syndrome, type 2	*MITF*
Waardenburg syndrome, type 4	*SOX10, EDNRB, EDN3*
Stickler syndrome	*COL2A1, COL11A1, COL11A2*
OSMED	*COL11A1, COL11A2*
Pendred syndrome	*SLC26A4*
Alport syndrome	*COL4A3, COL4A4, COL4A5*
Charcot–Marie–Tooth syndrome	*GJB1*
Renal tubular acidosis and hearing loss	*ATP6B1*
Jervell/Lange–Nielsen syndrome	*KCNQ1, KCNE1*
DFN1	*TIMM8A*
Neurofibromatosis, type 2	*NF2*
Osteogenesis imperfecta	*COL1A1, COL1A2*
Muenke syndrome	*FGFR3*
Branchio-oto-renal syndrome	*EYA1*
Treacher Collins syndrome	*TCOF1*
Townes–Brocks syndrome	*SALL1*

[a]Although this table is representative, it is far from exhaustive.

Based in part on Steel and Kros, 2001 (95), Petit, 2004 (84), and Yang et al., 2005 (107a).

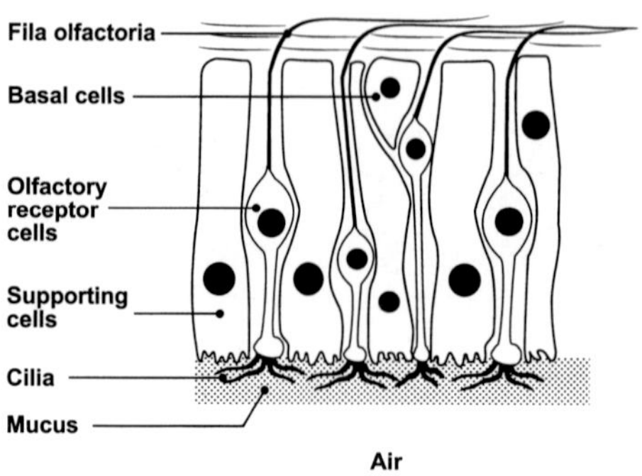

Figure 4-17 Olfactory epithelium.

act as convergence centers to integrate and relay signals from the same sensory neuron type. The respective receptor protein also determines the connection specific sensory neurons will have with a specific glomerulus. Further connections involve periglomerular cells and granular cells, which inhibit mitral and bristle cells, but higher centers can relieve this inhibition (31, 64).

The signals of mitral cell axons reach the anterior olfactory nucleus. Its neurons cross over in the anterior commissure of the contralateral

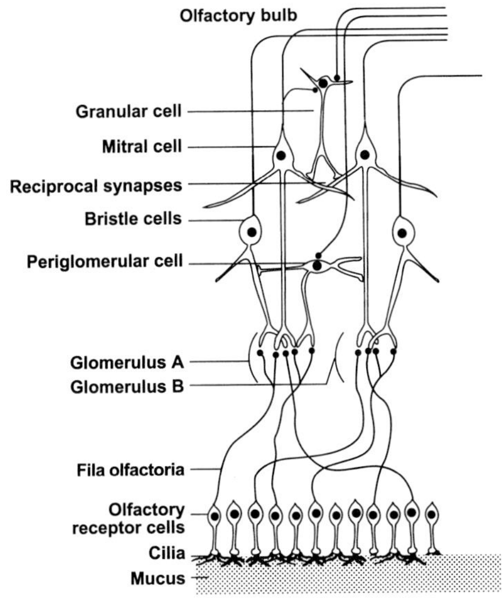

Figure 4-18 Olfactory pathway. Glomerulus A and glomerulus B act as convergence centers to integrate and relay signals from the same sensory neuron type. See text.

olfactory bulb, form the olfactory tracts, which project to the primary olfactory cortex, where the olfactory input is processed, and are relayed to the hypothalamus, limbic system, reticular formation, and neocortex directly or via the thalamus (31, 39) (Fig. 4-19).

Molecular Biology

The molecular basis of olfaction has been discussed by many authors (58, 66, 75, 87, 88). The olfactory gene family is the largest in the human genome, surpassing the immunoglobulin and T-cell antigen receptor family of genes. About 1,000 genes are known. Humans rely less on their sense of smell than other vertebrates. About 30% of human olfactory genes are functional; the other 70% are pseudogenes. In contrast, mice have as few as 5–10% pseudogenes. Olfactory receptor genes are found on all human chromosomes, except 20 and Y. They predominate on chromosomes 7, 16, and 17 (58, 75, 88).

Olfactory receptor genes can be divided into two main groups. Class I genes are found in fish. Class II genes are found in most mammals. Frogs express both types: Class I in tadpoles with gills and Class II in frogs with lungs. In humans, Class II genes predominate, but more than 10% of functional olfactory genes are of the Class I fish-like type (58).

Odorants possess the following molecular properties: some degree of water solubility, sufficiently high vapor pressure, low polarity, lipophilicity, and surface activity. The cilia of the sensory neurons are immersed in a layer of mucus. Odorant molecules dissolve in the mucus and bind to G protein-coupled receptors, which are 7-pass transmembrane proteins (64). Each

Figure 4-19 The rhinencephalon consists of the olfactory nerves, bulbs, tracts, and striae, a portion of the anterior perforated substance, primary olfactory cortex, entorhinal cortex of the parahippocampal gyrus, and corticomedial portion of the amygdala. The structures of the olfactory system are shaded and labeled. On the left side, the temporal lobe has been pulled to the left to expose more of the brain. The black line indicates projections of the mitral cells of the olfactory bulb. From Gilman and Newman, 1988 (39).

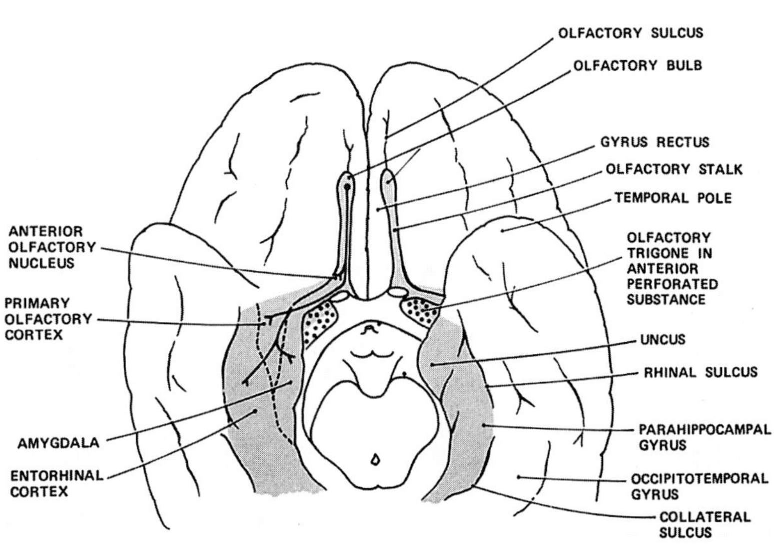

olfactory sensory neuron expresses only one receptor. Several models of odorant receptor transcriptional regulation have been proposed. In the short promoter model, the sequences immediately upstream of olfactory receptor genes contain transcription factor binding sites sufficient to regulate receptor transmission. Which gene is expressed depends on the transcription factor(s) expressed in the olfactory sensory neuron. In the locus control region model, proximal promoters contain transcription factor binding sites necessary to drive specific receptor transcription, but are not available until factors that bind to a distal locus control region make the region accessible to transcription. In the recombination model, olfactory receptor genes are translocated by recombination or copied by gene conversion into a single active locus for expression (58).

Trigeminal Sensory Nerve Endings

Trigeminal sensory nerve endings are found in the nasal mucosa. Although several times less sensitive than olfactory sensory neurons, trigeminal nerve endings do react to strong odors, such as ammonia, menthol, camphor, mustard, onion, or chili powder (64).

Vomeronasal Organ

Vertebrates have evolved two separate chemosensory systems: olfactory sensory neurons (*vide supra*) and the vomeronasal sensory neurons, which are located in the lining of the nasal septum. They have microvilli, in contrast to the cilia associated with olfactory neurons. The traditional distinction that common odors are routed through the olfactory system, whereas pheromones are routed through the vomeronasal system is now obsolete. Each pathway must be evaluated for any putative pheromone (4, 13, 54, 91a).

Receptor families expressed in the vomeronasal organ include the V1R family with about 35 members, V2R family with about 150 members, and V3R family. Within the vomeronasal organ itself, an upper zone expresses V1Rs and the G protein Gai2, and a lower zone that expresses V2Rs and the G protein GaO. The two zones differ in their projection to the accessory olfactory bulb. Gai2-positive neurons project to the rostral half and GaO-positive neurons project to the caudal half (4, 13, 54, 81).

It is now known that the neuroendocrine system can be controlled by common odors and pheromones, both acting through the main olfactory pathway. A new class of pheromones—non-volatile MHC class I peptides—activate both olfactory and vomeronasal sensory neurons. The Bruce effect, mediated by the vomeronasal system, is a form of pregnancy disruption in mice; postimplantation fails on exposure to an unknown male. However, the main olfactory pathway is required for mating preference (91a).

Humans do not have a functional vomeronasal system, but depend on the olfactory pathway. Menstrual cycle synchrony has been demonstrated in female populations in constant close contact, such as in a women's college dormitory (4, 13, 54, 81).

Follow Your Nose: Studies of Odors during Infancy, Childhood, and Adulthood

Twenty-five newborn infants were tested for their facial response to various odors prior to their first feeding. Photographs and movies were taken during the study. Records were made of the unstimulated face in response to a swab containing an odorless solvent used to dilute the test odorants: butter, banana, vanilla, shrimp, and rotten eggs. These stimuli were artificial food flavors approved by the U.S. Food and Drug Administration. Photographs of facial responses were coded as facial acceptance, facial indifference, or facial rejection by a panel of judges, including three physicians. The results are shown in Table 4-9. The smell of butter and banana had the highest acceptance rates. The smell of shrimp and particularly rotten eggs had the highest rejection rates (97).

Table 4-9 Facial Response of Neonates to Odors[a]

Odor Stimulus	Acceptance	Indifference	Rejection
Butter	62%	32%	6%
Banana	55%	25%	20%
Vanilla	46%	46%	8%
Shrimp	22%	25%	53%
Rotten eggs	0%	0%	100%

[a]Based on the facial responses of 25 neonates by 6 judges.

Modified from Steiner, 1977 (97).

In another study, 150 children 3 to 6 years of age were given plastic squeeze bottles containing several odors: bubble gum, sour milk, beer, and mineral oil as a neutral smell. If children liked the smell, they passed the bottle to a stuffed Big Bird toy. If they disliked the smell, they passed the bottle to an Oscar-the-Grouch toy for disposal. Separately, the children's parents answered questions about alcohol use. The smell of bubble gum was liked best (86%). The smell of sour milk was disliked the most (89%). About half the children liked the smell of beer (53%). However, when the data were grouped according to parental drinking habits, only 27% of children with parents who drank to escape liked the smell of beer, in contrast to 66% of children with nonproblem-drinking parents who liked the smell of beer (62).

In still another study, women were asked to judge odors from a wide range of men. Smells were coded as familiar, intense, pleasant, or spicy. Women were found to prefer the odor of males whose HLA most closely matched their own, and the HLA alleles in question were inherited from their fathers. This paternal effect is most readily explained by genomic imprinting (49, 50). Clearly, "Father nose best!"

Olfactory thresholds increase with age. The threshold for smelling foods, such as cherries, grapes, and lemons, is 11 times higher, on the average, for older persons than for younger persons (90).

Noses Nobody Knows: The Pathology of Sniffing

In olfactory pathology, the sense of smell can be absent (anosmia), diminished (hyposmia), increased (hyperosmia), or distorted (dysosmia). Table 4-10 summarizes conditions with abnor-

Table 4-10 Conditions with Abnormal Olfaction

Dysmorphic and genetic disorders	*Tumors*
Holoprosencephaly	Nasal polyposis
Arhinencephaly[a] (isolated, Kallmann syndrome, holoprosencephaly, trisomy 13 syndrome)	Nasopharyngeal angiofibroma
Absent nose[b]	Glioma (frontal lobe)
Basal encephalocele	Meningioma (olfactory groove)
Campomelic dysplasia[c] (*SOX9* mutations, 17q24.3-q25)	*Trauma*
Kallmann syndrome (*KAL1* mutations, Xp22.3; *FGFR1* loss-of-function mutations in *KAL2* (8p11.1-p11.2)	Head trauma
Cystic fibrosis (*CFTR* mutations, 7q31.2)	Fracture of cribriform plate
Nervous system disorders	Partial ethmoidectomy for hypertelorism correction[d]
Alzheimer disease	Laryngectomy
Epilepsy	*Endocrine disorders*
Huntington disease (*Huntingtin* gene, 4p16.3)	Cushing syndrome
Multiple sclerosis	Diabetes mellitus (many genes, many loci)
Parkinson disease	Hypothyroidism
Schizophrenia	Pseudohypoparathyroidism (*GNAS1* mutations, 20q13.2)
Infectious, viral, and allergic disorders	Turner syndrome (monosomy X, Xp22.33)
Common cold	*Chemicals*
Allergic rhinitis	Amphetamines
Chronic rhinitis	Cocaine
Adenoid hypertrophy	Some industrial chemicals
Sinusitis	Heavy smoking
Osteomyelitis of ethmoid	*Other conditions*
Meningitis	Cirrhosis of the liver
Abscesses of frontal lobe	Chronic renal failure
Bronchial asthma	Leprosy
Viral hepatitis	Sjögren syndrome

[a]Arhinencephaly is defined as absent olfactory tracts and bulbs. It can be a minimal form of holoprosencephaly in some cases, but not in others [Cohen, 1989 (18), 2001 (22)]. See Chapter 5 for etiology of holoprosencephaly.

[b]Absent nose is rare, but over 20 cases have been reported [Cohen, 1993 (20)].

[c]About 18% of cases have arhinencephaly [Cohen, 1989 (18), 2001 (22)].

[d]In this operation, the cribriform plate is preserved, but occasionally olfaction may be diminished.

Data from Brodal, 1981 (11), Cohen, 1989 (18), 1993 (20), 2001 (22), 2002 (23), Cohen and Sulik, 1992 (24), Deems et al., 1991(28), Gilman and Newman, 1988 (39), Kopala et al., 1998 (57), Moberg et al., 1999 (73), Schiffman, 1983 (90), Soussi–Yanicostas et al., 2002 (94).

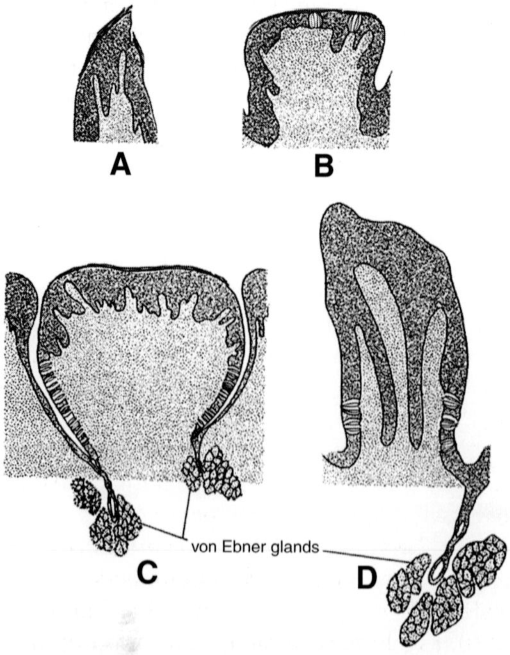

Figure 4-20 Lingual papillae. A, Filiform papilla. B, Fungiform papilla. C, Vallate papilla with associated serous minor salivary glands (von Ebner glands). D, Foliate papilla with associated von Ebner gland. Modified from Moss-Salentijn, 1985 (78).

mal olfaction under the following headings: dysmorphic and genetic disorders, nervous system disorders; infectious, viral, and allergic disorders; tumors; trauma; endocrine disorders; chemicals; and other conditions. The pathology of many conditions in Table 4-10 appears to be related to a decreased rate of olfactory neuron replacement; normally, replenishment occurs every 30–60 days (31, 89).

The most common causes of a diminished or absent sense of smell are upper respiratory infections, head trauma, and chronic nasal and paranasal sinus disease (28, 89). Arhinencephaly is simply absence of the olfactory tracts and bulbs; it is part of the holoprosencephalic spectrum in some cases, but not in others (22). Anosmin-1 is defective in the X-linked form of Kallmann syndrome, type 1 (*KAL1* mutations) (94). *FGFR1* loss-of-function mutations cause Kallmann syndrome, type 2 (KAL2) (81). Olfactory hallucinations occur in epilepsy and are virtually always unpleasant. Olfactory deficits are found in schizophrenia (57, 73).

Body odors caused by metabolic products and bacterial decomposition are of interest because of their diagnostic importance. Historical examples include, among others, typhoid fever (fresh-baked brown bread), isovaleric acidemia (odor of sweaty feet), diabetic coma (fruity odor), yellow fever (odor of a butcher shop), and maple syrup–urine disease (90).

GUSTATION

Anatomy and Histology

The tongue has four types of papillae: filiform, fungiform, foliate, and circumvallate (Fig. 4-20, Table 4-11). The filiform papillae are located on the anterior and posterior surface of the tongue. They are the most numerous papillae and they contain no taste buds. The fungiform papillae are confined to the anterior two-thirds of the tongue, numbering about 200. They contain one

Table 4-11 Lingual Papillae

Types of Papillae	Shapes	Locations	Special Considerations[a]
Filiform	Jagged mountain-shaped	Over tongue surface	Most abundant Mechanical function No taste buds
Fungiform	Small mushroom-shaped	Among filiform papillae	Red color because epithelium is thin. Have several taste buds
Circumvallate	Large mushroom-shaped	V-shaped groove at border between body and base of tongue	8–12 in number sunken below surface Associated serous minor salivary glands Have taste buds
Foliate	Large ridge-shaped	Sides of tongue near base	Associated serous minor salivary glands Have taste buds

[a]Taste buds are also found in the palate, epiglottis, pharynx, and upper esophagus.

or more taste buds. The foliate papillae are located on the lateral posterior borders of the tongue. They contain taste buds and are associated with serous minor salivary glands (von Ebner glands). The circumvallate papillae, 8–12 in number, are located at the junction of the oral and pharyngeal parts of the tongue. Taste buds are located on the sides of the papillae and are recessed below the tongue surface. Each papilla is surrounded by a moat bathed in saliva from von Ebner glands (9, 31).

Taste buds are found not only on the tongue, but also on the palate, epiglottis, pharynx, and upper esophagus. Each taste bud has a taste pore, which is a small opening on the epithelial surface. Taste cells are short lived and, like surrounding epithelial cells, are rapidly renewed in two-week cycles (9, 31). ATP is the key neurotransmitter for communication from the taste buds to the gustatory nerves (34a).

Innervation

The fungiform papillae are innervated by the sensory neurons of the geniculate ganglion, whose peripheral branches travel in the chorda tympani branch of the facial nerve (VII). The circumvallate papillae are innervated by sensory neurons in the petrosal ganglion, which travel in the lingual branch of the glossopharyngeal nerve (IX). Taste buds in the palate are innervated by the greater superficial branch of the facial nerve (VII). Taste buds of the epiglottis and esophagus are innervated by the superficial laryngeal branch of the vagus nerve (X). The endings of the facial (VII), glossopharyngeal (IX), and vagus (X) nerves are relayed by the nucleus tractus solitarius and converge on the postcentral gyrus via the thalamus, and the hypothalamus and limbic system via the pons (9, 31).

Physiology

Five taste qualities are known: sweet, salty, sour, bitter, and umami (Table 4-12). The concentrations of gustatory stimuli determine whether tastes are considered pleasant or unpleasant. Higher concentrations of salty, sour, and bitter tastants are commonly considered unpleasant, whereas sweet tastants tend to be more easily tolerated at higher concentrations (31). In contrast to the large number of odorant molecules recognized by smelling, fewer numbers of molecules are recognized by tasting. Furthermore, the threshold concentrations for detecting tastants is much higher than the threshold for detecting odorants (33).

The perception of flavor results primarily from gustatory and olfactory sensations together with tactile and thermal sensations (Table 4-5). Saliva is also essential for taste because it acts as a solvent for the chemical stimuli in food and transports them to the taste buds. The loss of olfaction results in almost complete loss of food enjoyment. The loss of salivation can also result in loss of food enjoyment (9). Combined olfaction and gustation for food enjoyment was recognized in 1826 by a French gourmet, Brillat-

Table 4-12 Taste Modalities

Taste Qualities	Examples of Taste Stimuli	Mechanisms of Transduction	Receptors	Effectors
Salty	Salt	Na^+, K^+, and H^+ acting directly on ion channels		
Sour	Citric acid	Na^+, K^+, and H^+ acting directly on ion channels		
Sweet	Sucrose	G protein-coupled receptors	Combined T1R2 + T1R3	TRPM5, PLCβ2
Bitter	Quinine, phenylthiocarbamide (PTC) G-n-propyl-2-thiouracil (PROP)	G protein-coupled receptors	Many T2Rs	TRPM5, PLCβ2
Umami	L-amino acids monosodium-L-glutamate (MSG)	G protein-coupled receptors	Combined T1R1 + T1R3 mGluR4	TRPM5, PLCβ2

Data from Bradley, 1995 (9); Adler et al., 2000 (1a); Chandrashekar et al., 2000 (16); Chaudhara et al., 2000 (17); Dulac, 2000 (33), Mombaerts, 2000 (76); Caicedo and Roper, 2001 (14); Max et al., 2001 (72); Montmayeur et al., 2001 (77); Nelson et al., 2002 (79); Amrein and Bray, 2003 (2); Damak et al., 2003 (26); Despopoulos and Silbernagl, 2003 (31); Kim et al., 2003 (55); and Zhang et al., 2003 (109).

Savarin, who said "... smell and taste are in fact but a single sense, whose laboratory is the mouth and whose chimney is the nose ..." (10).

Molecular Biology

Two modalities—sour and salty tastes—are mediated directly by the interaction of cations (Na^+, K^+, H^+) with ion channels. The other three modalities–sweet, bitter, and umami tastes– rely on G protein-coupled receptors, which are seven transmembrane proteins (76) (Table 4-12). These receptors contain a G protein α-subunit, gusductin (16). Multigene families responsible for the transduction of sweet, bitter, and umami tastes have been identified by many authors (1a, 2, 16, 17, 26, 71, 72, 77, 79, 86, 109).

Sweet and amino acid (umami) tastes are mediated by G protein-coupled receptors that combine to generate at least two heteromeric receptors. T1R2 and T1R3 associate to function as a broadly tuned sweet receptor. T1R1 and T1R3 form a universal L-amino acid sensor (79). It has also been shown that mGluR4, a metabotropic glutamate receptor variant, functions as a taste receptor for umami (17). Receptors for the bitter modality are discussed in a separate section below. Responses to sweet, bitter, and umami stimuli require two additional signaling molecules, TRPM5 and PLCβ2 (109). Heat activation of TRPM5 underlies thermal sensitivity of sweet taste (98a) (Table 4-12).

The Bitter End

Betty Botter bought some butter
but, she said, the butter's bitter;
if I put it in my batter
it will make my batter bitter,
but a bit of better butter
will make my batter better
So she bought a bit of butter
better than her bitter butter,
and she put it in her batter
and the batter was not bitter.
So 'twas better Betty Botter
bought a bit of better butter.

Bitter taste perception has a protective function in animals, including humans. A bitter taste elicits an aversive response to noxious and toxic stimuli, such as poisonous substances. Many bitter tastes are known, including, among others, quinine, G-n-propyl-2-thiouracil (PROP), and phenylthiocarbamide (PTC). For some individuals, PROP and PTC taste bitter, whereas for others, they do not (55, 86) (Table 4-12).

Bitter tastants are detected by about 30 T2Rs (Table 4-12) that map to 5p15, 7q31, and 12p13. A large number of T2R pseudogenes are also known to occur (1a, 16, 86). T2R genes are expressed in about 15% of taste cells (16). It has been suggested that bitter sensation does not occur by activation of a single, homogeneous population of bitter-sensitive taste cells, but by activation of multiple heterogeneous taste cells (14).

Gustofacial Responses of Neonates

Distinguishable gustofacial responses can be observed in neonates when sweet, sour, and bitter solutions are tasted. With sweet solutions, retraction at the corners of the mouth resembling a "smile" is accompanied by sucking and licking the upper lip. When an infant's bottle contains water with sugar in it, pain is diminished in response, for example, to immunization. With sour solutions, the lips are pursed and the nose wrinkles together with quick repeated blinking and flushing. With bitter solutions, the lips are "arched" with lowering at the corners of the mouth, protrusion of the tongue, salivation, spitting, and a facial expression showing "anger" and "dislike." If the solution is a bit stronger, vomiting may occur (29, 96). No characteristic gustofacial response is observed to salty solutions up to 7 months of age. However, different sucking responses show that neonates can taste and differentiate various concentrations of sodium chloride (36, 52).

When mothers are exposed to carrot juice during pregnancy or during breast feeding, their infants, at 7 months of age, show a distinct preference for carrot porridge rather than plain porridge. In contrast, when mothers are exposed to water during pregnancy and during breast feeding, their infants, at 7 months of age, did not show this preference (45).

Pathology

Gustatory disorders may be classified by the type of sensory loss: ageusia, or complete loss of taste; hypoageusia, or partial loss of taste; hyperageusia, or increased sensitivity to taste; and dysgeusia, or aberrant taste. Disorders of taste are much less common than disorders of olfaction (9).

Taste thresholds for sweeteners, salt, and amino acids are 2-2 1/2 times higher in the elderly, on the average, than in the young (90). Disorders of the nervous system, cancer, chronic renal failure, cirrhosis of the liver, and various endocrinopathies may be associated with gustatory abnormalities (89).

Loss of taste may occur from radiation treatment for head and neck tumors, which affects the salivary glands and results in loss of saliva essential as a transport medium for taste stimuli. Xerostomia may also occur with the salivary pathology of Sjögren syndrome. Burning mouth can be found in some postmenopausal women; taste perception can be altered (9) (see Chapter 7).

Gustatory hallucinations occur in some patients with epilepsy. Patients with familial dysautonomia, an autosomal recessive disorder, show a marked reduction in taste sensitivity. They also have absence of the red-colored fungiform papillae from the dorsum of the tongue, which is diagnostic of the disorder. Neurological lesions affecting cranial nerves VII, IX, and X, the nucleus tractus solitarius, or the pathway to the postcentral gyrus may result in altered taste perception. Lesions affecting the pathways from the salivary glands through cranial nerves VII and IX to the (superior and inferior) salivatory nucleus may also result in altered taste perception (39, 40).

OTHER SENSES

Mechanosensors

The sense of touch provided by mechanosensors is essential for perceiving the form, shape, and spatial nature of various objectives. Such stereognostic perception requires central nervous system integration from adjacent receptors and their coordination with tactile motor function. Tactile sensors are especially prominent in the palms, fingers, and orofacial areas (31).

Mechanoreceptors can generate action potentials during the dynamic phase of stimulus application as *rapidly adapting receptors (RA)* or during both dynamic and static phases as *slowly adapting receptors (SA)*. Both RA and SA can be subdivided into two types. The rapidly adapting type I (RA I) and the slowly adapting type I (SA I) have small receptor fields, whereas rapidly adapting type II (RA II) and slowly adapting type II (SA II) have large receptor fields (9).

Hairless areas of the skin contain the following mechanoreceptors. Meisser corpuscles, located in the papillary ridges of the dermis, are RA I type mechanoreceptors; hair follicle receptors serve the same function in hairy areas of the skin. Merkel cell neurite complexes, consisting of epithelial cells with a disk-like expansion of nerve fiber terminals, are SA I type mechanoreceptors. SA II responses originate in Ruffini corpuscles, which are long spindle-shaped structures made up of nerve fibers branching inside the receptor capsule. RA II receptors are Panician corpuscles of the hypodermis composed of layers of Schwann cell lamellae surrounding a central-receptor axon terminal (9, 31).

RA I type mechanoreceptors are located in facial skin, but responses from RA II units typical of Panician corpuscles are not found. Mechanoreceptors are present on the lips, tongue, and oral cavity. Receptors of the Ruffini type are found in the periodontal ligament (9).

The posterior funiculus–lemniscus system is the principal route by which the postcentral gyrus of the parietal lobe receives input from mechanosensors, thermosensors (*vide infra*), propriosensors (*vide infra*), and nocisensors (*vide infra*). Secondary afferent trigeminal fibers end in the posteromedial ventral nucleus of the thalamus. Tertiary afferent fibers end in the somatosensory cortex (31) (Table 4-5).

Thermosensors

Thermal sensations on the body surface, face, and oral cavity have dual expressions—areas of warm or cold stimulation separated by areas insensitive to thermal stimulation. Cold-sensitive spots are more numerous than warm-sensitive spots. The highest densities of both kinds of skin receptors are found on the face. Thermal receptors are important in the maintenance of body temperature. Depending on the environmental temperature, reflex sweating or shivering can be initiated, particularly in response to sudden changes. Heat sensors are also used for the perception of pungent substances such as capsaicin, the active ingredient in hot chili peppers (9). Central pathways involving thermosensors have already been discussed in the section on mechanosensors (*vide supra*) (Table 4-5).

Propriosensors

Proprioception, the sensing of muscular strength, position, and movement, is important in jaw muscle closure. Cutaneous mechanosensors assist propriosensors in muscle spindles, joints, and tendons (31).

Muscle spindles regulate muscular length. Two types of infrafusal muscle fibers are chain fibers and nuclear bag fibers. Type Ia[e] afferent neurons coil around both types, whereas type II[e] neurons wind around nuclear chain fibers only. These endings detect longitudinal stretching of infrafusal muscle fibers, determining their length by types Ia and II afferents, and changes in length by type Ia afferents. The efferent γ-motor neurons innervate both infrafusal fiber types, permitting variation in their length and stretch sensitivity. Golgi tendon organs, important in the control of muscle tension, are found at the junction of a muscle and its tendon. Their impulses are conveyed by type Ib[e] afferents (31). Central pathways involving propriosensors have already been discussed in the section on mechanosensors (*vide supra*) (Table 4-5).

Muscle spindles present in jaw elevator muscles—masseter,temporal,andmedialpterygoid—are concentrated in the deep and anterior parts of the muscles. Although muscle spindles are absent from the tongue in most species, they are present in primates. Jaw depressor muscles lack spindles. Small numbers of Golgi tendons are found deep in the masseter and temporalis at the point of insertion into the mandible. Receptor endings are found in the temporomandibular joint and are only important in extreme open rotation of the joint (9).

[e]Ia, Skeletal muscle efferent and afferents in muscle spindles. Ib, Skeletal muscle efferent and afferents in Golgi tendon organs. II, Mechano-afferents of the skin.

Nocisensors

Sensors for pain, present in all tissues except the brain and the liver, are bead-like endings of peripheral axons, the somata of which are located in the dorsal root ganglia and in the nuclei of the trigeminal nerve. Nociception may involve craniofacial skin, muscles, bone, oral and nasopharyngeal mucosae, teeth, and periodontium (9, 31).

Two groups of nocisensors are the myelinated, faster conducting Aδ type, activated by intense mechanical stimuli and the unmyelinated, slower conducting C type, activated by intense mechanical, thermal, and chemical stimuli. Pain can also be elicited from areas of the skin devoid of sensors because free nerve endings can act as nocisensors (9, 31). Central pathways involving nocisensors are discussed in the section on mechanosensors (*vide supra*) (Table 4-5).

MUSCLES OF FACIAL EXPRESSION

The muscles of facial expression are cutaneous and lie within the layers of the superficial fascia (Fig. 4-21). Generally, they arise from fascia or from bone and insert into the skin. Individual muscles seldom remain distinct throughout their length because they tend to merge with neighboring muscles at their terminations or attachments. Table 4-13 lists the muscles of facial expression, their anatomic locations, their associated VIIth nerve branch, and their actions. In this table, the muscles are arranged under the following headings: scalp, ear, eye, nose, mouth, and neck (41).

Upper motor neuron cell bodies for the facial region are located in the precentral gyrus of the frontal lobe. Axons pass through the corona radiata, internal capsule, crus cerebri of the midbrain, and dispersed pyramidal tract bundles of the pons. In the lower part of the pons, mostly crossed but some uncrossed fibers (a) pass into the facial nucleus that supplies the lower facial muscles and (b) pass into the reticular formation

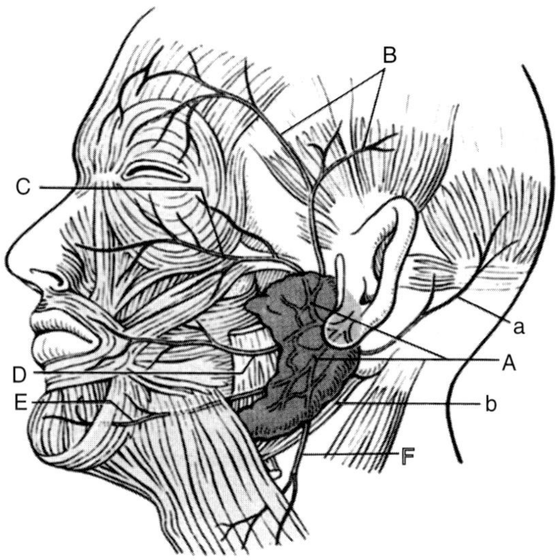

Figure 4-21 Muscles of facial expression and parotid gland. A, Parotid plexus of VII. B, Temporal branches of VII. C, Zygomatic branches of VII. D, Buccal branches of VII. E, Mandibular branches of VII. F, Cervical branch of VII for the platysma muscle. a, Posterior auricular nerve from VII before it enters the parotid gland. b, Preparotid branches of VII to the anterior belly of the digastric muscle. From Kahle, 1986 (53).

near that part of the nucleus that supplies the upper facial muscles. Axons involved in lower facial movement synapse directly with neurons in the facial nucleus. Axons involved in upper facial movement terminating in the reticular formation synapse with *interneurons* that project into the nucleus (105) (Fig. 4-22).

The difference in the termination of the upper motor neurons explains the contralateral lower facial weakness associated with upper motor neuron lesions. In the older concept, upper motor neuron facial weakness was attributed to bilateral upper motor neuron input into that part of the facial nucleus, which supplied the upper facial muscles, in contrast to contralateral input into that part of the facial nucleus, which supplied the lower facial muscles (Fig. 4-22). Studies have not verified the older concept (51, 105).

In addition to the motor component of VII for the muscles of facial expression, a sensory component carries gustatory impulses from the anterior two-thirds of the tongue and proprioceptive impulses from the face as well as an autonomic–parasympathetic component. These latter fibers arise in the superior salivary nucleus and provide secretory and vasodilatory function (a) to the submandibular and sublingual glands via the intermediate nerve, chorda tympani, and lingual nerve, and (b) to the lacrimal gland via the intermediate nerve, greater superficial petrosal nerve, vidian nerve, sphenopalatine ganglion, and zygomaticotemporal nerve. Proximal to the geniculate ganglion, all these components are in congruity (40).

A lesion at the level of the stylomastoid foramen results in complete paralysis of the facial muscles on the ipsilateral side. Although corneal sensation persists, the corneal reflex is lost on the affected side because the motor fibers for this reflex do not function (39, 40). Idiopathic swelling of the facial nerve in the distal part of the facial canal results in Bell's palsy (21). When a lesion affects the facial nerve in the internal acoustic canal or in the proximal part of the facial canal, the facial nerve fibers to the stapedius muscle and intermediate nerve fibers will be involved. Thus, in addition to unilateral facial paralysis, other ipsilateral effects include increased sensation to sound (hyperacusis), absence of taste sensation (ageusia) on the anterior two-thirds of the tongue, and disturbed secretion of tears and saliva. If a lesion occurs proximal to the geniculate ganglion, then during regeneration, fibers destined for the submandibular and sublingual salivary glands may become interchanged with those for the lacrimal gland. Thus, when a gustatory stimulus is evoked, lacrimation is produced (39, 40) (Fig. 4-23).

MUSCLES OF MASTICATION

The muscles of mastication proper and their accessory muscles together with their associated cranial nerves and actions are summarized in Table 4-14. The muscles of mastication proper are illustrated in Figure 4-24.

The temporal muscle arises from the temporal fossa, passes deep to the zygomatic arch, and inserts into the coronoid process of the mandible.

Table 4-13 Muscles of Facial Expression

Muscles	Locations	Seventh Cranial Nerve Branches	Actions
Scalp			
Occipitofrontalis	Forehead and occiput; frontal and occipital bellies connected by by galea aponeurotica	Posterior auricular (occipital belly), temporal (frontal belly)	Occipital and frontal bellies acting together tighten scalp, raise eyebrows, and wrinkle forehead; Frontal bellies acting alone raise eyebrows either on one or both sides
Ear			
Auricularis anterior	In front of helix	Temporal	Draws auricle forward and upward
Auricularis superior	Above auricle	Temporal	Draws auricle upward
Auricularis posterior	Behind auricle	Posterior auricular[a]	Draws auricle backward
Eye			
Corrugator	Medial end of eyebrow beneath frontalis and orbicularis oculi	Temporal	Draws eyebrows downward and medialward, resulting in vertical forehead wrinkles
Orbicularis oculi	Circumscribes orbit	Temporal, zygomatic	Closes eyelids
Nose			
Procerus	Lower part of nasal bone and upper part of lateral nasal cartilage	Buccal	Depresses eyebrows medially and produces transverse wrinkling over bridge of the nose
Nasalis	Cartilaginous nasal bridge and greater alar cartilage	Buccal	Depresses cartilaginous part of nose and draws ala toward the septum
Depressor septi	Lateral to philtrum	Buccal	Constricts aperture of nares
Dilator naris posterior	Partly beneath levator labii superioris	Buccal	Enlarges aperture of nares
Dilator naris anterior	Greater alar cartilage and integument near margin of nostril	Buccal	Enlarges aperture of nares
Mouth			
Levator labii superioris	Upper lip	Buccal	Elevates upper lip
Levator labii superioris alaeque nasi	Upper lip and side of nose	Buccal	Dilates nares and elevates upper lip
Levator anguli oris	Corner of mouth	Buccal	Lifts corner of mouth
Zygomaticus major	Cheek and corner of mouth	Buccal	Lifts corner of mouth
Zygomaticus minor	Cheek and corner of mouth	Buccal	Elevates upper lip
Risorius	Cheek	Buccal	Retracts corners of mouth
Orbicularis oris	Circumscribes mouth	Buccal	Closes, purses, and protrudes lips
Buccinator	Cheek	Buccal	Compresses cheek
Depressor labii inferioris	Lower lip	Mandibular	Depresses lower lip
Depressor anguli oris	Corner of mouth	Mandibular	Depresses corner of mouth
Mentalis	Chin	Mandibular	Raises and protrudes lower lip and wrinkles skin of chin
Neck			
Platysma	Neck, chin	Cervical	Draws outer part of lower lip downward and backward, widening corners of mouth; also assists in depressing the mandible

[a]Before the facial nerve enters the parotid gland to form a plexus, it gives off the posterior auricular branch.

Data from Goss, 1961, (41).

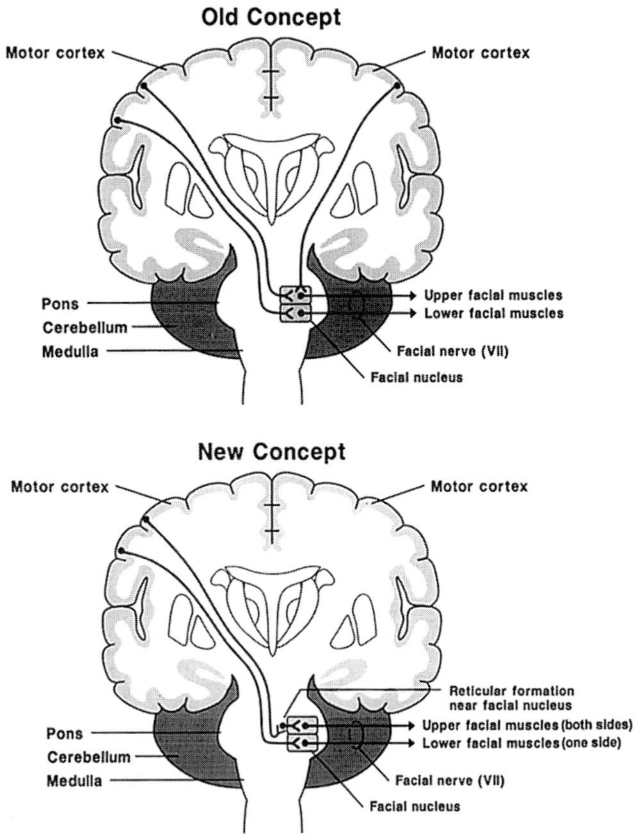

Figure 4-22 Upper motor neuron facial weakness. Old and new concepts. From Watson, 1995 (105).

The masseter muscle arises from the zygomatic process of the maxilla and the zygomatic arch itself, and inserts into the lateral portion of the mandibular ramus. The medial pterygoid muscle arises from the lateral pterygoid plate, pyramidal process of the palatine bone, and maxillary tuberosity, and inserts into the medial surface of the mandibular ramus. The lateral pterygoid muscle arises from the great wing of the sphenoid and infratemporal crest, and inserts into the neck of the mandibular condyle and the articular disk of the temporomandibular joint (41).

The temporal, masseter, and medial pterygoid muscles close the jaws. The lateral pterygoid muscle together with the anterior belly of the digastric muscle, and to a lesser extent, by the platysma muscle, open the jaws. The buccinator muscles, which compress the cheeks, help guide food onto the occlusal plane for crushing and grinding (41) (Table 4-14).

TRIGEMINAL NERVE

The trigeminal nerve (Table 4-3) is the largest of all cranial nerves. It is the sensory nerve of the face, mucous membranes, and other internal structures of the head. It is the motor nerve of the muscles of mastication. Thus, functional components include general somatic afferents (facial skin, eyeglobe and adnexa, oral and nasal mucosae), special visceral efferents (muscles of mastication), and general somatic efferents (proprioception for muscles of mastication) (41, 105).

The trigeminal nerve gives rise to the ophthalmic, maxillary, and mandibular divisions

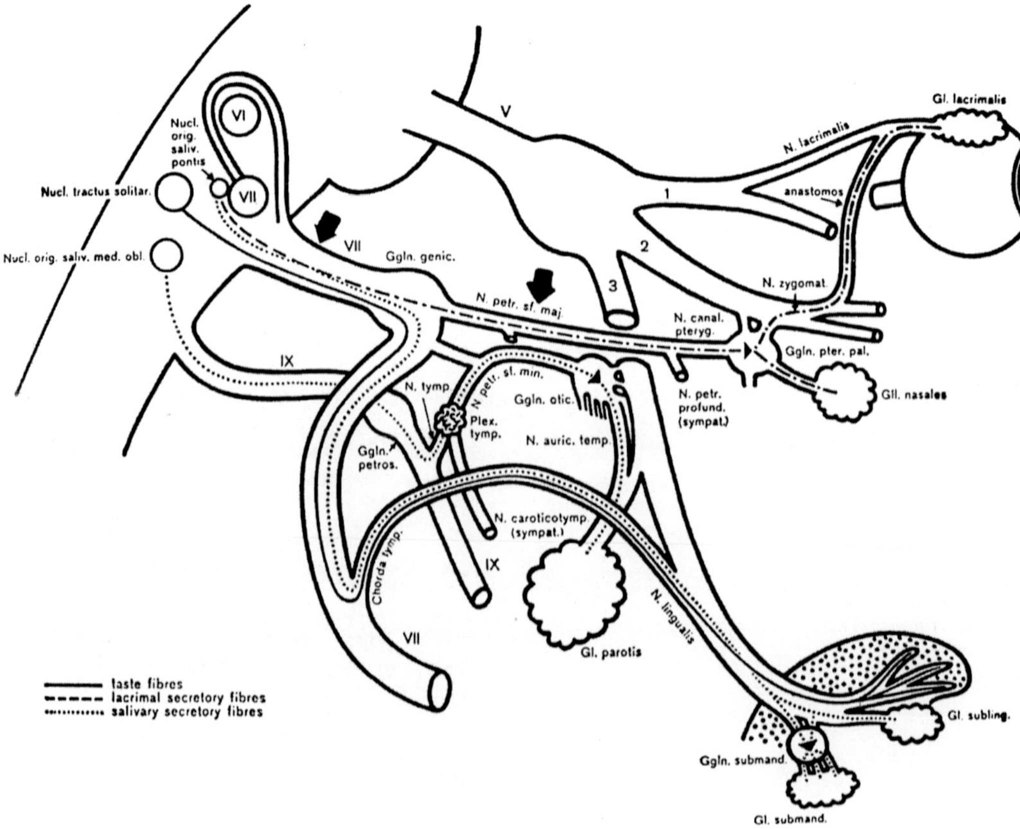

Figure 4-23 Gustatory lacrimation. Diagram shows two possible mechanisms. Most often, the lesion is proximal to the geniculate ganglion and in regeneration, fibers destined for the submandibular and sublingual salivary glands may become interchanged with those from the lacrimal gland. When the lesion is distal to the geniculate ganglion, interchange of fibers of the greater and lesser petrosal nerves become involved. From Gorlin et al., 2001 (40).

(Fig. 4-25). The ophthalmic division is sensory and supplies the eyeglobes, conjunctiva, lacrimal glands, part of the nasal mucosa and paranasal sinuses, and skin of the forehead, eyelids, and nose (41) (Fig. 4-25).

The maxillary division is sensory and supplies the skin of the middle part of the face, lower eyelids, sides of the nose, maxillary sinus, tonsils, hard and soft palate, upper gingivae and alveolae, and upper teeth (41) (Fig. 4-25).

The mandibular division is both sensory and motor. The sensory fibers supply the skin of the temporal region, ears, external auditory meatus, cheeks, lower lip, lower part of the face, oral mucous membranes, tongue, lower gingivae and alveolae, lower teeth, mandible, temporomandibular joint, mastoid air cells, and part of the dura mater and skull (41) (Fig. 4-25).

The motor fibers of the mandibular division supply the muscles of mastication (masseter, temporal, medial pterygoid, and lateral pterygoid muscles), myelohyoid muscle, anterior belly of the digastric muscle, tensor tympani, and tensor veli palatini (41) (Fig. 4-25).

Table 4-14 Muscles of Mastication

Muscles	Cranial Nerves (Branches)	Actions
Muscles Proper		
Temporal	V (Mandibular)	Closes jaws Retracts mandible
Masseter	V (Mandibular)	Closes jaws
Medial pterygoid	V (Mandibular)	Closes jaws
Lateral pterygoid	V (Mandibular)	Opens jaws Protrudes mandible Moves mandible side to side
Accessory Muscles		
Digastric		Assists in jaw opening
Anterior belly	V (Mandibular)	
Posterior belly	VII (Preparotid[a])	
Buccinator	VII (Buccal)	Compresses cheeks and is therefore an accessory muscle of mastication
Platysma	VII (Cervical)	Assists in jaw opening

[a]Branch of the VIIth cranial nerve before it enters the parotid gland to form the parotid plexus.

Data from Goss, 1961 (41).

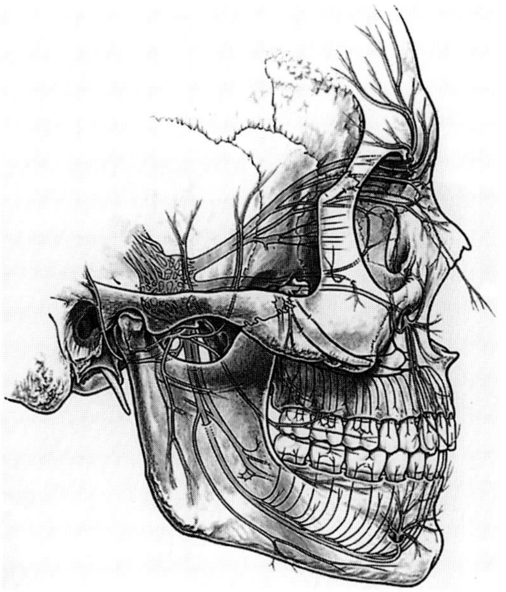

Figure 4-25 Trigeminal nerve showing ophthalmic, maxillary, and mandibular divisions. From Evers and Haegerstam, 1990 (34).

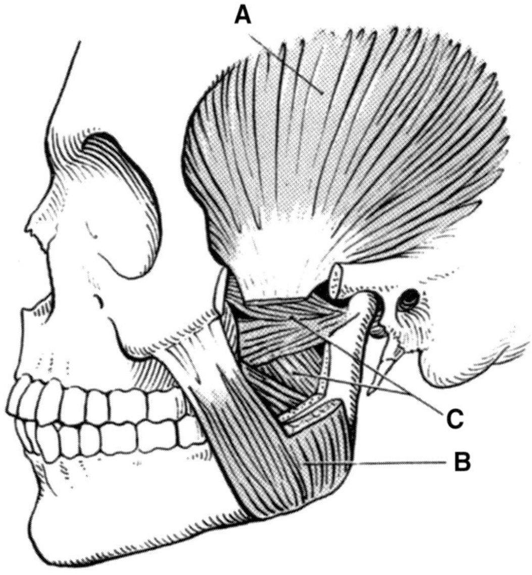

Figure 4-24 Muscles of mastication. A, Temporalis muscle. B, Masseter muscle. C, Lateral and medial pterygoid muscles. Modified from Kahle, 1986 (53).

SALIVARY GLANDS

There are three pairs of major salivary glands: parotid, submandibular, and sublingual. The parotid gland is the largest (14–20 g) followed by the submandibular gland (10–15 g). The sublingual gland is the smallest (~ 2 g) (19). The major salivary glands and their anatomic locations, secretions, ducts, nerve supply, and arterial supply are summarized in Table 4-15. The anatomic position of the parotid gland is shown in Figure 4-21.

The secretory system and duct system of the major salivary glands are shown in Figure 4-26. The parotid gland contains only serous acini; submandibular gland, mostly mucous tubules with serous demilunes, but also some pure mucous tubules and serous acini; and sublingual gland, mostly mucous tubules, some serous demilunes, and rare serous acini. Parotid saliva is completely serous; submandibular gland, mixed mucous and serous; and sublingual saliva, predominantly mucous (19).

Minor salivary glands located in the oral mucosa include labial glands, buccal glands, palatine glands, and lingual glands. In fact, they are found

Table 4-15 The Major Salivary Glands

Location	Secretion	Duct	Nerve Supply	Arterial Supply
Parotid gland				
Parotid region, interval between posterior border of the ramus and upper anterior border of the sternocleidomastoid muscle	Serous	Parotid duct travels anteriorly along the lateral border of the masseter muscle, rolls over the anterior border of the masseter to pierce the buccinator, and empties into the oral cavity at the occlusal level of the second maxillary molar	Preganglionic, lesser petrosal nerve, a branch of IX. Synapse: optic ganglion. Postganglionic fibers leave the otic ganglion and travel with the auriculotemporal branch of V-3 to the parotid region	Glandular branches of the superficial temporal artery and the transverse facial artery
Submandibular gland				
Superficial portion, submandibular region between the lateral aspect of the myelohyoid muscle and the submandibular fossa of the mandible. Deep portion, floor of the mouth between the base of the tongue and the sublingual gland	Serous and mucous	Submandibular duct continues forward from the deep portion of the gland and empties into the oral cavity behind the lower central incisors	Preganglionic, chorda tympani nerve, a branch VII. Synapse: submandibular ganglion. Postganglionic fibers leave the ganglion and pass to the gland	Glandular branches of the facial and lingual arteries
Sublingual gland				
Floor of the mouth, medial to the sublingual fossa of the mandible, above the myelohyoid muscle	Mucous	Sublingual ducts open directly into the oral cavity through openings in the sublingual fold, anterior ducts open medially into the submandibular duct	Preganglionic, chorda tympani nerve, a branch of VII. Synapse: submandibular ganglion. Postganglionic fibers leave the ganglion and pass to the gland	Glandular branches of the lingual artery

Modified from Liebgott, 1986 (67).

throughout the oral region, except the most anterior portion of the hard palate and gingivae. Almost all are mucous glands; the only serous glands are found in association with the circumvallate and foliate papillae of the tongue (19).

About 600–800 mls. of saliva are produced every 24 hours. The contributions of each gland to whole resting saliva are as follows: submandibular gland (67%), parotid gland (23%), and sublingual and minor glands (10%). When salivation is stimulated, the contribution of the parotid gland rises to 50%. Central nervous system control of salivary secretion and the nerve supply to the major salivary glands are summarized in Table 4-15 and illustrated in Figure 4-27. Salivation is affected in various pathological states: excess salivation (sialorrhea), pseudo-excess salivation (pseudo-sialorrhea), deficient salivation (xerostomia), and pseudodeficient salivation (pseudoxerostomia) (19) (Table 4-16).

The main electrolytes in saliva are sodium, potassium, calcium, chloride, bicarbonate, and inorganic phosphate. Although the salivary duct system can reabsorb ions, the amount depends on the salivary flow rate. In the parotid gland, for example, as salivary flow increases, sodium, chloride, and bicarbonate ions increase, whereas potassium ions and inorganic phosphate decrease. Iodine metabolism involves the salivary glands as well as the thyroid gland and gastric glands. They all concentrate iodide to many times the plasma level. Iodide secreted by the salivary glands and gastric glands is reabsorbed in the small intestine (19).

Saliva serves many functions: lubrication, which permits food particles to be tasted and aids in mastication, food bolus formation, and deglutition; preliminary digestion of starches by salivary amylase; protection by providing a mucosal barrier and producing various antimicro-

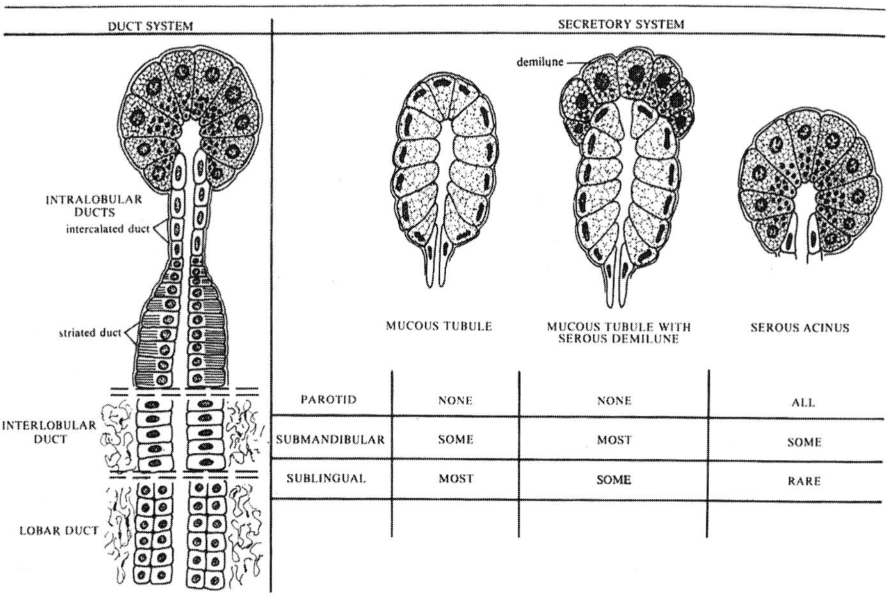

	MUCOUS TUBULE	MUCOUS TUBULE WITH SEROUS DEMILUNE	SEROUS ACINUS
PAROTID	NONE	NONE	ALL
SUBMANDIBULAR	SOME	MOST	SOME
SUBLINGUAL	MOST	SOME	RARE

Figure 4-26 Secretory units and duct system of the three major salivary glands. The table lists the relative proportions of the three types of secretory units in each salivary gland. From Borysenko and Beringer, 1984 (8).

bial factors (lysozyme, secretory IgA, and antibodies); and buffering capacity by increased bicarbonate ion concentration against acid produced by bacteria. Salivary glands also produce clotting factors, which aid in oral healing, and secrete physiologically active substances, such as nerve growth factor, vasoactive peptides, and regulatory peptides (19).

DENTITION

The dentition consists of two sets of teeth: 20 deciduous and 32 permanent (see Table 4-1 for details). Teeth have crowns projecting above the gingiva and one to three roots (depending on the particular tooth) attached to the alveolar bone by the periodontal ligament.

Tooth Development

The embryonic enamel organs of the primary dentition develop directly from downgrowths of the dental lamina from the oral epithelium. The enamel organs of the permanent dentition arise from successional dental lamina lingual to those of the primary teeth, although the three permanent molars develop from a backward extension of the dental lamina. Enamel, a product of ameloblasts, is of epithelial origin. Dentin, a product of the odontoblasts, and cementum, a product of cementoblasts, are of mesenchymal origin. The periodontal ligament and alveolar bone are also of mesenchymal origin (78).

Calcified Dental Tissue

Three types of mature calcified dental tissue are found. Enamel is composed of about 96% inorganic matter. Dentin is composed of about 80% inorganic matter. Cementum is composed of about 65% inorganic matter similar to that found in bone (8, 78). The chronologies of the deciduous and permanent dentitions are summarized in Tables 4-17 and 4-18 and illustrated in Figures 4-28 and 4-29 (69, 70, 91).

Dental Pathology

Caries is caused by the action of acidogenic bacteria on carbohydrates. Bacterial plaque on the

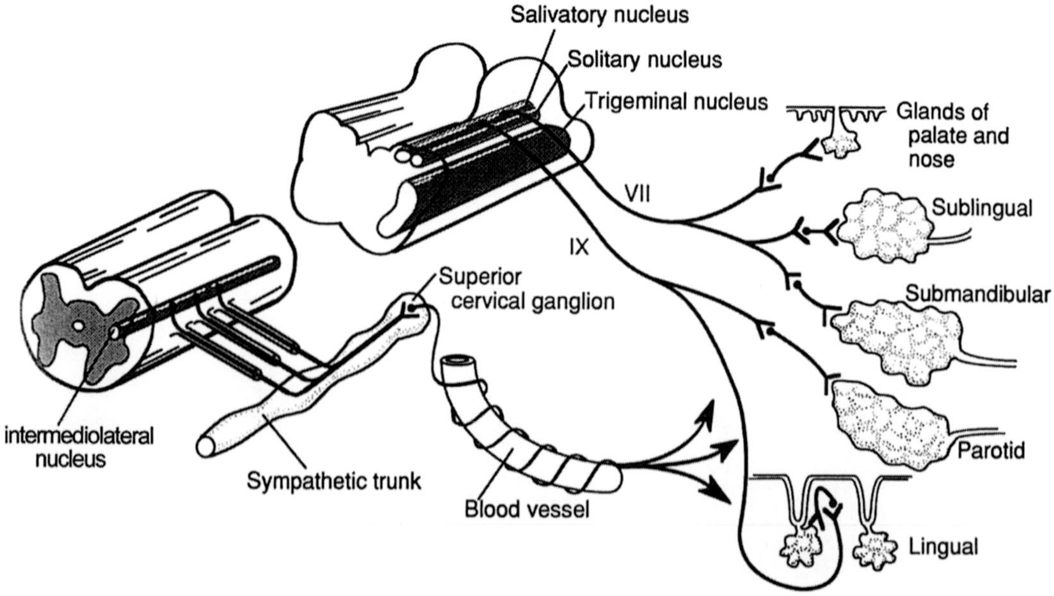

Figure 4-27 Central nervous system control of salivary secretion. In the medulla, the salivatory nucleus contains cell bodies of efferent preganglionic secretomotor fibers that are distributed to the salivary glands with the facial (VII) and glossopharyngeal (IX) nerves. These efferent fibers synapse in peripheral parasympathetic ganglia. Postganglionic fibers are distributed to the salivary glands. Cell bodies of preganglionic sympathetic neurons are in the intermediolateral nucleus of the thoracic spinal cord. Preganglionic fibers are distributed to the superior cervical ganglion of the sympathetic trunk; they synapse with the cell bodies of postganglionic neurons. Postganglionic sympathetic secretomotor fibers are distributed to the salivary glands with blood vessels. Modified from Bradley, 1995 (9).

teeth results in decalcification. If the process extends to the pulp, inflammation can cause severe pain, and if it continues through the apical foramen, a dentoalveolar abscess may occur. Improper oral hygiene with food deposits in the gingival crevices may result in gingivitis. If the process extends further, periodontitis with alveolar bone loss may result in loosening and loss of teeth.

In Class I occlusion, the relationship between the maxilla and the mandible is normal. However, individual tooth discrepancies can occur, resulting in a Class I malocclusion. In Class II malocclusion, mandibular retrusion occurs. A Class III malocclusion is characterized by mandibular prognathism.

Other dental abnormalities are listed in Table 4-19. More complete discussion is available elsewhere (40). Many odontogenic tumors are known, including, among others, ameloblas-

Table 4-16 Clinical Correlates of Salivary Flow

Excess (sialorrhea)
Acute oral inflammations
Drugs
Neurological disturbances
Pseudo-excess (pseudosialorrhea)
Esophageal atresia
Paralysis of tongue or pharyngeal muscles
Deficient (xerostomia or dry mouth)[a]
Inflammation of salivary glands (parotitis)
Dehydration
Drugs
Anxiety states
Radiation
Sjögren syndrome
Congenitally absent salivary glands
Significant loss of parotid acini with fatty replacement in old age[b]
Pseudodeficient (pseudoxerostomia)
Mouth breathing

[a]Dry mouth predisposes to dental caries.

[b]Can result in dry mouth, despite assertions to the contrary (9).

Table 4-17 Normal Chronological Development of the Primary Teeth

Tooth	Initiation, Week in Utero	Calcification Begins, Week in Utero	Crown Completed, Month	Eruption, Month	Root Completed, Year	Root Resorption Begins, Year	Tooth Shed, Year
Central incisor	7	14 (13–16)	1–3	6–9	1½	5–6	7–8
Lateral incisor	7	16 (14½–16½)	2–3	7–10	1½–2	5–6	7–9
Canine	7½	17 (15–18)	9	16–20	2½–3¼	6–7	10–12
First molar	8	15½ (14½–17)	6	12–16	2–2½	4–5	9–11
Second molar	10	18½ (16–23½)	10–12	20–30	3	4–5	11–12

Data from Logan and Kronfeld, 1933 (69); Lunt and Law, 1974 (70); and Schour and Massler, 1940 (91).

Table 4-18 Normal Chronological Development of the Permanent Teeth

Tooth	Initiation, Month	Calcification Begins	Crown Completed, Year	Eruption, Year	Root Completed, Year
Maxilla					
Central incisor	5–5¼ in utero	3–4 months	4–5	7–8	10
Lateral incisor	5–5¼ in utero	1 year	4–5	8–9	11
Canine	5½–6 in utero	4–5 months	6–7	11–12	13–15
First premolar	Birth	1½–1¾ years	5–6	10–11	12–13
Second premolar	7½–8	2–2½ years	6–7	10–12	12–14
First molar	3½–4 in utero	Birth	2½–3	6–7	9–10
Second molar	8½–9	2½–3 years	7–8	12–13	14–16
Third molar	3½–4 (years)	7–9 years	12–16	17–25	18–25
Mandible					
Central incisor	5–5¼ in utero	3–4 months	4–5	6–7	9
Lateral incisor	5–5¼ in utero	3–4 months	4–5	7–8	10
Canine	5½–6 in utero	4–5 months	6–7	9–11	12–14
First premolar	Birth	1¾–2 years	5–6	10–12	12–13
Second premolar	7½–8	2¼–2½ years	6–7	11–12	13–14
First molar	3½–4 in utero	Birth	2½–3	6–7	9–10
Second molar	8½–9	2½–3 years	7–8	11–13	14–15
Third molar	3½–4 (years)	8–10 years	12–16	17–25	18–25

Data from Logan and Kronfeld, 1933 (69) and Schour and Massler, 1940 (91).

toma, odontoma, adenomatoid odontogenic tumor, calcifying epithelial odontogenic tumor, and four types of cementomas (85).

MASTICATION, DEGLUTITION, AND VOMITING

Mastication

The chewing cycle is divided into four phases. At the point of minimal opening, the jaw partially opens at a slow rate followed by a faster opening phase. In jaw closure, a fast phase is followed by a final slow phase. A masticatory sequence consists of a number of chewing cycles from ingestion to swallowing. The muscles of mastication contract asynchronously with wide variability in the time of onset, peak of activity, and decline of activity. During chewing, the tongue and the buccinator muscles control the movement of food by guiding it to the occlusal table for crushing, grinding, and forming a bolus of food for swallowing. Motor and sensory nuclei in the brainstem play an essential role in controlling-mastication. Integrated activity of a number of muscles involve the trigeminal, hypoglossal, and facial nuclei. Coordination relies on sensory input from the oral cavity, terminating primarily in the trigeminal nucleus and solitary tract nucleus (9).

Deglutition

The oral phase of swallowing is under voluntary control, whereas the pharyngeal and esophageal

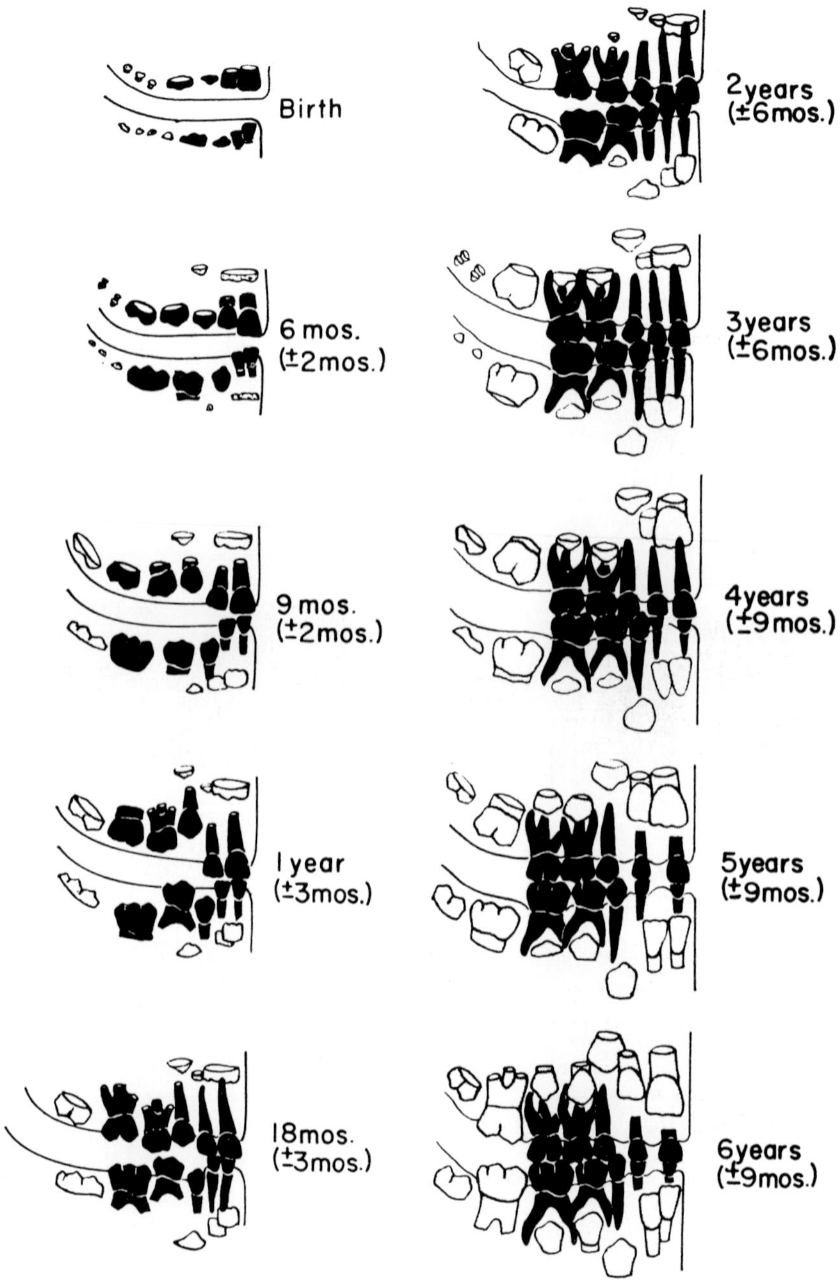

Figure 4-28 Development and eruption of the primary dentition and early development of the permanent dentition. From Schour and Massler, 1940 (91).

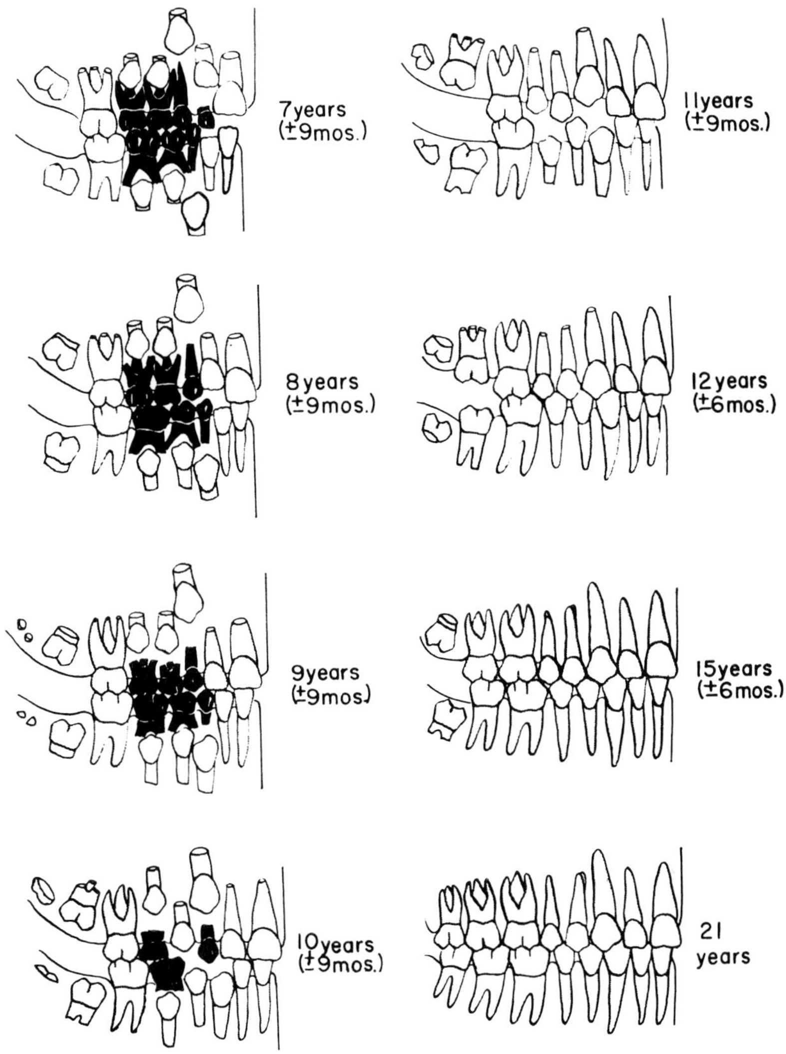

Figure 4-29 Further development and eruption of the permanent dentition. From Schour and Massler, 1940 (91).

phases are involuntary. As the tongue pushes a bolus of food into the oropharynx and then into the hypopharynx, the nasopharynx is reflexively blocked, respiration is inhibited, the vocal cords close, the epiglottis seals off the trachea, and the upper esophageal sphincter opens. At the beginning of deglutition, the lower esophageal sphincter also opens, caused by a vagovagal reflex, which is mediated by vasoactive intestinal peptide- and nitric acid-releasing neurons (9, 31).

Vomiting

Vomiting serves as a protective reflex for various toxins. It is also an important symptom in many other conditions, including, among others, intracranial bleeding, tumors, morning sickness of pregnancy, motion sickness, surgical anesthesia, radiation treatment, overextension of the stomach, delayed gastric emptying, inflammation of abdominal organs, and pyloric stenosis. Chronic vomiting can occur secondary to malnutrition (9, 31).

Table 4-19 Dental Abnormalities

Abnormalities	Comments
Delayed eruption	Untreated hypothyroidism, untreated hypopituitarism, trisomy 21 syndrome
Natal and neonatal teeth	Isolated (sometimes autosomal dominant), Hallermann–Streiff syndrome, Ellis–van Creveld syndrome (*EVC* gene), pachyonychia congenita, cyclopia
Premature exfoliation of deciduous teeth	Traumatic avulsion, hypophosphatasia, cyclic neutropenia, acrodynia, acatalasia, Langerhans cell histiocytosis, Papillon–Lefèvre syndrome: *CTSC* mutations (11q14), juvenile periodontitis: *CTSC* mutations (11q14), Haim–Munk syndrome: *CTSC* mutations (11q14)
Hypodontia	Absent or hypoplastic maxillary lateral incisors (2% of the general population), hypodontia: *MSX1* mutations (4p16), Witkop syndrome: *MSX1* mutations (4p16), *PAX9* (14q12-q13), hypodontia/clefting: *MSX1* mutations (4p16), hypodontia and hypoplastic teeth in hypohidrotic ectodermal dysplasia: [(*EDA1* mutations in X-linked form (Xq12.2-q13.1); *EDAR* mutations in autosomal dominant form (2q11-q13); and *EDARADD* mutations in autosomal recessive form (1q42.2-q43)], hypodontia in Rieger syndrome: *RIEG1* (4q25-q26) and *RIEG2* (13q14), *AXIN2* mutations (16p13.3) in the combination of hypodontia and colorectal cancer
Supernumerary teeth	Isolated, Gardner form of multiple adenomatous polyposis: *APC* mutations (5q21-q22)
Abnormal tooth position	Transposition, ectopic teeth
Abnormal tooth size	Small tooth size in rubella syndrome, large teeth with XYY karyotype, globodontia in otodental syndrome, canine radiculomegaly in oculofaciocardiodental syndrome: *BCOR* mutations (Xp11.4)
Tooth anomalies	Gemination (one tooth bud and one tooth with partially duplicated crown), twinning (one tooth bud with two teeth), fusion (two tooth buds and two teeth with dentin union), concresence (two tooth buds and two teeth with cementum union), dens invaginatus (invagination in the center of the tooth crown during development, giving a "tooth within a tooth" or "dens in dente" appearance)
Enamel hypoplasia	Various prenatal and neonatal infections; tetracycline can cause pigmented hypoplasia
Amelogenesis imperfecta	Amelogenesis imperfecta (hypoplastic local type): *ENAM* mutations (4q21), amelogenesis imperfecta (hypoplastic type): *ENAM* mutations (4q21), amelogenesis imperfecta (hypomature type): *KLK4* mutations (19q13.3-q13.4), amelogenesis imperfecta (hypoplastic type): *AMGN* mutations (Xp22.1-p22.3)
Dentin defects	Dentinogenesis imperfecta 1: *DSPP* mutations (4q21.3), dentin dysplasia, type 2: *DSPP* mutations (4q21.3), osteogenesis imperfecta with dentinogenesis imperfecta: *COL1A1* and *COL1A2* mutations (7q22.1, 17q21.31-q22)
Taurodontism	Isolated, amelogenesis imperfecta/taurodontism (type I and type II), Klinefelter syndrome and X-aneuploidy states (XXY, XXXY, XXXXY), trisomy 21 syndrome, Ackerman syndrome, tricho-dento-osseous syndrome: *DLX3* mutations (17q21.3-q22), amelogenesis imperfecta and taurodontism only: *DLX3* mutation (17q21.3-q22)
Miscellaneous tooth defects	Dystrophic epidermolysis bullosa (enamel pitting), Morquio syndrome (hypoplastic enamel), pseudohypoparathyroidism (males have thin enamel with small pits; females have normal enamel thickness with large pits), tuberous sclerosis (pit-shaped enamel defects), vitamin D-resistant rickets (large pulp chambers with high pulp horns)

Vomiting is heralded by nausea and retching. It is accompanied by widespread autonomic effects, such as salivation, pupillary dilation, sweating, and pallor. During the act of vomiting, the diaphragm is in the inspiratory position and the abdominal muscles contract, exerting pressure on the stomach. Contraction of the duodenum blocks the way to the gut and the lower esophageal sphincter relaxes, resulting in the ejection of stomach contents (31).

The vomiting center is located in the reticular formation of the medulla oblongata. It is controlled mainly by fourth ventricle chemosensors in the area postrema that monitor blood content without interference of the blood–brain barrier (31).

REFERENCES

1. Abelson JF, Kwan KY, O'Roak BJ, Baek DY, Stillman AA, Morgan TM, Mathews CA, Pauls DL, Rašin M-R, Gunel M, Davis NR, Ercan-Sencicek AG, Guez DH, Spertus JA, Leckman JF, Dure LS IV, Kurlan R, Singer HS, Gilbert

DL, Farhi A, Louvi A, Lifton RP, Sestan N, State MW: Sequence variants in *SLITRK1* are associated with Tourette's syndrome. Science 310: 317–320, 2005.

1a. Adler E, Hoon MA, Mueller KL, Chandrashekar J, Ryba NJP, Zuker CS: A novel family of mammalian receptors. Cell 100:693–702, 2000.

2. Amrein H, Bray S: Bitter-sweet solution in taste transduction. Cell 112:283–287, 2003.
3. Ashley–Koch A, Wolpert CM, Menold MM, Zaeem I, Basu S, Donnelly SL, Ravan SA, Powell CM, Qumsiyeh MB, Aylsworth AS, Vance JM, Gilbert JR, Wright HH, Abramson RK, DeLong GR, Cuccaro ML, Pericak–Vance MA: Genetic studies of autistic disorder and chromosome 7. Genomics 61:227–236, 1999.
4. Bargmann CI: Olfactory receptors, vomeronasal receptors, and the organization of olfactory information. Cell 90:585–587, 1997.
5. Bishop DV, North T, Donlan C: Genetic basis of specific language impairment: Evidence from a twin study. Dev Med Child Neurol 37:56–71, 1995.
6. Bishop DVM: Putting language genes in perspective. Trends Genet 18:57–59, 2002.
7. Boghosian–Sell L, Comings DE, Overhauser J: Tourette syndrome in a pedigree with a 7;18 translocation: Identification of a YAC spanning the translocation breakpoint at 18q22.3. Am J Hum Genet 59:999–1005, 1996.
8. Borysenko M, Beringer T: Functional Histology. Second Edition. Boston: Little, Brown & Company, 1984.
9. Bradley RM: Essentials of Oral Physiology. St. Louis: Mosby-Year Book, 1995.
10. Brillat–Savarin J: La Physiologie du Goût, ou Meditations de Gastronomie Transcendante, Dedie aux Gastronomes Parisiens par un Professeur. Paris: A. Sautelet et Cie, Vol. 8, 1826.
11. Brodal A: Neurological Anatomy. Third Edition. New York: Oxford University Press, 1981.
12. Bruce HA, Margolis RL: FOXP2: Novel exons, splice variants, and CAG repeat length stability. Hum Genet 111:136–144, 2002.
13. Buck LB: The molecular architecture of odor and pheromone sensing in mammals. Cell 100: 611–618, 2000.
14. Caicedo A, Roper SD: Taste receptor cells that discriminate between bitter stimuli. Science 291: 1157–1560, 2001.
15. Calder AJ, Keane J, Manes F, Antoun N, Young AW: Impaired recognition and experience of disgust following brain injury. Nat Neurosci 3: 1077–1078, 2000.
16. Chandrashekar J, Mueller KL, Hoon MA, Adler E, Feng L, Guo W, Zuker CS, Ryba NJP: T2Rs function as bitter taste receptors. Cell 100:703–711, 2000.
17. Chaudhara N, Landin AM, Roper SD: A metabotropic glutamate receptor variant functions as a taste receptor. Nat Neurosci 3:113–119, 2000.
18. Cohen MM Jr: Perspectives on holoprosencephaly. Part III. Spectra, distinctions, continuities, and discontinuities. Am J Med Genet 34:271–289, 1989.
19. Cohen MM Jr: Saliva and Salivary Glands. Teaching handout. 1990–2000.
20. Cohen MM Jr: The nose. In Human Malformations and Related Anomalies. R.E. Stevenson, J.G. Hall, and R.M. Goodman, Editors. New York: Oxford University Press. Vol. II, pp. 323–339, 1993.
21. Cohen MM Jr: Perspectives on craniofacial asymmetry. III. Common and/or well-known causes of asymmetry. Int J Oral Maxillofac Surg 24:127–133, 1995.
22. Cohen MM Jr: Problems in the definition of holoprosencephaly. Am J Med Genet 103:183–187, 2001.
23. Cohen MM Jr: Malformations of the craniofacial region: Evolutionary, embryonic, genetic, and clinical perspectives. Am J Med Genet (Semin Med Genet) 115:245–268, 2002.
24. Cohen MM Jr, Sulik KK: Perspectives on holoprosencephaly. Part II. Central nervous system, craniofacial anatomy, syndrome commentary, diagnostic approach, and experimental studies. J Craniofac Genet Dev Biol 12:196–244, 1992.
25. Dale PS, Simonoff E, Bishop DV, Eley TC, Oliver B, Price TS, Purcell S, Stevenson J, Plomin R: Genetic influence on language delay in two-year-old children. Nat Neurosci 1:324–328, 1998.
26. Damak S, Rong M, Yasumatsu K, Kokrashvili Z, Varadarajan V, Zou S, Jiang P, Ninomiya Y, Margolskee RF: Detection of sweet and umami taste in the absence of taste receptor T1r3. Science 301:850–853, 2003.
27. Davidson RJ, Ekman P, Saron CD, Senulis JA, Friesen WV: Approach–withdrawal and cerebral asymmetry: Emotional expression and brain physiology I. J Personality Soc Psychol 58:330–341, 1990.
28. Deems DA, Doty RL, Settle G, Moore–Gillon V, Shaman P, Mester AF, Kimmelman CP, Brightman VJ, Snow JB: Smell and taste disorders, a study of 750 patients from the University of Pennsylvania Smell and Taste Center. Arch Otolaryngol Head Neck Surg 117:519–528, 1991.
29. Desor JA, Maller O, Andrews K: Ingestive response of human newborns to salty, sour, and bitter stimuli. J Comp Physiol Psychol 89:966–970, 1975.
30. Despopoulos A, Silbernagl S: Color Atlas of Physiology. Third Edition. New York: Georg Thieme Verlag, 1986.
31. Despopoulos A, Silbernagl S: Color Atlas of Physiology. Fifth Edition. New York: Georg Thieme Verlag, 2003.
32. Dodson KM, Sismanis A, Nance WE: Superficial siderosis: A potentially important cause of ge-

netic as well as nongenetic deafness. Am J Med Genet 130A:22–25, 2004.
33. Dulac C: The physiology of taste, vintage 2000. Cell 100:607–610, 2000.
33a. Enlow DH: Facial Growth. Third Edition. Philadelphia: W.B. Saunders, 1990.
34. Evers H, Haegerstam G: Introduction to Dental Local Anaesthesia. Fribourg: Megaglobe SA, 1990.
34a. Finger TE, Danilova V, Barrows J, Bartel DL, Vigers AJ, Stone L, Hellekant G, Kinnamon SC: ATP signaling is crucial for communication from taste buds to gustatory nerves. Science 310: 1495–1499, 2005.
35. Fisher SE, Vargha–Khadem F, Watkins KE, Monaco AP, Pembrey ME: Localisation of a gene implicated in a severe speech and language disorder. Nat Genet 18:168–170, 1998.
36. Fomon SJ, Thomas LN, Filer LJ Jr: Acceptance of unsalted strained foods by normal infants. J Pediatr 76:242–246, 1970.
37. Foster D, Danker WH: The Nature of Stimuli: Basic Principles of Sensory Evaluation. Philadelphia: American Society for Testing and Materials, 1980.
38. Geschwind N: Specialization of the human brain. In The Brain: A Scientific American Book. San Francisco: W.H. Freeman, pp. 108–117, 1979.
39. Gilman S, Newman SW: Manter and Gatz's Essentials of Clinical Neuroanatomy and Neurophysiology. Seventh Edition. Philadelphia: F.A. Davis, 1988.
40. Gorlin RJ, Cohen MM Jr, Hennekam RCM: Syndromes of the Head and Neck. Fourth Edition. New York: Oxford University Press, 2001.
41. Goss CM: Gray's Anatomy. Twenty-seventh Edition. Philadelphia: Lea & Febiger, 1961.
42. Gouras P: Color vision (online, 2004). http://webvision.med.utah.edu/color.html
43. Graham A: Hear, hear for the inner ear. Science 290:1904–1905, 2000.
44. Hauser MD, Chomsky N, Fitch WT: The faculty of language: What is it, who has it, and how did it evolve? Science 298:1569–1579, 2002.
45. Holden C: Random samples. Science 289:387, 2000.
46. Holden C: The origin of speech. Science 308: 1316–1319, 2004.
47. Holowka S, Petitto LA: Left hemisphere cerebral specialization for babies while babbling. Science 297:1515, 2002.
48. Hubel DH, Wiesel TN: Brain mechanisms of vision. In The Brain: A Scientific American Book. San Francisco: W.H. Freeman, pp. 84–96, 1979.
49. Isles A: Father nose best. Trends Genet 18:182, 2002.
50. Jacob S, McClintock MZ, Zolano B, Ober C: Paternally inherited HLA alleles are associated with women's choice of male odor. Nat Genet 30:175–179, 2002.
51. Jenny AB, Sapir CB: Organization of the facial nucleus and corticofacial projection in the monkey: A reconsideration of the upper motor neuron facial palsy. Neurology 37:930–939, 1987.
52. Jensen K: Differential reactions to taste and temperature stimuli in newborn infants. Genet Psychol Monogr 12:361–479, 1932.
53. Kahle W: Color Atlas and Textbook of Human Anatomy. Volume 3. Nervous System and Sensory Organs. Third Edition. New York: Georg Thieme Verlag, 1986.
53a. Kennedy HJ, Crawford AC, Fettiplace R: Force generation by mammalian hair bundles supports a role in cochlear amplification. Nature 433:880–883, 2005.
54. Keverne EB: Pheromones, vomeronasal function, and gender-specific behavior. Cell 108:735–738, 2002.
55. Kim U-k, Jorgenson E, Coon H, Leppert M, Risch N, Dragna D: Positional cloning of the human quantitative trait locus underling taste sensitivity to phenylthiocarbamide. Science 299: 1221–1225, 2003.
56. Kimura D: A note on cerebral dominance in hearing. Acta Otolaryngol 56:617–618, 1963.
57. Kopala LC, Good KP, Torrey EF, Honer WG: Olfactory function in monozygotic twins discordant for schizophrenia. Am J Psychiat 155:134–136, 1998.
58. Kratz E, Dugas JC, Ngai J: Odorant receptor gene regulation: Implications from genomic organization. Trends Genet 18:29–34, 2002.
59. Ladher RK, Anakwe KU, Gurney AL, Schoenwolf GC, Francis–West PH: Identification of synergistic signals initiating inner ear movement. Science 290:1965–1967, 2000.
60. Lai CS, Fisher SE, Hurst JA, Levy ER, Hodgson S, Fox M, Jeremiah S, Povey S, Jamison DC, Green ED, Vargha–Khadem F, Monaco AP: The SPCHI region on human 7q31: Genomic characterization of the critical interval and localization of translocations associated with speech and language disorder. Am J Hum Genet 67:357–368, 2000.
61. Lai CS, Fisher SE, Hurst JA, Vargha–Khadem F, Monaco AP: A forkhead-domain gene is mutated in a severe speech and language disorder. Nature 413:519–523, 2001.
62. Larkin M: Beer smell repels children of drinkers. Lancet 356:743, 2000.
63. Law J, Boyle J, Harris F, Harkness A, Nye C: Prevalence and natural history of primary speech and language delay: Findings from a systematic review of the literature. Int J Lang Commun Disord 35:165–188, 2000.
64. Leffingwell JC: Olfaction (online, 2004): http://www.leffingwell.com/olfaction.htm.
65. Lewis BA, Thompson LA: A study of developmental speech and language disorders in twins. J Speech Hear Res 35:1086–1094, 1992.

66. Li J, Ishii T, Feinstein P, Mombaerts P: Odorant receptor gene choice is reset by nuclear transfer from mouse olfactory sensory neurons. Nature 428:393–399, 2004.
67. Liebgott B: The Anatomical Basis of Dentistry. Toronto: B.C. Decker, 1986.
68. Lodish H, Baltimore D, Berk A, Zipursky SL, Matsudaira P, Darnell J: Molecular Cell Biology. Third Edition. New York: W.H. Freeman and Company, 1995.
69. Logan WHG, Kronfeld R: Development of the human jaws and surrounding structures from birth to the age of 15 years. J Am Dent Assoc 20: 379–427, 1933.
69a. Lowrey GH: Growth and Development of Children. Sixth Edition. Chicago: Year Book Medical Publishers, 1973.
70. Lunt RC, Law DB: A review of the chronology of calcification of the deciduous teeth. J Am Dent Assoc 89:599–606, 1974.
71. Matsunami H, Montmayeur JP, Buck LB: A family of candidate taste receptors in human and mouse. Nature 404:601–604, 2000.
72. Max M, Shanker YG, Huang L, Rong M, Liu Z, Campagne F, Weinstein H, Damak S, Margolskee RF: *Tas1r3*, encoding a new candidate taste receptor, is allelic to the sweet responsiveness locus *Sac*. Nat Genet 28:58–63, 2001.
73. Moberg PJ, Agrin R, Gur RE, Gur RC, Turetsky BI, Doty RL: Olfactory dysfunction in schizophrenia: A qualitative and quantitative review. Neuropharmacol 21:324–340, 1999.
74. Moffett BC Jr: Syllabus for course in fundamentals of skeletal biology related to clinical dentistry. Unpublished manuscript, University of Washington, 1973.
75. Mombaerts P: Odorant receptor genes in humans. Curr Opin Genet Dev 9:315–320, 1999.
76. Mombaerts P: Better taste through chemistry. Nat Genet 25:130–132, 2000.
77. Montmayeur J-P, Liberles SD, Matsunami H, Buck LB: A candidate taste receptor gene near a sweet taste locus. Nat Neurosci 4:492–498, 2001.
78. Moss–Salentijn L, Hendricks–Klyvert M: Dental and Oral Tissues. Second Edition. Philadelphia: Lea and Febiger, 1985.
79. Nelson G, Chandrashekar J, Hoon MA, Feng L, Zhao G, Ryba NJP, Zuker CS: An amino-acid taste receptor. Nature 416:199–202, 2002.
80. O'Brien EK, Zhang X, Nishimura C, Tomblin JB, Murray JC: Association of specific language impairment (SLI) to the region of 7q31. Am J Hum Genet 72:1536–1543, 2003.
81. OMIM: On-Line Mendelian Inheritance in Man, 2004.
81a. Panda S, Nayak SK, Campo B, Walker JR, Hogenesch JB, Jegla T: Illumination of the melanopsin signaling pathway. Science 307:600–604, 2005.
82. Patton KT, Thibodeau GA: Mosby's Handbook of Anatomy and Physiology. St. Louis: C.V. Mosby, 2000.
83. Petek E, Windpassinger C, Vincent JB, Cheung J, Boright AP, Scherer SW, Kroisel PM, Wagner K: Disruption of a novel gene (*IMMP2L*) by a breakpoint in 7q31 associated with Tourette syndrome. Am J Hum Genet 68:848–858, 2001.
84. Petit C: Memorial lecture. Hereditary sensory defects: From genes to pathogenesis. Am J Med Genet 130A:3–7, 2004.
85. Pindborg JJ, Kramer IRH, Torloni H: Histological Typing of Odontogenic Tumours, Jaw Cysts, and Allied Conditions. Geneva: World Health Organization, 1971.
86. Reed DR, Nanthakumar E, North M, Bell C, Bartoshuk LM, Price RA: Localization of a gene for bitter-taste perception to human chromosome 5p15. Am H Hum Genet 64:1478–1480, 1999.
87. Reed RR: After the holy grail: Establishing a molecular basis for mammalian olfaction. Cell 116: 329–336, 2004.
88. Rouquier S, Taviaux S, Trask BJ, Brand–Arpon V, van den Engh G, Demaille J, Giorgi D: Distribution of olfactory receptor genes in the human genome. Nat Genet 18:243–250, 1998.
89. Schiffman S: Taste and smell in disease. Part I. New Engl J Med 308:1275–1279, 1983.
90. Schiffman S: Taste and smell in disease. Part II. New Engl J Med 308:1337–1343, 1983.
91. Schour I, Massler M: Studies in tooth development. The growth pattern of human teeth. J Am Dent Assoc 27:1918–1931, 1940.
91a. Shepherd GM: Smells, brains and hormones. Nature 439:149–151, 2005.
92. Shu W, Yang H, Zhang L, Lu MM, Morrisey EE: Characterization of a new subfamily of winged-helix/forkhead (Fox) genes that are expressed in the lung and act as transcriptional repressors. J Biol Chem 276:27488–27497, 2001.
93. Sininger YS, Cone–Wesson B: Asymmetric cochlear processing mimics hemisphere specialization. Science 305:1581, 2004.
94. Soussi–Yanicostas N, de Castro F, Julliard AK, Perfettini I, Chédotal A, Petit C: Anosmin-1, defective in the X-linked form of Kallmann syndrome, promotes axonal branch formation from olfactory bulb output neurons. Cell 109:217–228, 2002.
95. Steel KP, Kros CJ: A genetic approach to understanding auditory function. Nat Genet 27:143–149, 2001.
96. Steiner JE: The gustofacial response: Observations on normal and anencephalic newborn infants. In Oral Sensation and Perception: Development in the Fetus and Infant. J.F. Bosma, Editor. Bethesda: U.S. Department of Health, Education, and Welfare. DHEW Publication No. (NIH) 73–5460. pp. 254–278, 1973.
97. Steiner JE: Facial expressions of the neonate infant indicating the hedonics of food-related

chemical stimuli. In Taste and Development: The Genesis of Sweet Preference. J.E. Weiffenbach, Editor. Bethesda: U.S. Department of Health, Education, and Welfare. DHEW Publication No. (NIH) 77–1068. pp. 173–189, 1977.

98. Sturm RA, Frudakis TN: Eye colour: Portals into pigmentation genes and ancestry. Trends Genet 20:327–332, 2004.

98a. Talavera K, Yasumatsu K, Voets T, Droogmans G, Shigemura N, Ninomiya Y, Margolskee RF, Nilius B: Heat activation of TRPM5 underlies thermal sensitivity of sweet taste. Nature 438: 1022–1025, 2005.

99. Thompson MD, Comings DE, Abu-Ghazalah R, Jereseh Y, Lin L, Wade J, Sakurai T, Tokita S, Yoshida T, Tanaka H, Yanagisawa M, McIntyre Burnham W, Moldofsky H: Variants of the *Orexin2/hcrt2* receptor gene identified in patients with excessive daytime sleepiness and patients with Tourette's syndrome comorbidity. Am J Med Genet Part B (Neuropsychiatric Genetics) 129B:69–75, 2004.

100. Tomblin JB, Records NL, Buckwalter P, Zhang X, Smith E, O'Brien M: Prevalence of specific language impairment in kindergarten children. J Speech Lang Hear Res 40:1245–1260, 1997.

101. Vannini V, Pogliani G: The New Atlas of the Human Body. Milan: Corgi, 1980.

102. Walters BA, Raff ML, Ver Hoeve J, Tesser R, Langar LO, France TD, Glass IA, Pauli RM: Spondylometaphyseal dysplasia with cone–rod dystrophy. Am J Med Genet 129:265–276, 2004.

103. Warburton P, Baird G, Chen W, Morris K, Jacobs BW, Hodgson S, Docherty Z: Support for linkage of autism and specific language impairment to 7q3 from two chromosome rearrangements involving band 7q31. Am J Med Genet 96:228–234, 2000.

104. Wassink TH, Piven J, Vieland VJ, Huang J, Swiderski RE, Pietila J, Braun T, Beck G, Folstein SE, Haines JL, Sheffield VC: Evidence supporting WNT2 as an autism susceptibility gene. Am J Med Genet 105:406–413, 2001.

105. Watson CW: Basic Human Neuroanatomy. Fifth Edition. Philadelphia: Lippincott, Williams and Wilkins, 1995.

106. Wilson JH, Wensel TG: The nature of dominant mutations of rhodopsin and implications for gene therapy. Mol Neurobiol 28:149–158, 2003.

107. Witten PE, Hall BK. Seasonal changes in the lower jaw skeleton in male Atlantic salmon (*Salmo salar L.*): Remodelling and regression of the kype after spawning. J Anat 203:435–450, 2003.

107a. Yang T, Pfister M, Blin N, Zenner HP, Pusch CM, Smith RJH: Genetic heterogeneity of deafness phenotypes linked to DFNA4. Am J Med Genet 139A:9–12, 2005.

108. Zatorre RJ, Belin P, Penhune VB: Structure and functions of auditory cortex: Music and speech. Trends Cogn Sci 6:37–46, 2002.

109. Zhang Y, Hoon MA, Chandrashekar J, Mueller KL, Cook B, Wu D, Zuker CS, Ryba NJP: Coding of sweet, bitter, and umami tastes: Different receptor cells sharing similar signaling pathways. Cell 112:293–301, 2003.

5

Dysmorphic and Genetic Perspectives

ANOMALIES, COMPLEX ANOMALIES, AND SYNDROMES

Anomalies may be simple, as in cleft lip–palate; complex, as in holoprosencephaly; or syndromic, as in Apert syndrome. There are many different ways to classify syndromes. Geneticists, craniofacial surgeons (see Tables 6-2 to 6-4, 6-7), and pediatricians all categorize syndromes in different ways. There are probably as many methods of systematizing these disorders as there are clinicians and craniofacial biologists working in the field. Any attempt to bring order to a group of disparate conditions is helpful, even if various classifications crosscut each other. A morphologic–etiologic classification of craniofacial syndromes is shown in Table 5-1. Selected molecular disorders with craniofacial anomalies are summarized in Table 5-2.

TYPES OF ANOMALIES

Most anomalies observed at birth can be sorted into one of three basic categories—malformations, deformations, or disruptions (Fig. 5-1). There are practical reasons for distinguishing these because the clinical implications of each category are different (22).

A *malformation may be defined as a morphologic defect of an organ, part of an organ, or a larger area of the body resulting from an intrinsically abnormal developmental process*. Polydactyly, cleft lip, and renal agenesis serve as examples. Malformations may be relatively simple or complex. The later the defect is initiated, the simpler the malformation. Malformations initiated earlier during organogenesis tend to have more far-reaching consequences (22).

A *deformation may be defined as an abnormal form or position of a part of the body caused by nondisruptive mechanical forces*. Important deformities include clubfoot, congenital hip dislocation, and congenital postural scoliosis. Deformations arise most frequently during late fetal life. Since the most common cause is intrauterine molding by mechanical forces, the musculoskeletal system is usually affected. The most important factor contributing to deformations is lack of fetal movement, whatever the cause. Deformations may result from mechanical, malformational, or functional causes (22).

First pregnancies tend to be associated with unstretched uterine and abdominal muscles, which may result in uteroplacental insufficiency and, in turn, may lead to oligohydramnios. Breech presentation is common since the uterus is too compressed to allow the fetus to rotate into the cephalic position. Uterine constraints on fetal movement allow mild but persistent extrinsic forces to deform the fetus (22).

During the first few days after birth, infants with deformities can usually be folded into their atypical prenatal postures. Radiographically, the close correspondence between the abnormal posture of the fetus before delivery and the pos-

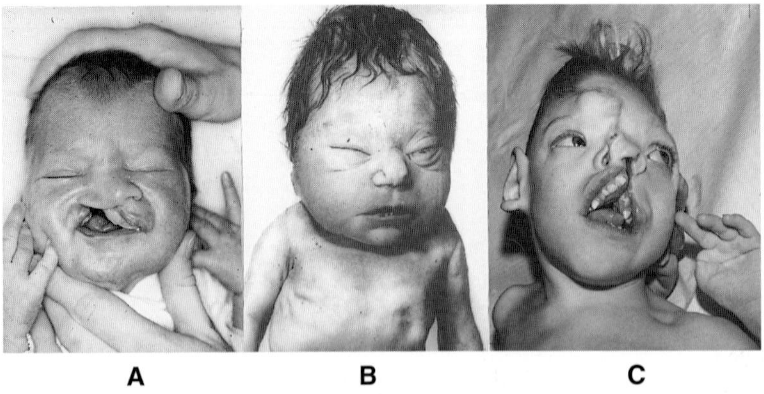

Figure 5-1 A, Malformation: cleft lip-palate. B, Deformation: Potter-type face, resulting from oligohydramnios. C, Disruption: amniotic bands of face, cranium, and hand. From Cohen, 1997 (22).

Table 5-1 Classification of Craniofacial Syndromes

Morphologic
Clinical description
Developmental pathogenesis
Dysmorphic growth and development
Etiologic
Teratogenic
Chromosomal
Monogenic
Mendelian inheritance pattern
Molecular delineation

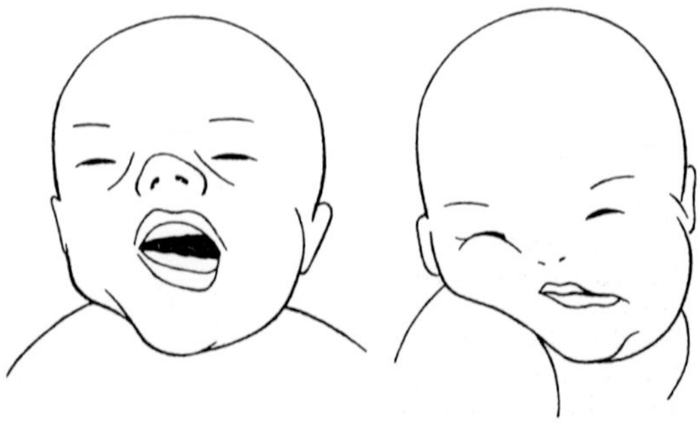

Figure 5-2 Mandibular deformation, resulting from a sharply lateroflexed positioning of the head *in utero* with the shoulder pressing against the mandible for a long period of time. Courtesy of the Mead Johnson and Co.

ture of the infant after birth has been observed repeatedly. *Such posture has been termed the position of comfort.* For example, a mandibular deformation may result from a sharply lateroflexed position of the head with the shoulder pressed against the mandible for a long period of time late during intrauterine life (22) (Fig. 5-2).

A *disruption is a morphologic defect of an organ, part of an organ, or a larger region of the body resulting from a breakdown of, or interference with, an originally normal developmental process*. An amputation of a digit *in utero* caused by an amnionic band serves as an example as do other band-caused disruptions such as ring constriction and distal pseudosyndactyly. In some cases, amnionic bands can produce dramatic consequences, such as asymmetric encephaloceles, bizarre facial clefting, and limb amputations (22).

INTERRELATIONSHIPS BETWEEN TYPES OF ANOMALIES

It should be recognized that although the distinctions between the three types of anomalies—malformations, deformations, and disruptions—are useful for clinical purposes, particularly in the newborn period (Table 5-3), the three classes of anomalies are interrelated in some instances during growth after birth. For example, cleft lip–palate, a malformation, develops secondary facial deformation during postnatal growth (22).

Malformations can also result in deformations appearing at birth, such as bilateral renal agenesis, producing oligohydramnios and the facial and limb deformities of Potter sequence. Some anomalies, such as Robin sequence, may be malformational or deformational, depending on the initiating factor–intrinsic mandibular hypoplasia (malformational) or extrinsic mandibular constraint (deformational) (22).

Table 5-2 Selected Genetic Disorders with Craniofacial Anomalies[a]

Name	Chromosome Map Locations	Mutation-Bearing Genes
Pycnodysostosis	1q21	*CTSK*
Papillary thyroid/renal neoplasia	1q21	*PTCRN*
Lipoid proteinosis	1q21	*EMC1*
Hutchinson–Gilford progeria	1q21.2	*LMNA*
Mandibuloacral dysplasia	1q21.2	*LMNA*
Hyperparathyroidism/jaw tumor syndrome	1q25–q31	*HRPT2*
van der Woude syndrome	1q32–q41	*IRF6*
Popliteal pterygium syndrome	1q32–q41	*IRF6*
Hypohidrotic ectodermal dysplasia, type 2b	1q42.2–q43	*EDARADD*
Holoprosencephaly	2p21	*SIX3*
Hereditary gingival fibromatosis, type 1	2p21–p22	*SOS1*
Hypohidrotic ectodermal dysplasia, type 2a	2q11–q13	*EDAR*
Hypohidrotic ectodermal dysplasia, type 3	2q11–q13	*EDAR*
Holoprosencephaly	2q14	*GLI2*
Waardenburg syndrome, type 1	2q37	*PAX3*
Waardenburg syndrome, type 2	3p12–p14	*MITF*
Septo-optic dysplasia	3p21.1	*HESX1*
Jansen metaphyseal chondrodysplasia	3p21.1–p22	*PTHR1*
Mucopolysaccharidosis IVB	3p21.33	*GLBI*
Holoprosencephaly	3p21–p23	*TDGF1*
EEC syndrome	3q27	*TP63*
Hay–Wells syndrome	3q27	*TP63*
Ellis–van Creveld syndrome	4p16	*EVC*
Hypodontia	4p16	*MSX1*
Hypodontia/clefting	4p16	*MSX1*
Witkop syndrome	4p16	*MSX1*
Mucopolysaccharidosis IH	4p16.3	*IDA*
Mucopolysaccharidosis IS	4p16.3	*IDA*
Achondroplasia	4p16.3	*FGFR3*
Hypochondroplasia	4p16.3	*FGFR3*
Thanatophoric dysplasia, type 1	4p16.3	*FGFR3*
Thanatophoric dysplasia, type 2	4p16.3	*FGFR3*
SADDAN	4p16.3	*FGFR3*
Crouzonodermoskeletal syndrome	4p16.3	*FGFR3*
Muenke syndrome	4p16.3	*FGFR3*
Cherubism	4p16.3	*SH3BP2*
Amelogenesis imperfecta 2 (hypoplastic local type)	4q21	*ENAM*
Amelogenesis imperfecta (hypoplastic type)	4q21	*ENAM*
Dentinogenesis imperfecta 1	4q21.3	*DSPP*
Dentin dysplasia, type 2	4q21.3	*DSPP*
Mucolipidosis II	4q21–q23	*GNPTA*
Mucolipidosis III	4q21–q23	*GNPTA*
Rieger syndrome, type 1	4q25–q26	*RIEG1*
Mucopolysaccharidosis VI	5q13.3	*ARSB*
Familial aldenomatous polyposis/Gardner syndrome	5q21–q22	*APC*
Diastrophic dysplasia	5q31–q34	*DTDST*
Achondrogenesis, type 1B	5q31–q34	*DTDST*
Atelosteogenesis, type II	5q31–q34	*DTDST*
Treacher Collins syndrome	5q33–q33.1	*TCOF1*
Boston–type craniosynostosis	5q34–q35	*MSX2*
Parietal foramina	5q34–q35	*MSX2*
Sotos syndrome	5q35	*NSD1*
Cleidocranial dysplasia	6p21	*RUNX2*
Stickler syndrome	6p21	*COL11A1*
Oculodentodigital dysplasia	6q22–q23	*GJA1*
Chondrodysplasia, rhizomelic type	6q22–q24	*PEX*
Greig cephalopolysyndactyly	7p13	*GLI3*
Pallister–Hall syndrome	7p13	*GLI3*
Saethre–Chotzen syndrome	7p21	*TWIST*
Mucopolysaccharidosis VII	7q21.11	*GUSB*
Osteogenesis imperfecta	7q22.1	*COL1A1, COL1A2*
Holoprosencephaly	7q36	*SHH*

(*continued*)

Table 5-2 (continued)

Name	Chromosome Map Locations	Mutation-Bearing Genes
Pfeiffer syndrome	8p11.2–p12	*FGFR1*
Branchio-oto-renal syndrome	8q13.3	*EYA1*
Cohen syndrome	8q22	*COH1*
Trichorhinophalangeal syndrome	8q24.12	*TRPS1*
Langer–Giedion syndrome	8q24.11–q24.13	*TRPS1* + *EXT1*
Holoprosenecephaly	8q24.3	*FAST1*
Nevoid basal cell carcinoma syndrome	9q22.3	*PTCH*
Holoprosencephaly	9q22.3	*PTCH*
Tuberous sclerosis	9q33–q34	*TSC1*
Multiple endocrine neoplasia, type 2B	10q11.2	*RET*
Bannayan–Riley–Ruvalcaba/Cowden syndrome	10q23	*PTEN*
Apert syndrome	10q25.3–q26	*FGFR2*
Crouzon syndrome	10q25.3–q26	*FGFR2*
Pfeiffer syndrome	10q25.3–q26	*FGFR2*
Jackson–Weiss syndrome	10q25.3–q26	*FGFR2*
Beare–Stevenson cutis gyrata syndrome	10q25.3–q26	*FGFR2*
Parietal foramina	11p11.2	*ALX4*
Beckwith–Wiedemann syndrome	11p15.5	*IGF2, H19, KCNQ1, KCNQ1OT1, CDKN1C*
Smith–Lemli–Opitz syndrome	11q12–q13	*DHCR7*
Papillon–Lefèvre syndrome	11q14	*CTSC*
Juvenile periodontosis	11q14	*CTSC*
Haim–Munk syndrome	11q14	*CTSC*
Cleft lip-palate, ectodermal dysplasia (AR)	11q23–q24	*PVRL1*
Cleft lip ± cleft palate	11q23–q24	*PVRL1*
Stickler syndrome	12q13.1–q13.3	*COL2A1*
Spondyloepiphyseal dysplasia congenita	12q13.1–q13.3	*COL2A1*
Achondrogenesis, type 2	12q13.1–q13.3	*COL2A1*
Mucopolysaccharidosis IIID	12q14	*GNS*
Rieger syndrome, type 2	13q14	*RIEG2*
Holoprosencephaly	13q32	*ZIC2*
Hypodontia	14q12–q13	*PAX9*
Marfan syndrome	15q21	*FBN1*
Bloom syndrome	15q26.1	*BLM*
Tuberous sclerosis	16p13.3	*TSC2*
Rubinstein–Taybi syndrome	16p13.3	*CBP*
Townes–Brocks syndrome	16q12.1	*SALL1*
Mucopolysaccharidosis IVA	16q24.3	*GALNS*
Neurofibromatosis, type 1	17q11.2	*NF1*
Tricho-dento-osseous syndrome	17q21	*DLX3*
Amelogenesis imperfecta and taurodontism only	17q21	*DLX3*
Osteogenesis imperfecta	17q21.31–q22	*COL1A1, COL1A2*
Oligodontia and colorectal cancer syndrome	17q24	*AXIN2*
Campomelic dysplasia	17q24.3	*SOX9*
Mucopolysaccharidosis IIIA	17q25.3	*HSS*
Holoprosencephaly	18p11.3	*TGIF*
Peutz–Jeghers syndrome	19p13.3	*STK11*
Amelogenesis imperfecta, hypomaturation type	19q13.3–q13.4	*KLK4*
Alagille syndrome	20p12	*JAG1*
Acromesomelic dysplasia	20q11.2	*CDMP1*
Fibrous dysplasia	20q13	*GNAS1*
McCune–Albright syndrome	20q13	*GNAS1*
Albright hereditary osteodystrophy	20q13	*GNAS1*
Neurofibromatosis, type 2	22q11.21–q13.1	*NF2*
Chondrodysplasia punctata, X-linked dominant Conradi–Hünermann type	Xp11.22–p11.23	*3β-hyroxysterol* Δ^8, Δ^7 *isomerase*
Oculofaciocardiodental syndrome	Xp11.4	*BCOR*
Opitz BBB/G syndrome	Xp22	*MIDI*
Amelogenesis imperfecta 1 (hypoplastic type)	Xp22.1–p22.3	*AMGN*
Chondrodysplasia, punctata, X-linked recessive type	Xp22.3	*ARSE*
Hypohidrotic ectodermal dysplasia	Xq12.2–q13.1	*ED1*
FG syndrome	Xq12–q21.31	*FGS1*
Cleft palate ± ankyloglossia	Xq12–q21	*TBX22*
Craniofrontonasal syndrome	Xq13.1	*EFNB1*
Simpson–Golabi–Behmel syndrome	Xq26	*GPC3*

Name	Chromosome Map Locations	Mutation-Bearing Genes
Fragile X syndrome	Xq27.3	*FMR1*
Mucopolysaccharidosis II	Xq27.3–q28	*IDS*
Otopalatodigital syndrome, type 1	Xq28	*FLN A*
Otopalatodigital syndrome, type 2	Xq28	*FLNA*
Frontometaphyseal dysplasia	Xq28	*FLNA*
Melnick–Needles syndrome	Xq28	*FLNA*

[a]Most disorders listed have other abnormalities below the craniofacial region, but several are confined only to the craniofacial area. For example, Hutchinson-Gilford progeria, Albright hereditary osteodystrophy, and tricho-dento-osseous syndrome involve abnormalities below the craniofacial region. Holoprosencephaly, Treacher Collins syndrome, and hypodontia/clefting are confined to the craniofacial area.

From OMIM, 2004 (90).

Table 5-3 General Comparison of Malformations, Deformations, and Disruptions

Features	Malformations	Deformations	Disruptions
Time of occurrence	Embryonic	Fetal	Embryonic/fetal
Level of disturbance	Organ	Region	Area
Perinatal mortality	+	−	+
Clinical variability of any given anomaly	Moderate	Mild	Extreme
Multiple causes of any given anomaly	Very frequent	Less common	Less common
Spontaneous correction	−	+	−
Correction by posture	−	+	−
Correction by surgery	+	±	+
Relative recurrence rate	Higher	Lower	Extremely low
Approximately frequency in newborns	2–3%	1–2%	1–2%

From Cohen, 1997 (22).

MINOR ANOMALIES AND THE FACE

Minor anomalies serve as diagnostic aids for many malformation syndromes because of their common occurrence (Table 5-4). For example, in Down syndrome, 79% of all malformations detectable by clinical examination are minor anomalies; in trisomy 18 syndrome, 38%; in trisomy 13 syndrome, 50%; and in Turner syndrome, 73% (106). The diagnosis of such syndromes as Down, de Lange, Williams, and Weaver is evident in all cases by the facial features alone (Fig. 5-3).

Of great interest is the occurrence of a major

Table 5-4 Minor Craniofacial Anomalies

Head	Ears	Oral regions
Aberrant scalp hair patterning	Primitive shape	Borderline small mandible
Flat occiput	Lack of helical fold	Incomplete form of cleft lip
Bony occipital spur	Asymmetric size	Bifid uvula
Third fontanel	Posterior angulation	Aberrant frenula
	Small ears	Enamel hypoplasia
Eyes	Protuberant ears	Malformed teeth
Epicanthic folds	Absent tragus	
Epicanthus inversus	Double lobule	*Neck*
Upslanting palpebral fissures	Auricular tag	Mild webbed neck
Downslanting palpebral fissures	Auricular pit	Branchial cleft fistula
Short palpebral fissures	Narrow external auditory meatus	
Dystopia canthorum		
Minor hypertelorism	*Nose*	
Minor hypotelorism	Small nares	
Minor ptosis	Notched alas	
Coloboma, lid		

Adapted from Marden et al., 1964 (70).

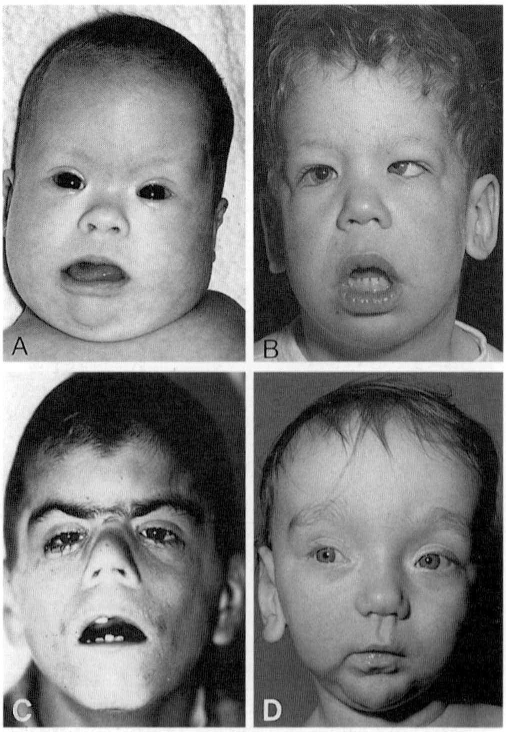

Figure 5-3 Four faces with minor anomalies. A, Down syndrome. Upslanting palpebral fissures, epicanthic folds, low nasal bridge, small nose, and protruding tongue. B, Williams syndrome. Strabismus, flat midface, anteverted nostrils, long philtrum, and thick lips. C, de Lange syndrome. Confluent eyebrows, long curly eyelashes, anteverted nostrils, and long philtrum. D, Weaver syndrome. Broad forehead, ocular hypertelorism, long philtrum, and micrognathia. A, C, D, From Cohen, 1997 (22). B, Courtesy of J.M. Opitz, Salt Lake City, Utah.

malformation in association with many newborns who have three or more minor anomalies (70). The implication is clear: Any newborn with three or more minor anomalies should be carefully evaluated for possible hidden major malformations, such as cardiac, renal, or vertebral defects. About 40% of patients with idiopathic mental retardation have three or more malformations, of which most are minor (106). Thus, the presence of minor anomalies may also be considered an aid in prognosis; they alert the clinician to monitor the child's mental development especially carefully (22) (Table 5-5). Of all the minor anomalies on the surface of the body, 71% occur in the craniofacial region and the hand (106) (Fig. 5-4).

Table 5-5 Significance of Minor Anomalies

Fact	Implication
In newborns with 3 or more minor anomalies, 90% have major anomaly	Search for major occult anomaly
Minor anomalies present in many multiple congenital anomaly syndromes	Aid in diagnosis
42% of idiopathic mental retardation cases have 3 or more anomalies of which 80% are minor anomalies	Aid in prognosis

From Cohen, 1997 (22).

Minor anomalies overlap with the general population. Single minor anomalies occur in approximately 15% of all newborns. The occurrence of two is less common and the presence of three or more is unusual, occurring in approximately 1% of all newborns (22).

Before ascribing significance to any minor anomaly, it should be ascertained if the anomaly is present in any other member of the family. Because minor anomalies are common in the general population, in some instances groupings of facial minor anomalies may represent variant additive patterns instead of true malformation syndromes. A variant additive pattern can be identified in the general population because the minor anomalies that make up the pattern in the proband may be observed separately in various relatives (22).

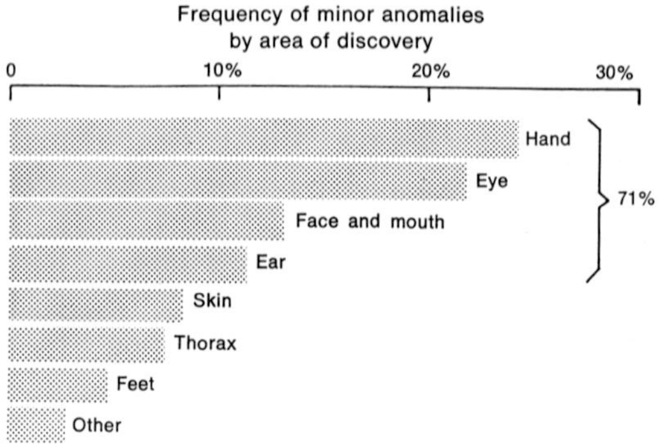

Figure 5-4 Frequency of minor anomalies by region. Of all minor anomalies, 71% occur in the head and neck region and the hand. From Smith and Bostian, 1964 (106).

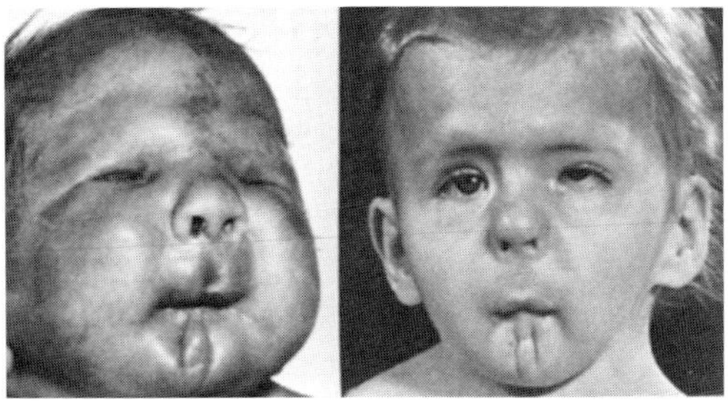

Figure 5-5 Whistling face syndrome (craniocarpotarsal dystrophy). Left, From Rintala, 1968 (96). Right, Courtesy of J. Külz, Rostock, Germany.

EVOLUTION OF FACIAL PHENOTYPES

Some phenotypes, such as the whistling face syndrome (craniocarpotarsal dystrophy), are both striking and constant (96) (Fig. 5-5). Other phenotypes may change with time. In the de Lange syndrome, the features of the face are usually constant enough at any age to impart the correct diagnostic impression, and the resemblance of affected individuals transcends the racial background of the patients (55). Occasionally, mild expression may render clinical diagnosis difficult, particularly at birth. In a few instances, the characteristic phenotype of the de Lange syndrome evolves with time (92) (Fig. 5-6). In some conditions, such as Proteus syndrome, the evolution of the phenotype is dramatic (36) (Fig. 5-7). In other conditions, such as the cri-du-chat syndrome, the phenotypic features tend to become less distinct with time (10).

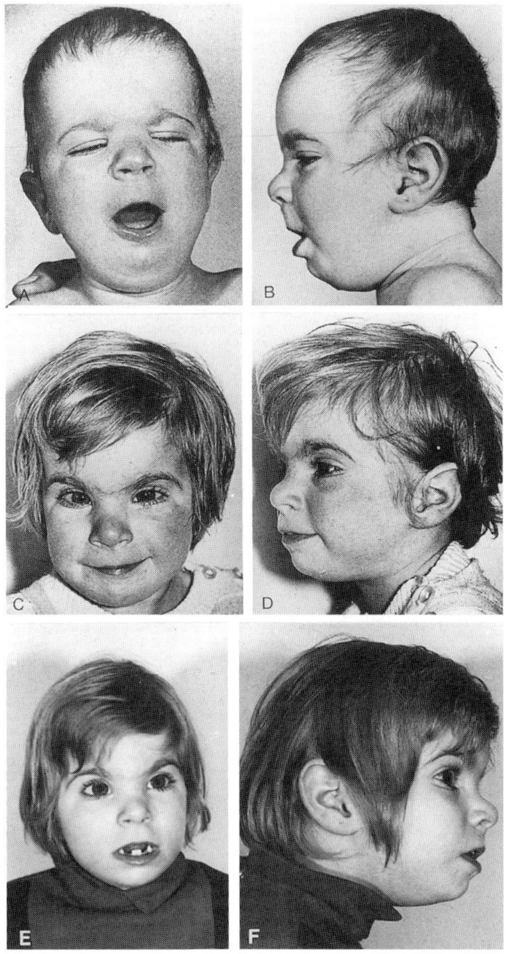

Figure 5-6 Photographic documentation showing evolution of de Lange phenotype. A, B, Patient at 8 months of age. Microcephaly, downslanting palpebral fissures, right epicanthic fold, convergent strabismus, small nose with anteverted nostrils, and slight posterior angulation of the ears were noted on physical examination at that time. C, D, Patient at 21 months of age with de Lange phenotype evident by that time. E, F, Patient at 3 years of age. A–D, From Passarge et al., 1971 (92). E, F, Courtesy of E. Passarge, Hamburg, Germany.

OROFACIAL CLEFTING

Various types of cleft lip and cleft palate are diagrammed in Figure 5-8, and three patients with cleft lip are shown in Figure 5-9.

Embryonic Basis of Orofacial Clefting

Exact timing and positioning of facial prominences during embryonic development are critical. Any alteration may result in a cleft lip. Genetic or environmental factors (or both) that inhibit the flow of neural crest cells or decrease their numbers may affect their masses so that contact between the facial prominences is inadequate or impossible. The epithelium that covers the mesenchyme may not undergo programmed cell death, so that fusion cannot take place. Any change in the position of the nasal placodes or abnormal directional growth of the facial prominences may result in cleft lip (47, 48).

Clefts of the lip and palate occur together in

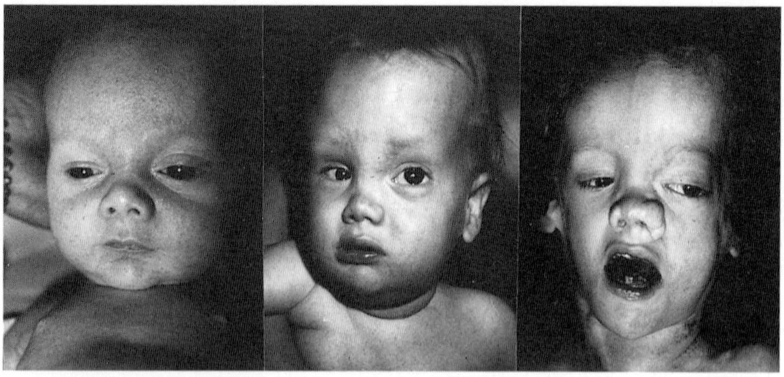

Figure 5-7 Proteus syndrome. Evolution of facial dysmorphism. Left, Age 1 month. Center, Age 5 months. Right, Age $5\frac{1}{2}$ years. Note the cerebriform connective tissue nevus on the left side of the nose. From Cohen et al., 2002 (36).

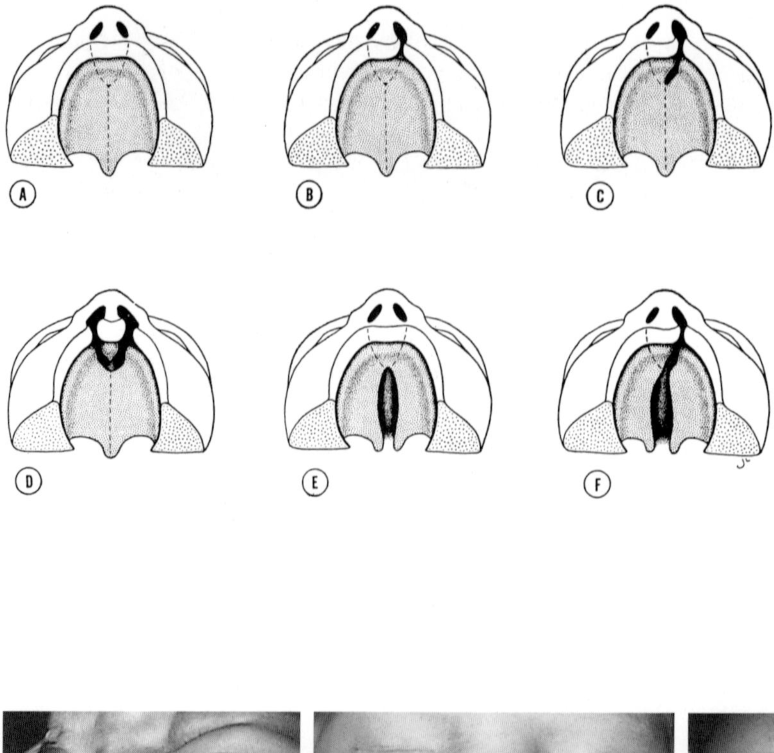

Figure 5-8 Ventral view of palate, alveolar ridge, lip, and nose. A, Normal. B, Unilateral cleft lip. C, More extensive unilateral cleft lip. D, Bilateral cleft lip. E, Cleft palate. F, Unilateral cleft lip and cleft palate. From Langman, 1983 (66).

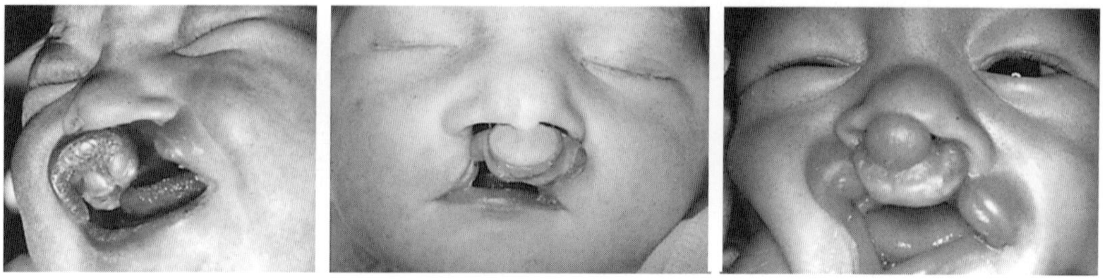

Figure 5-9 Cleft lip-palate. Left, Unilateral cleft lip–palate. Center, Right, Bilateral cleft lip of varying severity.

approximately 45% of cases. It has been suggested that the tip of the tongue becomes wedged in the labial cleft, thus not allowing the tongue to drop, which would inhibit palatal contact and fusion. This explanation is probably simplistic. It seems more likely that a reduction in the size of the labial maxillary prominences and the palatine process of the maxillary prominences is a more reasonable explanation (47, 48). Cleft palate may result from defective growth of the palatine shelves, failure of elevation of the shelves, failure of fusion of the shelves, or post-fusion rupture of the shelves. In some instances, micrognathia is a contributing factor that may inhibit elevation of the palatine shelves as, for example, in Robin sequence (48).

Causes of Cleft Lip and Cleft Palate

Cleft lip with or without cleft palate is etiologically distinct from isolated cleft palate. In a patient with cleft lip with or without cleft palate, if another family member is affected, he or she will have either isolated cleft lip or cleft lip together with cleft palate but not isolated cleft palate alone. Similarly, if a patient has isolated cleft palate, another affected family member can have only isolated cleft palate. There are two exceptions. First, in genetic isolates that are inbred, both types may occur by chance. Second, in many genetic syndromes with orofacial clefting, both types may be found. For example, in the autosomal dominantly inherited van der Woude syndrome, in which orofacial clefting is found together with lip pits, patients may have cleft lip, cleft lip–palate, or isolated cleft palate (17, 31).

Much evidence supports the view that genetic factors are associated with orofacial clefting. In twins with cleft lip–palate, concordance is far greater in monozygotic twins (40%) than in dizygotic twins (4.2%). In twins with isolated cleft palate, concordance is also higher in monozygotic twins (35%) than in dizygotic twins (7.8%) (121).

Many clefts occur sporadically. In some families, however, more than one individual is affected. In a few large families, Mendelian inheritance has been established (47, 48). Thus, orofacial clefting is heterogeneous, and variation in liability is probably determined by a number of major genes, minor genes, environmental factors, and a developmental threshold (45). Two alternative hypotheses for orofacial clefting have been advanced. One is the multifactorial threshold model. The other involves a single major gene with reduced penetrance. Both theories have their proponents (48).

Many candidate genes for orofacial clefting have been suggested by allelic association and by linkage analysis (2, 11, 43, 69, 75, 85, 94, 95, 100–102, 107, 121). Some syndromes in which specific gene mutations have been identified include, among others, van der Woude syndrome (*IRF6*) (58), popliteal pterygium syndrome (*IRF6*) (58), EEC syndrome (*TP63*) (6, 15), Hay–Wells syndrome (*TP63*) (73), autosomal recessive cleft palate/ectodermal dysplasia (*PVRL1*) (108), hypodontia/clefting (*MSX1*) (112), and cleft palate/ankyloglossia (*TBX22*)(69a). Candidate genes for isolated cleft lip and/or palate are summarized in Table 5-6. In the aggregate, these mutations explain about 5% of isolated cases with an additional 10–15% of cases explained by variations surrounding the *IRF6* gene (86).

Table 5-6 Candidate Genes for Isolated Cleft Lip and/or Palate

Gene	Meaning	Chromosome Map Location
MSX1	Muscle segment homeobox 1	4p16.1
MSX2	Muscle segment homeobox 2	5q34–q35
IRF6	Interferon regulatory factor 6	1q32–q41
FGFR1	Fibroblast growth factor receptor 1	8p12
PTCH	Patched	9q22.3
PVRL1	Poliovirus receptor-like 1	1q25–q31
PVRL2	Poliovirus receptor-like 2	19q13.2–q13.4
PRDM16	PR domain-containing protein 16	1p36.3
JAG2	Jagged 2	14q32
TBX10	T-box 10	11q13.1–q13.2
TBX22	T-box 22	Xq12–q21
TGFB3	Transforming growth factor β3	14q24
LHX8	Lim homeobox gene 8	4q25–q31
SATB2	Special AT-rich sequence-binding protein 2	2q33
SKI1	Skinny hedgehog 1	1q32

In the aggregate, these candidate genes explain about 5% of isolated cases with an additional 10–15% of cases explained by variations surrounding the *IRF6* gene (86). Some of these gene mutations can result in other disorders. For example, mutations in *PTCH* usually cause nevoid basal cell carcinoma syndrome, isolated basal cell carcinoma, and, rarely, holoprosencephaly (30).

Robin Sequence and Robin Complexes

Robin sequence is commonly defined as cleft palate, micrognathia, and glossoptosis (Fig. 5-10). However, Robin sequence has been defined in different ways by using different criteria (104, 105). In Table 5-7, various criteria for Robin sequence are summarized. Definitions I through IV are clearly Robin sequence, but definition V is non-Robin in type. The cleft palate is V-shaped and, by polysomnography, some respiratory compromise is evident. The mandible is normal or may be slightly small subjectively, but it is impossible to be certain. Thus, at the interface, it is not always possible to distinguish between Robin sequence and ordinary cleft palate (17, 23, 29).

Because of etiologic, pathogenetic, and phenotypic differences, Robin sequence or Robin complexes of various types can occur (23, 29). These changes are shown in Figure 5-11. Double lines indicate general causes: malformation, deformation, or connective tissue dysplasia. Thick arrows indicate Robin sequence. Thin lines show all possible ways in which Robin complexes can arise (23).

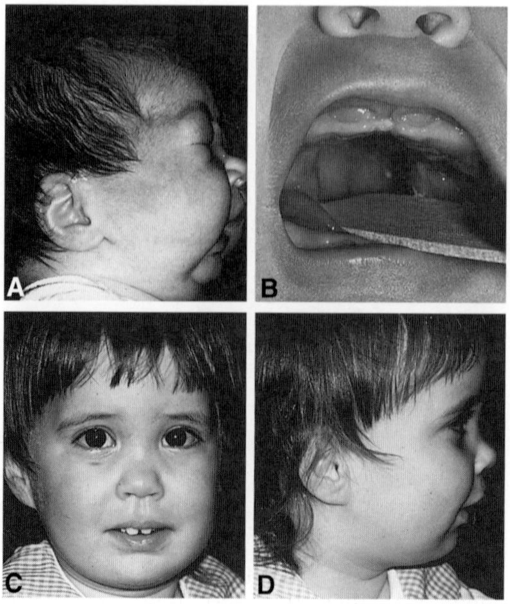

Figure 5-10 Robin sequence. Patient with Robin sequence showing micrognathia and U-shaped cleft palate (A, B), later showing catch-up growth (C, D). From Cohen, 1997 (22).

Table 5-7 Defining Robin Sequence Using Different Criteria

Definition	Mandibular Deficiency	Cleft Palate	Upper Airway Obstruction
I	+	+(U)	+
II	+	+(U or V)	+
III	+	+(U)	−
IV	+	±(U or V)	±
V (non-Robin)	−[a]	+(V)	+

U = U-shaped cleft palate. V = V-shaped cleft palate.

[a]Mandible is normal or may be slightly small subjectively, but it is impossible to be certain.

Modified from Cohen, 1999 (23), 2001 (29).

A distinction is made between micrognathia and retrognathia (Fig. 5-11). Micrognathia refers to size, retrognathia to position. In Treacher Collins syndrome, the mandible is short (61, 93). In deletion 22q11 syndrome, the mandible is essentially normal in size but retrognathic in position because the cranial base angle is larger than normal (3). Most Robin conditions are micrognathic or retrognathic, but not usually both together (microretrognathia) (22, 23, 29).

Different Robin complexes are exemplified by del(22q11) syndrome and spondyloepiphyseal dysplasia congenita. In del(22q11) syndrome, retrognathia, caused by an obtuse cranial base angle, and cleft palate, either submucous or true, are found. The flat cranial base angle and retrognathia do not contribute to pharyngeal obstruction, which results, in fact, from hypotonia (3). Robin sequence does not occur. Rather, in this form of Robin complex, all of the manifestations are causally, but not sequentially, related (23).

An example of multiple mechanisms of respiratory compromise can be found in spondyloepiphyseal dysplasia congenita: small mechanically abnormal chest, tracheobronchomalacia, and/or central apnea due to cervical or medullary compression caused by cervical instability (52). Another possibility is upper respiratory obstruction based on Robin sequence.

Median Clefts

True median cleft lip occurs with premaxillary agenesis and failure of completion of the nose. Median clefts may occur with holoprosence-

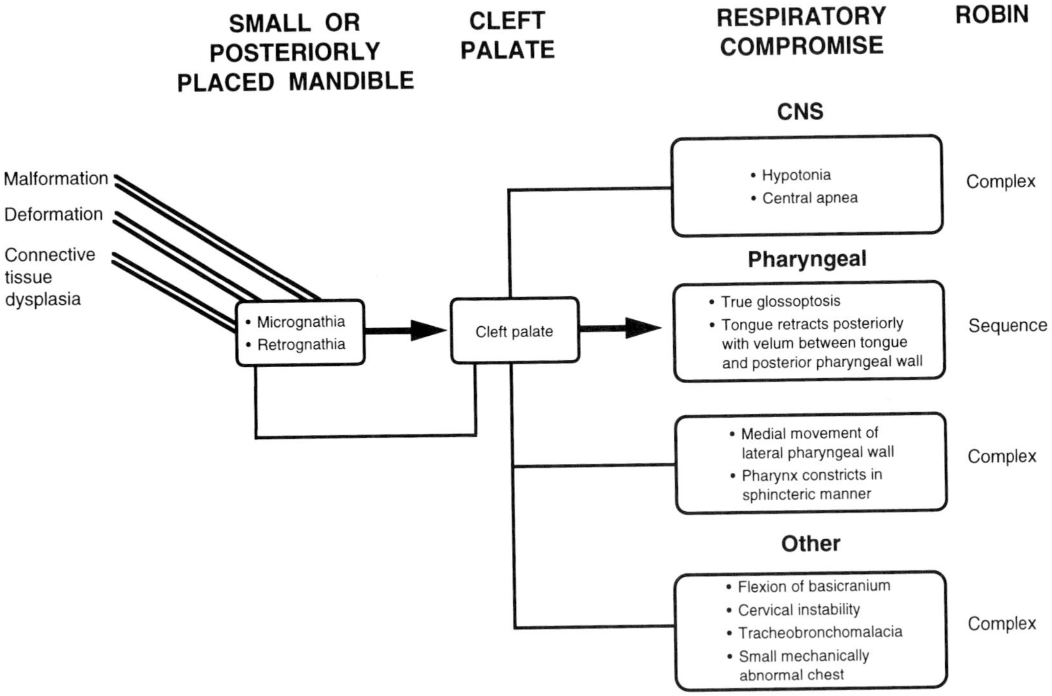

Figure 5-11 Robin sequence and complexes. Double lines indicate general causes: malformation, deformation, or connective tissue dysplasia. Thick arrows indicate Robin sequence. Thin lines show all possible ways that Robin complexes can occur. From Cohen, 1999 (23).

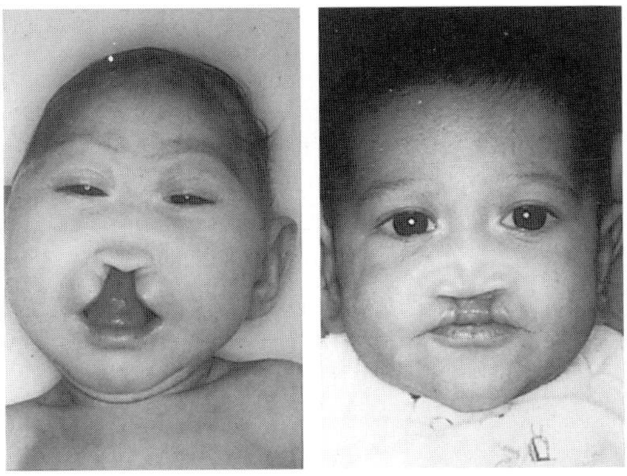

Figure 5-12 Median cleft lip. Left, With holoprosencephaly. Note associated microcephaly. Right, Without holoprosencephaly. Note normal head circumference. Courtesy of A. Richieri-Costa, Bauru, Brazil.

phaly (Fig. 5-12) or as an isolated malformation (22, 24).

Unusual Craniofacial Clefts and the Tessier Classification

Tessier (109) described an anatomic and descriptive classificatory system in which the various types of bony and soft tissue defects, which he called clefts, are situated along definite axes with numbers assigned to the sites of clefting, depending on their relationships to the sagittal midline (Fig. 5-13). Clefting may involve bone and/or soft tissue, but rarely to the same extent. From the sagittal midline to the infraorbital foramen, abnormalities of soft tissue predominate.

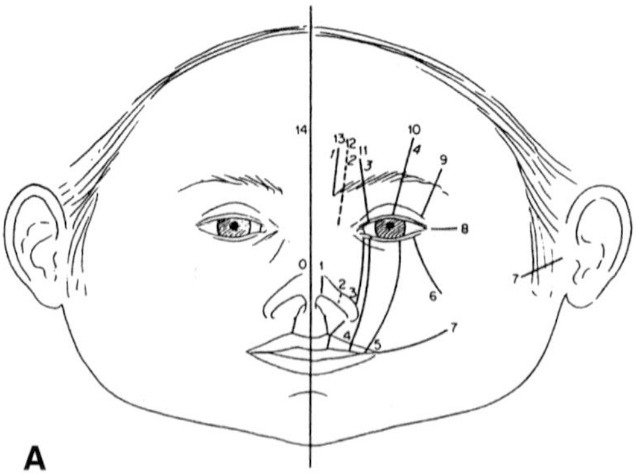

A

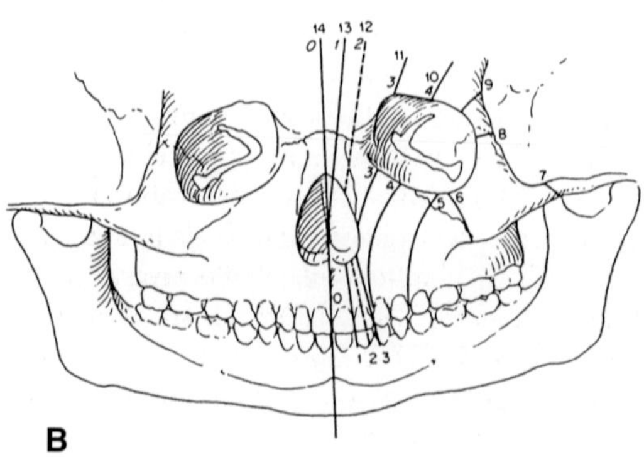

B

Figure 5-13 Tessier craniofacial clefting system. A, Soft tissue clefts. B, Bony clefts. Dotted lines represent uncertain localization or uncertain clefting. Note that northbound cranial lines have different numbers than their counterpart southbound facial lines. Thus, the system is descriptive and anatomic and avoids etiologic or pathogenetic speculation. For example, the cause of a number 10 cleft may possibly be different than the cause of a number 4 cleft. From Tessier, 1976 (109).

From the infraorbital foramen to the temporal bone, however, osseous defects are more severe than those of soft tissue, a notable exception being the ear. Clefts through the orbit use the lower eyelid as an equator. Cleft numbered lines may be either northbound (cranial) or southbound (facial). Cranial numbered lines have facial numbered counterparts, yet these numbers are different to avoid the implication that they necessarily have the same etiopathogenesis. Thus, the Tessier classification permits description of both the location and the extent of unusual facial clefts. Several different types of Tessier clefts may occur in the same patient.

In some instances, overlying soft tissue defects predict the possibility of underlying bony clefts. Elsewhere, I have referred to such features (e.g., colobomatous notching of the upper or lower eyelids or the nostrils, or interruption of the eyelashes or the eyebrows) as Tessier signs (22). Also included in this category are hairline indicators that may point to the cleft (76) (Fig. 5-14). A complex form of Tessier clefting is shown in Figure 5-15.

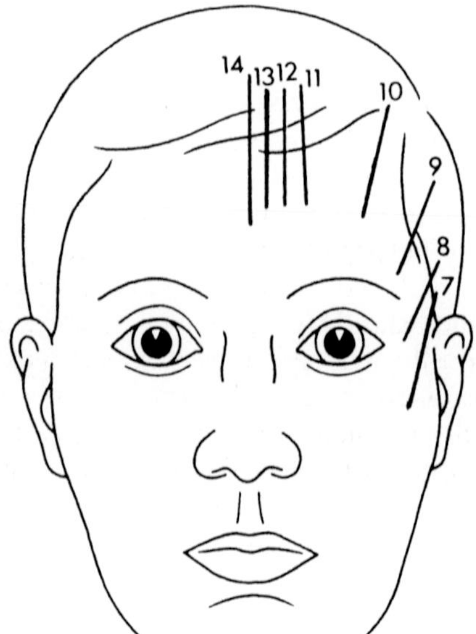

Figure 5-14 Diagram of hairline indicators that are superior and lateral extensions of Tessier clefts. See Figure 5-13. From Moore et al., 1988 (76).

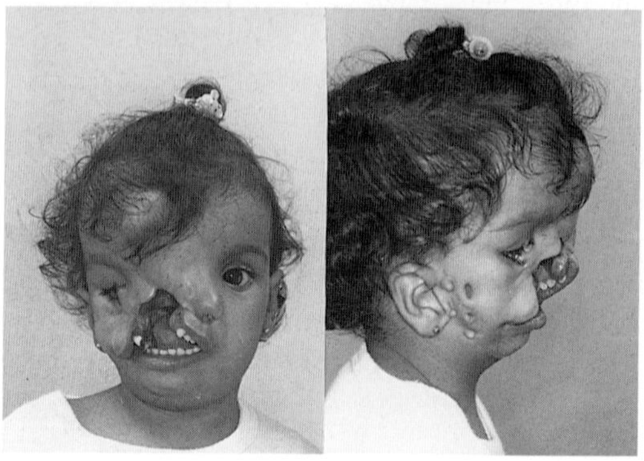

Figure 5-15 Complex orofacial clefting. Computed tomography scan showed anterior cranium bifidum occultum and a gap in the sphenoid bone. Note the ear tags. Hairline indicators point to cleft areas. Courtesy of A. Richieri-Costa, Bauru, Brazil.

The causes of most Tessier clefts are unknown. The overwhelming majority occur sporadically. One exception is Treacher Collins syndrome, which is autosomal dominantly inherited. Some Tessier clefts are malformations that can be explained by faulty embryogenesis. Others represent disruptions, however, such as those associated with amniotic bands (22).

ANENCEPHALY

Anencephaly is characterized by an open neural tube in the cephalic region, with an exposed mass of degenerating neural tissue on the skull floor. The cranial vault is absent, producing characteristic bulging of the eyes and absence of the neck (see Fig. 1-15). Both the membranous neurocranium and the chondrocranium are grossly malformed. Anencephaly is classified anatomically as meroacrania if the defect does not involve the foramen magnum, holoacrania if the defect extends through the foramen magnum, and holoacrania with rachischisis if spina bifida accompanies anencephaly (67).

HOLOPROSENCEPHALY

Central Nervous System

Holoprosencephaly is a developmental field defect of impaired midline cleavage of the embryonic forebrain. In alobar holoprosencephaly, the prosencephalon fails to cleave sagittally into cerebral hemispheres, transversely into telencephalon and diencephalon, and horizontally into olfactory tracts and bulbs.[a] Holoprosencephaly may be graded into alobar, semilobar, and lobar types. Although the classic definition is unambiguous, problems are encountered at the less severe end of the phenotypic spectrum. The classic anatomic definition of holoprosencephaly and the variable definitions used in genetics are both valid and depend on the context in which they are used (26).

Face

In a classic article, DeMyer et al. (42) discussed a graded series of facial anomalies that occur with holoprosencephaly (Fig. 5-16). The face predicts the brain approximately 80% of the time. The other 20% of the time, the facial changes are not diagnostic (19, 41). In cyclopia, the most extreme variant, a single median eye with varying degrees of doubling of the intrinsic ocular structures is associated with arrhinia and usually with proboscis formation. In ethmocephaly, two separate hypoteloric eyes are associated with arhinia and proboscis formation. In cebocephaly, hypotelorism is associated with a blind-ended, single-nostril nose. With median cleft lip, hypotelorism is associated with a flat nose and a median cleft because of agenesis of the primary palate. Less severe facial anomalies may include hypotelorism or hypertelorism, lateral cleft lip, or iris coloboma, or any combination of these features (39).

Numerous mild malformations have been found in holoprosencephaly (4, 9, 39, 51, 53, 71, 72). A single maxillary central incisor (9, 19) and, much less commonly, absence of the nasal septal cartilage (53) can be useful markers in autosomal dominant holoprosencephaly. Other identified defects have included stenosis of the nasal aperture (4), absence of the labial frenum (71), and absence of the philtral ridges (72).

Causes of Holoprosencephaly

Human holoprosencephaly is etiologically heterogeneous and pathogenetically variable. Identifiable genetic causes account for about 15–20% of all cases of holoprosencephaly. Monogenic inheritance has been reported, including autosomal dominant inheritance with wide expressivity and incomplete penetrance (8, 18), autosomal recessive inheritance (32, 33), and X-linked inheritance (54, 77).

Known human mutations for holoprosencephaly are summarized in Table 5-8. Because various pathways are linked to the sonic hedge-

[a] Although "cleave" is the commonly used term, no splitting actually occurs. In the sagittal plane, budding of the telencephalic vesicles takes place. In the transverse plane, "cleavage" is an arbitrary designation given to regions. In the horizontal plane, budding of the olfactory tracts takes place (26).

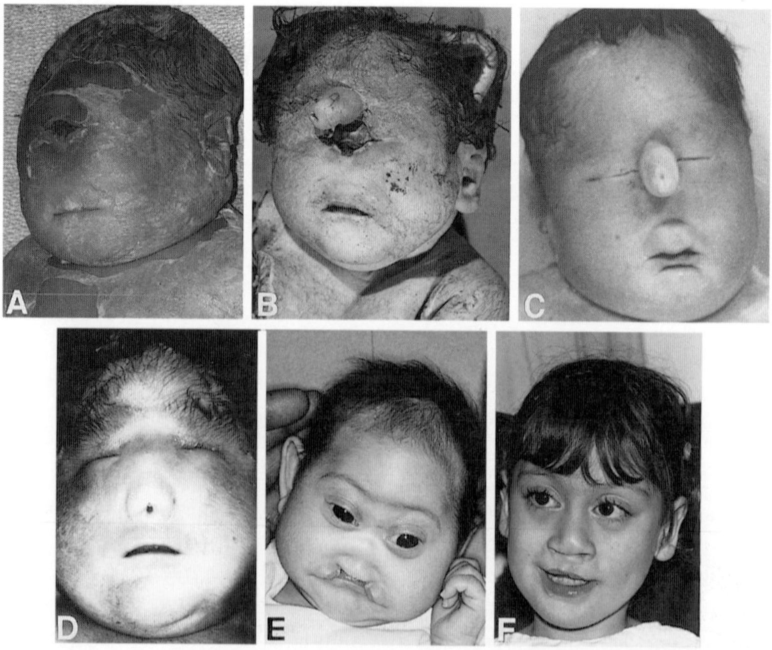

Figure 5-16 Spectrum of facial anomalies associated with varying degrees of holorosencephaly. A, Cyclopia without a proboscis. Note the single central eye. B, Cyclopia with a proboscis. C, Ethmocephaly. D, Cebocephaly. Ocular hypotelorism with a single-nostril nose. E, Median cleft lip, flat nose, and ocular hypotelorism. F, Ocular hypotelorism and a surgically repaired cleft lip. A–D, From Cohen et al., 1971 (33). E, From DeMeyer et al., 1964 (42).

Table 5-8 Known Mutated Genes in Holoprosencephaly[a]

Mutated Gene	Chromosome Map Location	Holoprosencephalic Phenotype	Frequency among All Cases of Holoprosencephaly
SHH	7q36	Great variability (severe to mild)	17% of familial cases; 3.7% of sporadic cases
ZIC2	13q32	Severe brain defect; normal or mildly dysmorphic face	3–4%
SIX3	2p21	Variable, usually severe	1.3%
TGIF	18p11.3	Variable	1.5%
PTCH	9q22.3	Variable	Rare (4 cases); *PTCH* mutations common in nevoid basal cell carcinoma syndrome and in isolated basal cell carcinomas
GLI2	2q14	Variable	1.8%
FAST1	8q24.3	Variable	20 missense variants in 100 holoprosencephaly cases and 350 patients with congenital heart defects; 3 mutations were found in both holoprosencephaly and in cardiac defects
TDGF1	3p21–p23	Variable	0.5%
DHCR7	11q12–q13	Holoprosencephalic faces	2–4% of Smith–Lemli–Opitz syndrome cases

[a]See text for references.

From Cohen, 2003 (30).

hog signaling network, the mutations are interrelated (78) (Fig. 5-17). *SHH* (7, 87, 97, 98), *PTCH* (74), and *GLI2* (99) are within the sonic hedgehog signaling network itself. *DHCR7* (57, 116, 120, 122) involves 7-dehydrocholesterol reductase in cholesterol biosynthesis. *ZIC2* (13, 14) and *GLI2* (99) are related; GLI proteins are translocated to cell nuclei by coexpressed ZIC proteins (12, 60). *TDGF1* (40), *TGIF* (50), and *FAST1* (91) involve *Nodal/TGFβ* signaling. *SIX3* (114) has not been linked to any of these interrelated pathways to date.

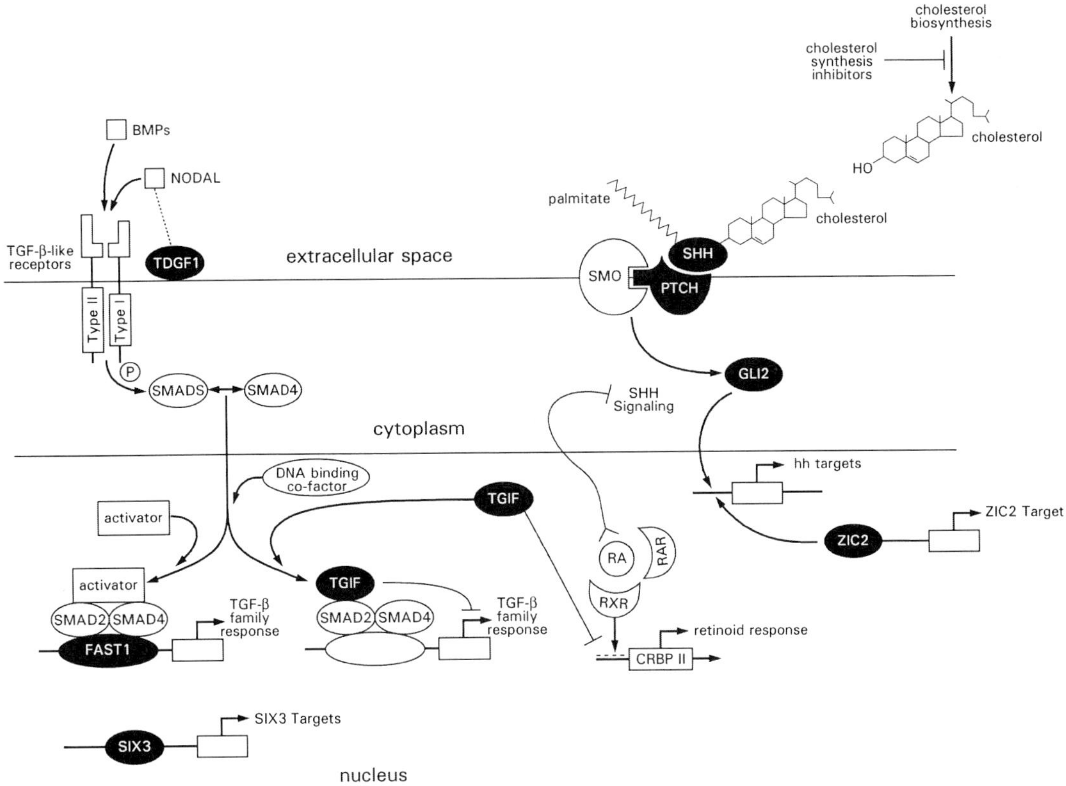

Figure 5-17 Mutations for holoprosencephaly in black: *SHH*, *PTCH*, *GLI2*, *ZIC2*, *TGIF*, *TDGF1*, *FAST1*, and *SIX3*. Another mutation, *DHCR7* involving 7-dehydrocholesterol reductase deficiency, is one of the cholesterol synthesis inhibitors (upper right). Because various pathways are linked to the sonic hedgehog signaling network, the mutations are interrelated: *SHH*, *PTCH*, and *GLI2* are within the network itself. *ZIC2* and *GLI2* are related: GLI proteins are translocated to cell nuclei by coexpressed ZIC proteins. *TDGF1*, *TGIF*, and *FAST1* involve Nodal/TGFβ signaling. *SIX3* has not been linked to any of these interrelated pathways to date. From Muenke and Beachy, 2001 (78). Modified and updated through the courtesy of M. Muenke, Bethesda, Maryland.

Teratogenic causes of holoprosencephaly are also known. About 1–2% of newborn infants of mothers with diabetes have holoprosencephaly (5). Holoprosencephaly may also be seen on occasion with fetal alcohol syndrome and retinoic acid embryopathy (38).

CRANIOSYNOSTOSIS

The term *craniostenosis* is used to indicate premature fusion of one or more sutures. Technically, *craniosynostosis* is the process of premature sutural fusion, and craniostenosis is the result. Actually, the terms have been used interchangeably, and craniosynostosis is replacing craniostenosis as the more common term (35).

Head shape depends on which sutures are prematurely synostosed (Figs. 5-18, 5-19), the order in which they synostose, and the timing at which they synostose. Craniosynostosis may be of prenatal or perinatal onset, or may occur during infancy or childhood. The earlier synostosis occurs, the more dramatic the effect on subsequent cranial growth and development. The later synostosis occurs, the less the effect on cranial growth and development. Synostosed skulls

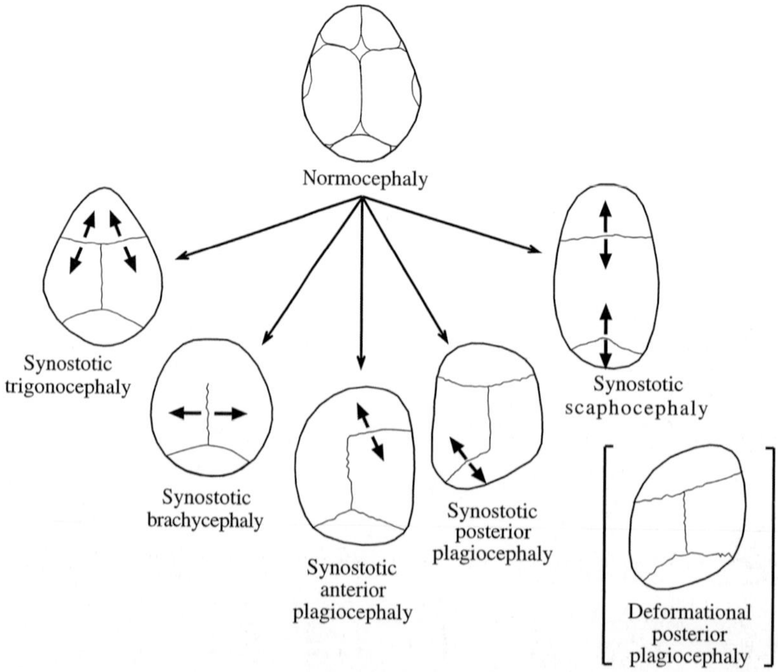

Figure 5-18 In synostotic scaphocephaly, the sagittal suture is prematurely closed. In synostotic brachycephaly and synostotic anterior plagiocephaly, the coronal suture is closed either bilaterally or unilaterally. In synostotic trigonocepahly, the metopic suture is prematurely closed. In the rarely occurring synostotic posterior plagiocephaly, the lambdoid suture is closed unilaterally. This condition is frequently confused with deformational posterior plagiocephaly (bracketed) in which the sutures are patent. The degree of alteration in cranial shape depends on the timing of synostosis. It may be congenital or develop during infancy, resulting in less severe cranial distortion. Furthermore, cranial shape can be modified when two or more sutures are involved. This figure specifically uses the adjective synostotic to describe the cranial shape because brachycephaly, dolichocephaly, and trigonocephaly are known to occur without synostosis. From Cohen and MacLean, 2000 (35).

with minimal dysmorphism have been observed (35).

Large neurosurgical services generate a characteristic pattern of sutural involvement (35, 103) (Fig. 5-20). Sagittal synostosis is the most common type, occurring more frequently in males than in females. Coronal synostosis is less common than sagittal synostosis; females slightly predominate or, in some surveys, males and females have the same frequency. Multiple synostoses of various types are less common than coronal synostosis. Both metopic and lambdoid synostosis occur with low frequency (35).

In sagittal synostosis, calvarial midline ridging usually occurs in the posterior half. Coronal synostosis may be bilateral, resulting in brachycephaly, or unilateral, resulting in synostotic frontal plagiocephaly. Metopic ridging is common during infancy and childhood, and is not associated with craniosynostosis. Such trigonocephaly self-corrects with time, resulting in a normal cosmetic appearance (35).

Plagiocephaly

Plagiocephaly is defined as an asymmetric skull. Well-known types include synostotic anterior

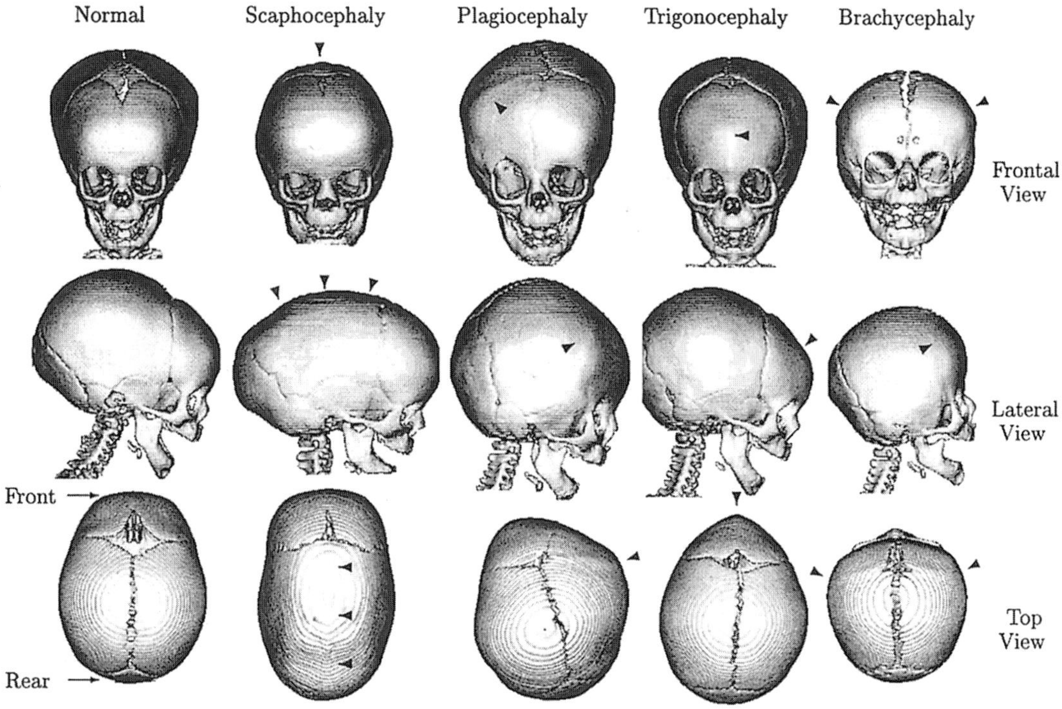

Figure 5-19 Three-dimensional surface views of five infants (less than 6 months of age) reconstructed from volumetric CT scans. These include a normal skull and the four most common cranial vault deformities: scaphocephaly, plagiocephaly, trigonocephaly, and brachycephaly. The triangles show the sites of the abnormality and the arrows show the principal direction of abnormal growth. Scaphocephaly is characterized by a high, narrow cranial vault that is elongated from front to back. Plagiocephaly is characterized by an asymmetric cranial vault shape. Trigonocephaly is characterized by a pointed forehead and depression of the temporal regions. Brachycephaly is characterized by a short, broad cranial vault shape. From Vannier, 2000 (113).

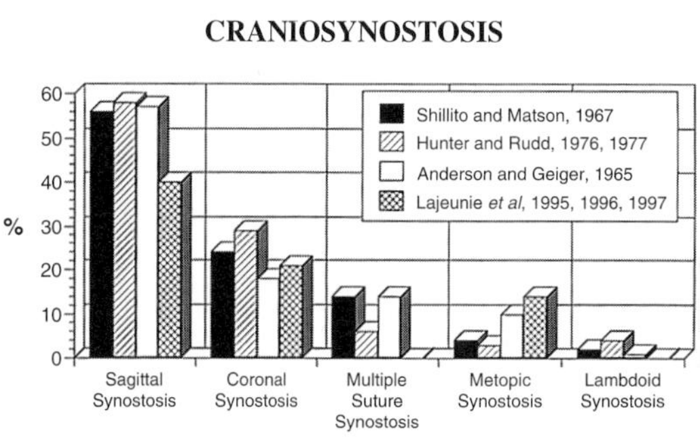

Figure 5-20 Characteristic pattern of craniosynostosis in four neurosurgical services. From Cohen and MacLean, 2000 (35).

plagiocephaly (unilateral coronal synostosis), synostotic posterior plagiocephaly (unilateral lambdoid synostosis), deformational anterior plagiocephaly, and deformational posterior plagiocephaly. Of these, deformational posterior plagiocephaly is common and lambdoid synostosis is rare (35) (Table 5-9).

Deformational posterior plagiocephaly, caused by intrauterine factors such as hypotonia, fetal positioning, and prematurity, can produce asymmetric flattening of the occiput that becomes favored by infants sleeping on their backs, which exaggerates the plagiocephaly. This type of plagiocephaly has increased dramatically since the

Table 5-9 Types of Plagiocephaly

Types	Sutures Synostosed	Relative Frequencies
Synostotic anterior plagiocephaly	Unilateral coronal	Less common than bilateral coronal synostosis
	Unilateral frontosphenoidal[a]	Very rare
	Unilateral frontozygomatic[a]	Very rare
Synostotic posterior plagiocephaly	Unilateral lambdoid	Rare
Deformational anterior plagiocephaly	No sutures synostosed	Formerly common; now uncommon
Deformational posterior plagiocephaly	No sutures synostosed	Common

[a]These have been reported in the absence of unilateral coronal synostosis.

Modified from Cohen and MacLean, 2000 (35).

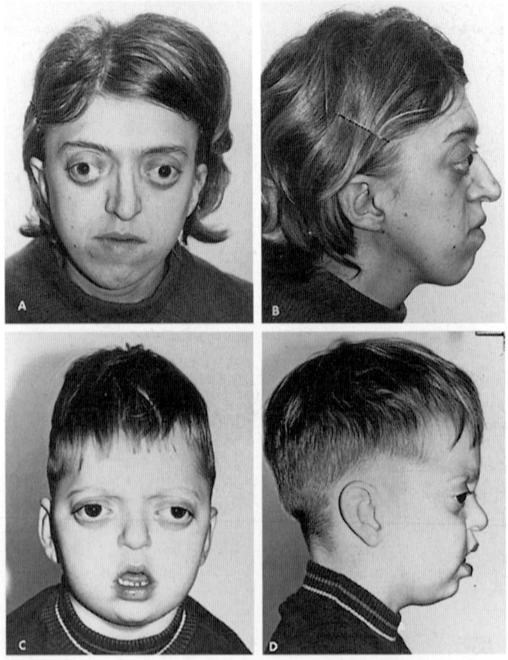

Figure 5-21 Crouzon syndrome in a mother and her son. A–D, Brachycephaly, ocular proptosis, and maxillary hypoplasia. From Cohen and MacLean, 2000 (35).

1992 recommendation of the American Academy of Pediatrics for supine infant sleeping to reduce the risk of sudden infant death syndrome (SIDS) (35).

Crouzon and Apert Syndromes

Crouzon syndrome is characterized by craniosynostosis, maxillary hypoplasia, shallow orbits, and ocular proptosis (Fig. 5-21). Inheritance is autosomal dominant, with 67% of cases being familial and 33% being sporadic (35).

Apert syndrome is characterized by craniosynostosis, midface hypoplasia, and symmetric syndactyly of the hands and feet (Figs. 5-22, 5-23). The facial phenotype is variable, from mild to severe (Fig. 5-24). Inheritance is autosomal dominant, with most cases representing new mutations (20, 35).

In Apert syndrome infants, a wide midline calvarial defect is present, extending from the glabella to the posterior fontanel (Figs. 5-25, 5-26). Bony islands, beginning anywhere within the defect, form and coalesce, resulting in complete obliteration of the defect during the first two years of life (Figs. 5-25, 5-26). No proper sagittal or metopic suture ever forms. Although the coronal suture is closed at birth, the anterolateral fontanels are abnormally large, and both the lambdoid and squamosal sutures are patent. Characteristics of suture default zones and suture formation are summarized in Table 5-10. Sutural interdigitations never form at the calvarial bony edges of the

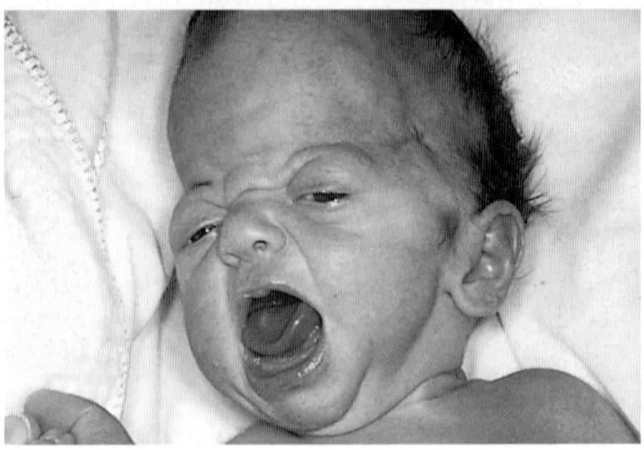

Figure 5-22 Infant with Apert syndrome. Note the high forehead, supraorbital grooves, and ocular proptosis. From Cohen and Kreiborg, 1993 (20).

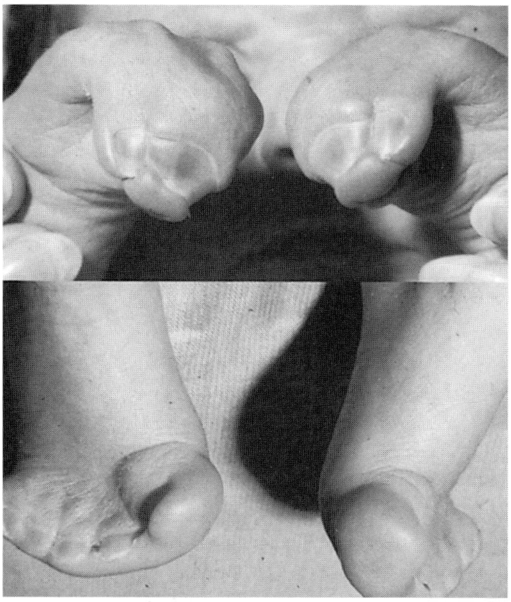

Figure 5-23 Hands and feet of an infant with Apert syndrome. Note the symmetric syndactyly. From Cohen and Kreiborg, 1993 (20).

midline defect. Therefore, the term *craniosynostosis* does not describe the bony islands that coalesce to obliterate the default zone. In Apert syndrome, sutures fail to form *ab initio* in the midline defect, resulting in a suture default zone. Thus, the appropriate term is *sutural agenesis*, not *craniosynostosis* (34, 35, 62).

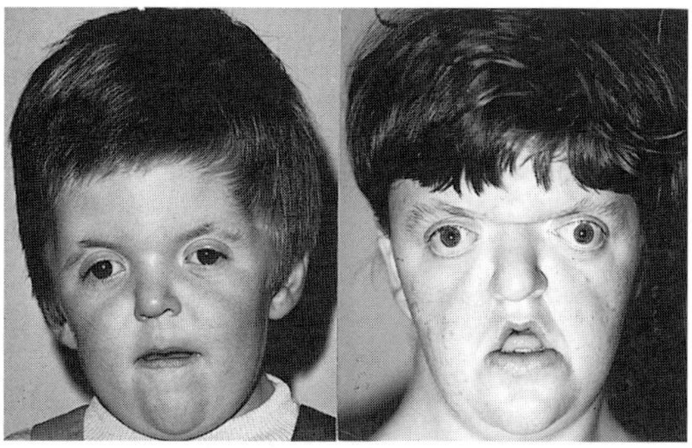

Figure 5-24 Facial characteristics of Apert syndrome. Left, Mild involvement. Note the break in the continuity of the eyebrows. Right, Severe involvement. Note the trapezoidal-shaped mouth. From Cohen and MacLean, 2000 (35).

Table 5-10 Sutures and Suture Zones in Apert Syndrome

Region	Characteristics
Metopic below glabella	Suture formation
Metopic above glabella	Suture default zone; part of midline calvarial defect; closure by coalescence of bony islands
Sagittal	Suture default zone; part of midline calvarial defect; closure by coalescence of bony islands
Coronal	Suture formation; prematurely synostosed at birth
Lambdoid	Suture formation; true Wormian bones on occasion
Squamosal	Suture formation
Sphenotemporal	Suture formation

From Cohen and Kreiborg, 1996 (34).

Etiologic Heterogeneity and Pathogenetic Variability

Craniosynostosis is etiologically heterogeneous and pathogenetically variable. Overall, 8% of all craniosynostosis pedigrees are familial (n = 725 pedigrees). Evidence generally supports autosomal dominant inheritance. For coronal synostosis, 14.4% of pedigrees are familial (n = 180 pedigrees); for sagittal synostosis, 6% of pedigrees are familial (n = 366 pedigrees); and for metopic synostosis, 5.6% of pedigrees are familial (n = 179 pedigrees) (63–65). Although some pedigrees show synostosis of a specific suture, such as the coronal, sagittal, or metopic suture, other pedigrees may show fusion of different sutures in affected relatives of the same family. Many syndromes have been identified, and well over 100 are known (35).

Different chromosomal aberrations have been associated with craniosynostosis. In some conditions, craniosynostosis is common [e.g., dup(3q), del(7p), del(9p), and del(11q)], and in others, craniosynostosis is an unusual occurrence [e.g., dup(5p), del(6)(q22.2-q23.1), del(8q), and dup(15q)] (35).

Primary craniosynostosis is common and includes single—and multiple—suture synostosis–for example, coronal synostosis or coronal and sagittal synostosis. Examples of secondary synostosis include hyperthyroidism and Hurler syndrome. Conditions with secondary synostosis are

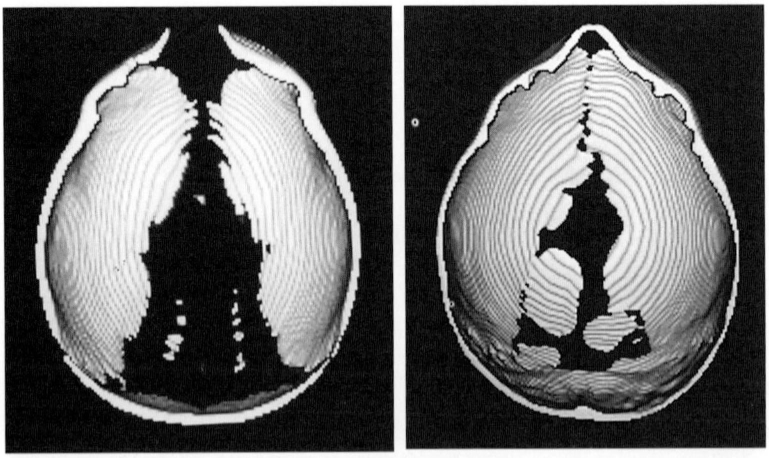

Figure 5-25 Three-dimensional reconstruction of a CT scan showing an endocranial view of the cranium. Left, Huge midline defect extending from the frontal bone to the posterior fontanel, a characteristic of the infant Apert skull. Age 1 month. Right, The same patient at age 8 months. Islands of bone are coalescing in the posterior region. Gap in the frontal bone is now bridged. Note the irregular narrowing of the midline defect compared to that of the figure on the left. From Kreiborg et al., 1993 (62).

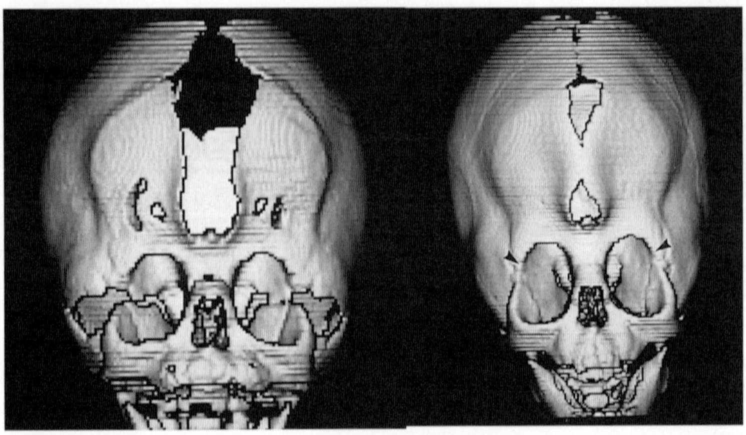

Figure 5-26 Three-dimensional reconstruction of CT scans in an Apert syndrome infant. Left, Extensive midline calvarial defect and large anterolateral fontanels extending into the orbits at 2 months of age. Right, Bony coalescence of the midline calvarial defect at 8 months of age. The break in the continuity of the eyebrows (see Figure 5-24, left) corresponds to underlying bony defects in the orbits (arrows). From Kreiborg et al., 1993 (62).

listed in Table 5-11. They include six teratogens that may be associated with craniosynostosis in some cases (35).

Koskinen–Moffett et al. (59) provided evidence for prenatal head constraint as a cause of craniosynostosis in mice. The evidence in humans is indirect; it is more compelling for sagittal synostosis than for coronal or metopic synostosis (35).

Mutations and Craniosynostosis Syndromes

Mutations have been identified in *FGFRs*, *TWIST*, *MSX2*, *EFNB1*, and *ALPL* (Table 5-12). Most mutations are found in *FGFR2*, although there are several in *FGFR3* and in *FGFR1* (30a, 35, 89, 111, 116a–118) (Fig. 5-27).

Mutations identified on *FGFRs* may be located extracellularly or intracellularly. Almost all nucleotide changes are of the missense or splice-site types. Occasionally, small inframe insertions and deletions are recorded. *FGFR* mutations involve gain-of-function. They may create or destroy cysteine residues, leaving an unpaired cysteine that can produce intermolecular disulfide bonding and constitutive receptor activation. Accentuated ligand binding, resulting in gain-of-function, occurs in mutations from the linker region of *FGFR1* (Pfeiffer syndrome), *FGFR2* (Apert syndrome), and *FGFR3* (Muenke syndrome). All human *FGFR* mutations known to

Table 5-11 Conditions with Secondary Craniosynostosis

Metabolic disorders
Hyperthyroidism
Rickets (various forms)
Mucopolysaccharidoses and related disorders
Hurler syndrome
Morquio syndrome
β-glucuronidase deficiency
Mucolipidosis III
α-D-mannosidase deficiency
Hematologic disorders
Thalassemia
Sickle cell anemia
Polycythemia vera
Congenital hemolytic icterus
Teratogens
Diphenylhydantoin
Retinoids
Valproate
Aminopterin
Fluconazole
Cyclophosphamide
Malformations
Holoprosencephaly
Microcephaly
Encephalocele
Iatrogenic disorders
Hydrocephalus with shunt

From Cohen and MacLean, 2000 (35).

Table 5-12 Genes Bearing Known Mutations for Craniosynostosis

FGFR1
Osteoglophonic dysplasia
Pfeiffer syndrome
FGFR2
Apert syndrome
Pfeiffer syndrome
Crouzon syndrome
Jackson–Weiss syndrome
Beare–Stevenson cutis gyrata syndrome
Nonclassifiable and variable craniosynostosis
FGFR3
Thanatophoric dysplasia, type I
Thanatophoric dysplasia, type II
Crouzonodermoskeletal syndrome
Muenke syndrome
TWIST
Saethre–Chotzen syndrome
MSX2
Boston–type craniosynostosis
EFNB1
Craniofrontonasal syndrome
ALPL
Hypophosphatasia, particularly infantile type

Modified from Cohen and MacLean, 2000 (35).

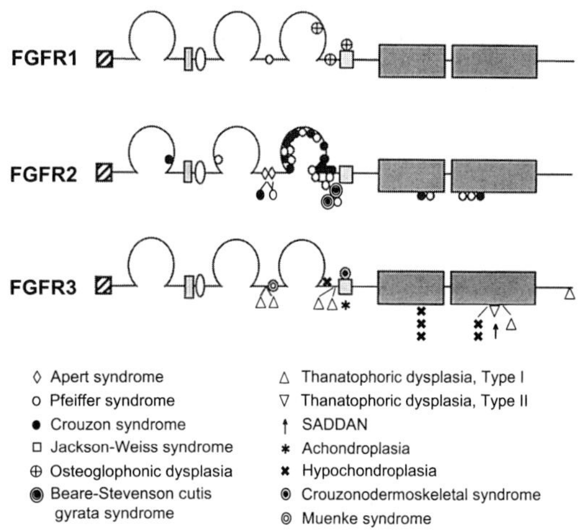

Figure 5-27 Most mutations for major craniosynostosis syndromes are on *FGFR2*, with several on *FGFR3* and on *FGFR1*. Note the mutations for short-limb skeletal dysplasias (achondroplasia, hypochondroplasia, thanatophoric dysplasia type I, and thanatophoric dysplasia type II) on *FGFR3*. Hatched square, signal peptide; solid oblong, acid box; solid oval, CAM homology domain; solid square, transmembrane domain; long solid oblongs, kinase domains 1 and 2. Three loops from left to right are immunoglobulin-like domains (IgI, IgII, IgIII). For clarity, only a few of the many mutations for Crouzon and Pfeiffer syndromes are shown on *FGFR2*. Modified from Cohen and MacLean, 2000 (35).

date allow early human development to proceed normally but interfere with later development, particularly of bone. When mesenchyme is converted to bone, the molecular crosstalk is highly complex and only partially understood (1, 30a, 35, 88).

Other gene mutations are also known (Table 5-12). In Saethre–Chotzen syndrome, mutations may be of the nonsense, missense, insertion, or deletion type and may be found outside or within the coding region. Most mutations truncate the *TWIST* protein, resulting in haploinsufficiency. Boston-type craniosynostosis results from a heterozygous gain-of-function mutation in *MSX2*. Craniofrontonasal syndrome is caused by heterozygous loss-of-function mutations in *EFNB1*. Craniosynostosis may be observed in hypophosphatasia, particularly the infantile type, caused by mutations in *ALPL* (16, 30a, 35, 56, 79, 90, 111, 115).

Table 5-13 Some Branchial Arch Syndromes

Syndrome	Major Features	Chromosome	Gene
Hemifacial microsomia/ Goldenhar syndrome	Hypoplasia and asymmetry involving mandible and ear, usually unilateral; if bilateral, more severe on one side; may be associated with various anomalies especially of the eyes (epibulbar dermoids), cardiovascular malformations, skeletal defects (especially vertebral), and CNS anomalies	?	?
Treacher Collins syndrome	Bilateral zygomatic hypoplasia, downslanting palpebral fissures, malformed ears, micrognathia	5q32–q33.1	*TCOF1*
Townes–Brocks syndrome	Dysplastic ears, ear tags, sensorineural deafness, bifid or supernumerary finger-like thumbs, hypoplastic kidneys, anal stenosis	16q12.1	*SALL1*
Branchio-oto-renal (BOR) syndrome	Cup-shaped ears, preauricular pits, branchial cleft fistulas, mixed deafness renal anomalies	8q13.3	*EYA1*

TREACHER COLLINS SYNDROME

Branchial arch syndromes make up an etiologically heterogeneous group of disorders. The best known of these are hemifacial microsomia, its variant Goldenhar syndrome, and Treacher Collins syndrome. Some branchial arch syndromes are summarized in Table 5-13.

Treacher Collins syndrome consists of bilateral zygomatic hypoplasia, downslanting palpebral fissures, malformed ears, and micrognathia (Fig. 5-28). Inheritance is autosomal dominant, and expressivity varies. The syndrome maps to 5q32-q33.1, and mutations in the *TCOF1* gene are of the nonsense, insertion, deletion, or splice-site types. The introduction of premature termination codons into the reading frame indicates haploinsufficiency (44, 46, 110).

CLEIDOCRANIAL DYSPLASIA

Cleidocranial dysplasia is an autosomal dominant skeletal disorder characterized by short stature, brachycephaly, delayed closure of fontanels and sutures, Wormian bones, midface hypoplasia, unerupted teeth, supernumerary permanent teeth, aplasia or hypoplasia of the clavicles, and other skeletal anomalies, such as hypoplastic iliac wings and brachydactyly (25, 27, 83) (Figs. 5-29 to 5-31).

Cleidocranial dysplasia is caused by mutations in *RUNX2*. These mutations result in haploinsufficiency and may be of the deletion, insertion, missense, or nonsense types (84). *RUNX2* is the master gene for bone (Fig. 5-32). In the mouse, cleidocranial dysplasia is caused when one allele is suppressed (*Runx2*$^{+/-}$). The null

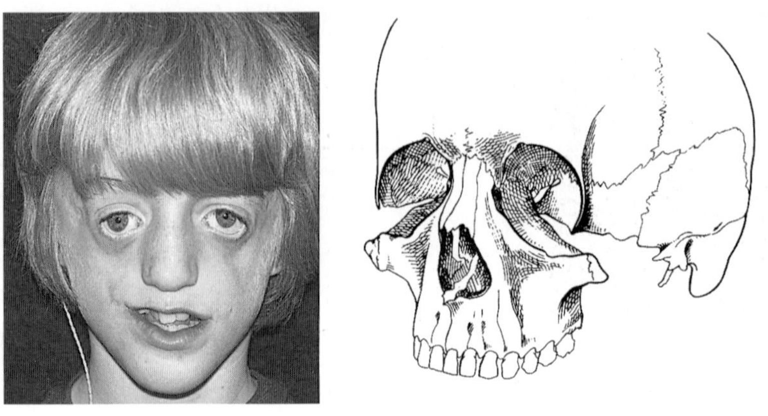

Figure 5-28 Treacher Collins syndrome. Left, Clinical appearance showing downslanting palpebral fissures and malar deficiency. Note the abnormal shape of the lower eyelids, which often have missing lashes medially. From Cohen, 1997 (22). Right, Bony clefting of ovoid-shaped orbits with absence of the zygomatic arches. From Tessier, 1976 (109). See Figure 5-13; these correspond to Tessier-type cleft numbers 6, 7, and 8.

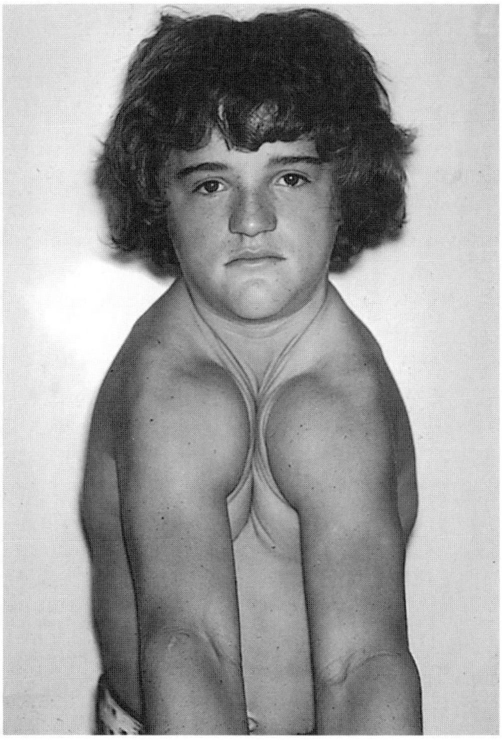

Figure 5-29 Cleidocranial dysplasia. Approximation of the shoulders in the midline, resulting from absence of the clavicles. From Cohen, 2000 (25).

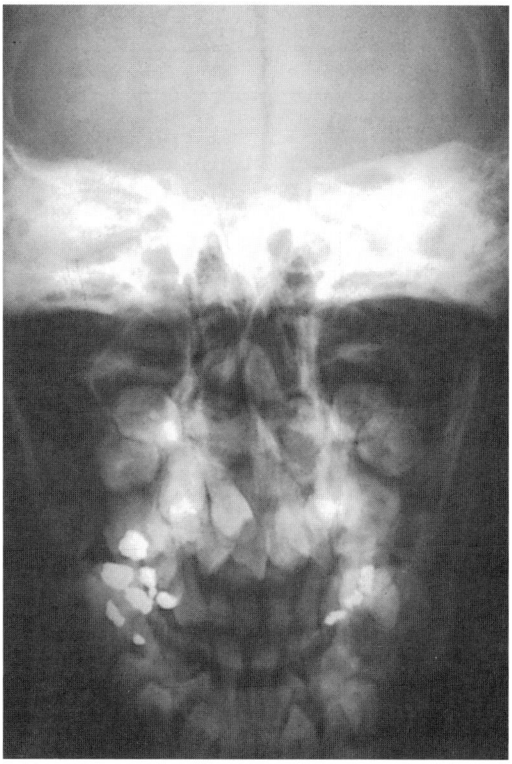

Figure 5-31 Cleidocranial dysplasia. Radiograph showing multiple unerupted supernumerary teeth. From Cohen, 2000 (25).

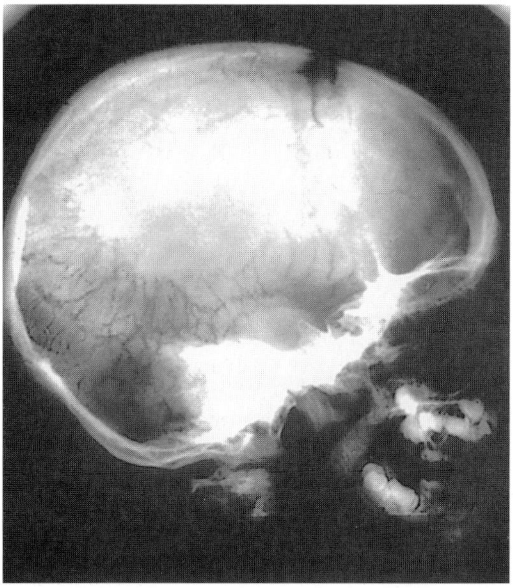

Figure 5-30 Cleidocranial dysplasia. Lateral radiograph. Patent anterior fontanel and Wormian bones. From Cohen, 2000 (25).

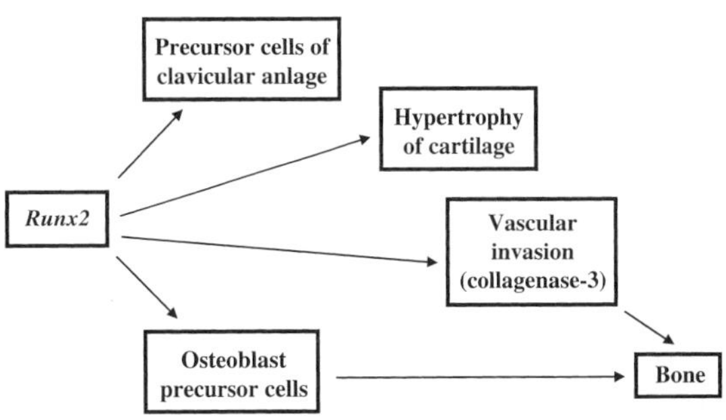

Figure 5-32 *Runx2* controls the differentiation of precursor cells into osteoblasts, regulates chondrocyte differentiation toward hypertrophy, may cause vascular invasion of cartilage by regulating collagenase-3, and controls the differentiation of precursor cells of the clavicular anlage. Modified and adapted from Mundlos, 1999 (83).

mutant, in which *Runx2* is suppressed at both alleles ($Runx2^{-/-}$), shows lack of ossification of both endochondral and membranous bone (83).

EHLERS–DANLOS SYNDROME, TYPE I

The Ehlers–Danlos syndromes are an etiologically heterogeneous group of generalized connective tissue disorders. Features include hyperextensible and doughy skin, connective tissue fragility, atrophic scars, and hypermobile joints. The autosomal dominant type I, the severe form, is caused by a COL5A1-linked defect in some families. Recorded mutations have included exon skipping, translocation, and cysteine substitution (48). Hyperextensible facial skin is demonstrated in Figure 5-33.

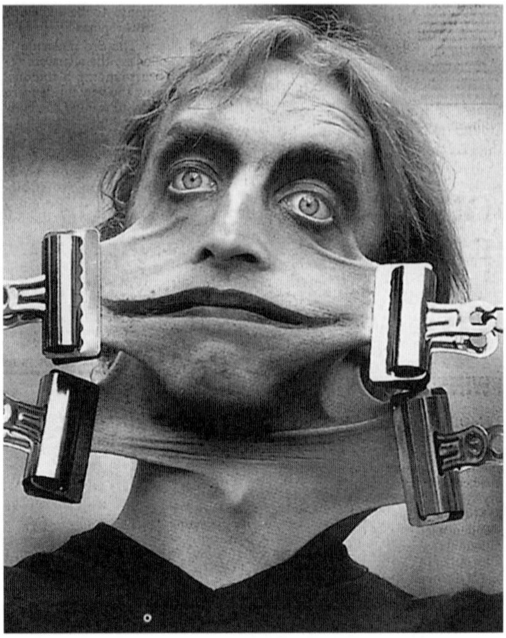

Figure 5-33 Ehlers-Danlos syndrome, type I. Demonstration of hyperextensible facial skin with clips. From Edinburgh Festival, 2001.

ASYMMETRIC DYSMORPHISM

Asymmetric dysmorphism has already been discussed under the headings of plagiocephaly, Apert syndrome, and craniofacial clefts of the Tessier type. In this section, hemihyperplasia, hemifacial microsomia, and hemiatrophy are considered.

Hemihyperplasia may affect either one tissue such as bone with secondary regional consequences or multiple tissues on a primary basis such as in "hemihypertrophy." An example of a bone abnormality with secondary regional consequences is unilateral enlargement of a mandibular condyle, which results in lower facial asymmetry (21).

By general usage, *hemihyperplasia* usually refers to multi-affected tissues. Although the term *hemihypertrophy* has commonly been used, it is inappropriate because the condition so obviously refers to hemihyperplasia (21) (Fig. 5-34).

Hemihypoplasia may affect one tissue such as bone with secondary regional consequences or multiple tissues on a primary basis. Condylar hypoplasia is an example of one affected tissue with secondary lower facial asymmetry. Hemifacial microsomia is an example of multi-affected tissues (21).

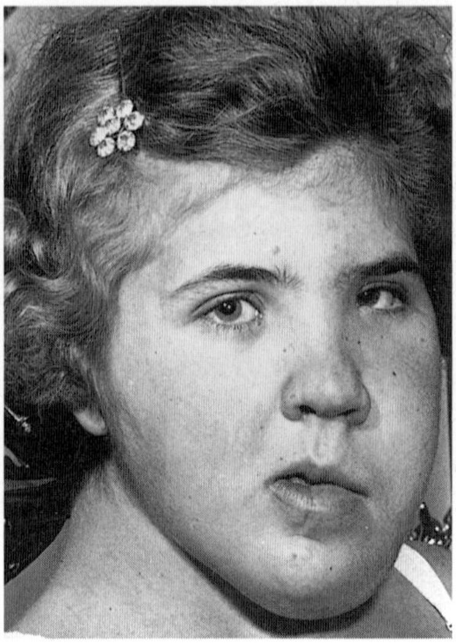

Figure 5-34 Facial hemihyperplasia. From Gorlin and Meskin, 1962 (49).

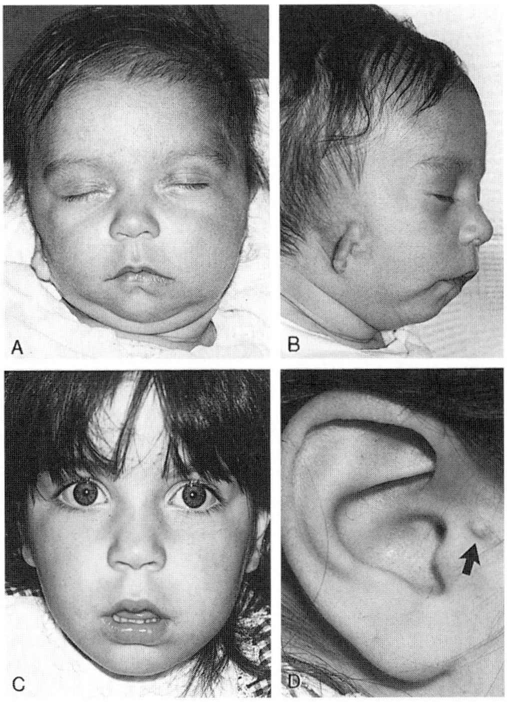

Figure 5-35 A, B, Hemifacial microsomia. C, D, Sister with ear tag only. From Cohen et al., 1989 (37).

Hemifacial microsomia involves aural, oral, and mandibular growth. The disorder may be mild or severe, and involvement is limited to one side of the face in most cases, but bilateral involvement, with more severe expression on one side, is also known to occur (Fig. 5-35). Goldenhar syndrome is a variant of hemifacial microsomia. There is a 3:2 predilection for right-sided involvement. Most cases are sporadic, but familial examples have been recorded (37).

The quintessential craniofacial disorder with *hemiatrophy* is Romberg syndrome, or progressive hemifacial atrophy (Fig. 5-36). Facial changes begin during the first decade and spread slowly and progressively to involve soft tissue, muscle, cartilage, and underlying bone. There is marked predilection for left-sided involvement. The overlying skin may become pigmented. After hemifacial atrophy progresses for several years, the process usually becomes arrested. Jacksonian epilepsy, which may occur in some cases, is often late in appearing (28, 68).

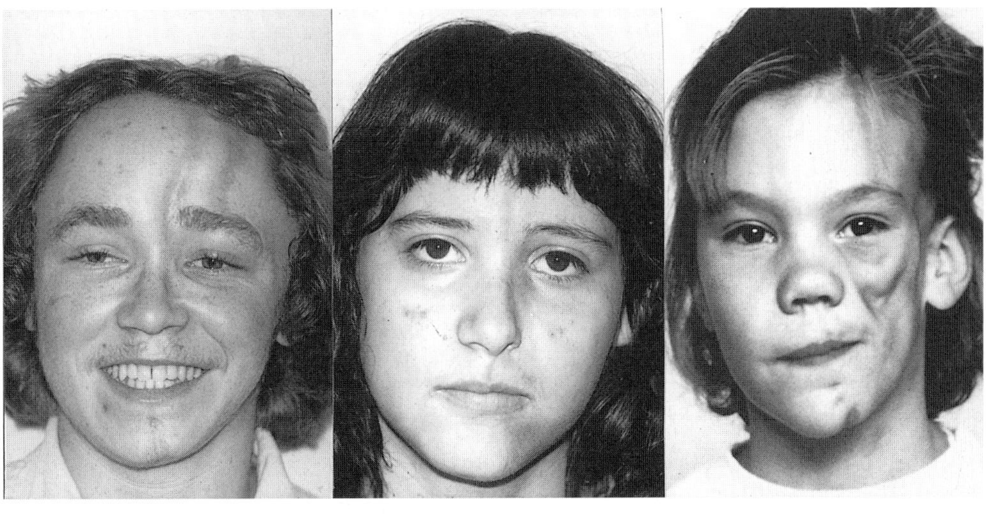

Figure 5-36 Romberg syndrome (hemifacial atrophy). Increasing severity from left to right. Left, Note linear depressions of the forehead and the asymmetric mandible. Center, Hemifacial involvement with an atrophic alar wing and loss of cartilage. Right, Severe involvement with loss of soft tissue, muscle, and bone. Note the sparing of the alar wing compared to the patient in the center. From Cohen, 2001 (28).

Table 5-14 Mulliken Classification

	Hemangioma	Vascular Malformation
Presentation	Usually not present at birth	Always present at birth but may not be apparent in some cases
Female:Male ratio	3:1 (higher in some series)	1:1
Cellular criteria	Endothelial hyperplasia	Flat endothelium
Course	Rapid postnatal growth followed by slow involution	Growth is commensurate with the child's growth

Adapted from Mulliken and Glowacki, 1982 (82).

Table 5-15 Biologic Classification of Vascular Markings

HEMANGIOMA	MALFORMATION
	Single channel
Proliferating phase	Capillary (CM)
Involuting phase	Lymphatic (LM)
	Venous (VM)
	Multiple channels
	CLM
	CVM
	CLVM
	LVM

Adapted from Mulliken, 1993 (80).

VASCULAR TUMORS AND MALFORMATIONS

Hemangiomas and Vascular Malformations

A distinction can be made between hemangiomas and vascular malformations based on cellular kinetics and clinical behavior (29a, 80–82) (Table 5-14). Hemangiomas have endothelial hyperplasia with rapid postnatal growth followed by slow involution (119) (Fig. 5-37). In contrast, vascular malformations are characterized by flat endothelium, and growth of the lesion is commensurate with growth of the child. In vascular malformations, a single type of channel may predominate or multiple channels of various types may occur (80–82) (Table 5-15, Figs. 5-38 to 5-42). Clinical conditions such as Sturge–Weber syndrome, Klippel–Trenaunay syndrome, and Parkes Weber syndrome have vascular malformations, not hemangiomas (36) (Table 5-16). Mulliken (80) noted that "standard" terminology used for vascular lesions has led to confusion, improper diagnosis, and misdirected research efforts (Fig. 5-43).

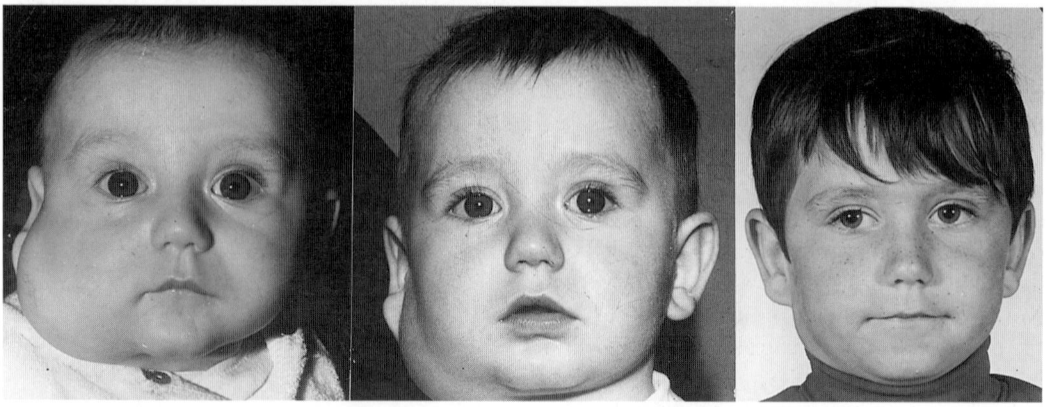

Figure 5-37 Proliferation and involution of a hemangioma with no surgical intervention. Left, A 3-month-old boy with a right parotid hemangioma. Note impingement on the ear and extension onto the neck. Center, Same child at 12 months of age with early signs of involution. Right, Same child at 6 years of age showing total resolution of the parotid hemangioma, but with excess wrinkling of the skin. From Williams, 1975 (119).

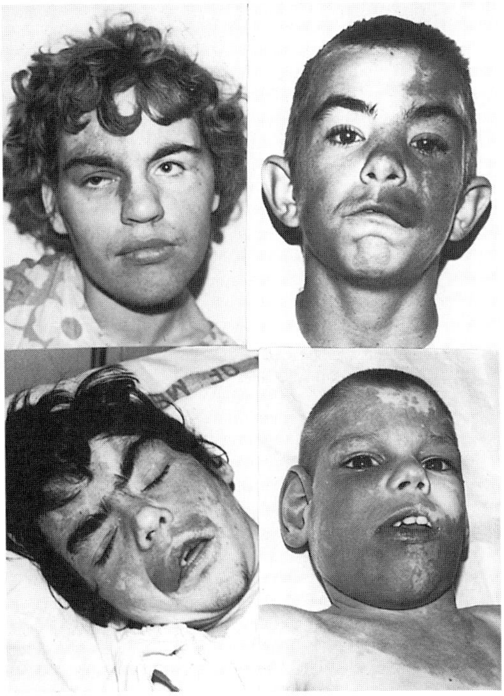

Figure 5-38 Sturge-Weber syndrome. Capillary malformation. Unilateral involvement (top). Bilateral involvement (bottom). Note extension onto the neck and chest of the patient on the bottom right. From Cohen et al., 2002 (36).

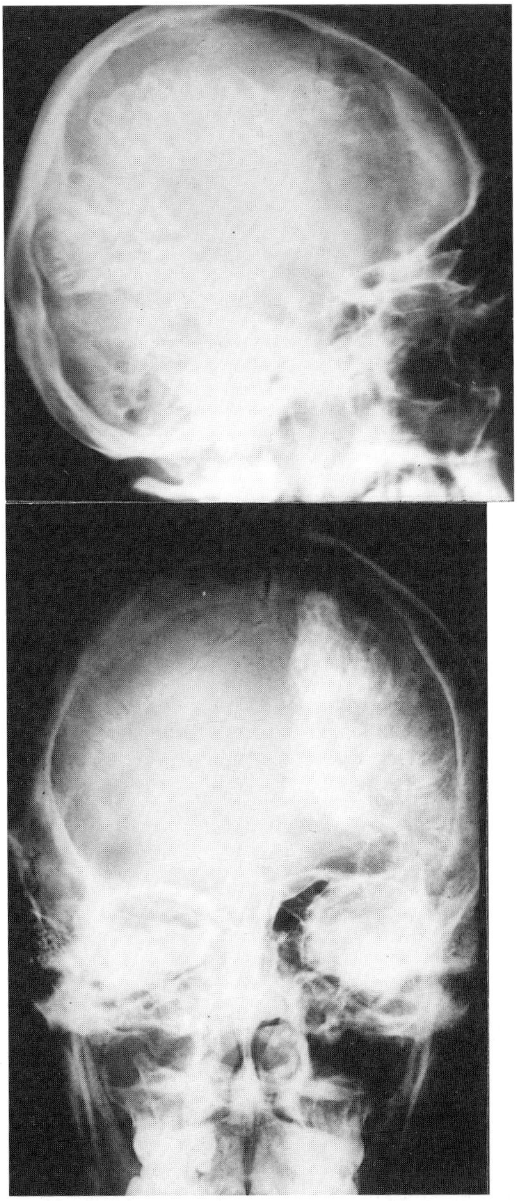

Figure 5-39 Sturge-Weber syndrome. Radiographs showing unilateral distribution of double contoured calcification of cerebral cortex. From Cohen et al., 2002 (36).

Sturge–Weber Syndrome

Sturge–Weber syndrome, a sporadically occurring disorder, is defined as a capillary malformation of the leptomeninges with or without choroid and facial V_1 or V_1–V_2 involvement (Table 5-16, Figs. 5-38 to 5-40). The syndrome can be explained by an embryonic defect with secondary consequences. During the sixth week of development, a vascular plexus forms around the cephalic portion of the neural tube and beneath the ectoderm destined to become facial skin. Normally, this vascular plexus regresses during the ninth week, but in Sturge–Weber syndrome, it persists, resulting in a capillary malformation of the leptomeninges overlying the cerebral cortex together with a facial "portwine" stain on the ipsilateral side. Variation in the persistence or regression of the vascular plexus accounts for cases with unilateral or bilateral involvement and also for cases with capillary malformation of the

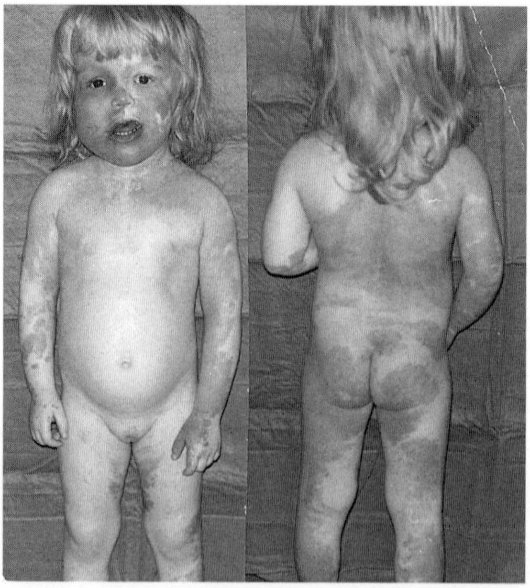

Figure 5-40 Sturge-Weber syndrome. Bilateral capillary malformations of the face, chest, back, buttocks, arms, and legs. This patient does *not* have Sturge-Weber syndrome *with Klippel-Trenaunay syndrome. There are capillary malformations only*. For Klippel-Trenaunay syndrome, additional features include lymphatic malformations, lymphatic vesicles, venous malformations of the lower limb, venous flares, and enlargement of a lower limb often involving the toes. From Cohen et al., 2002 (36).

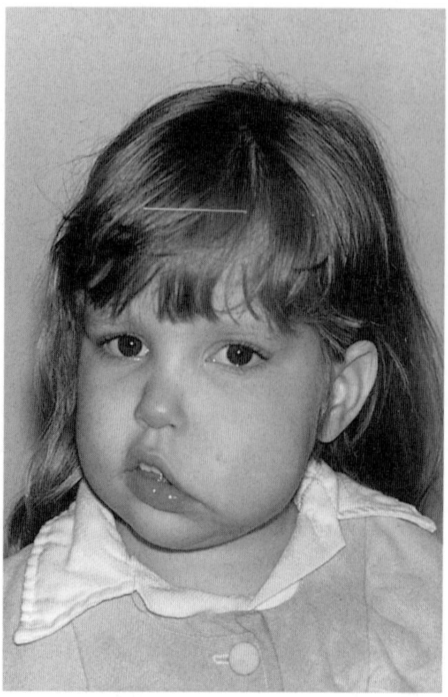

Figure 5-41 Lymphatic malformation of the face.

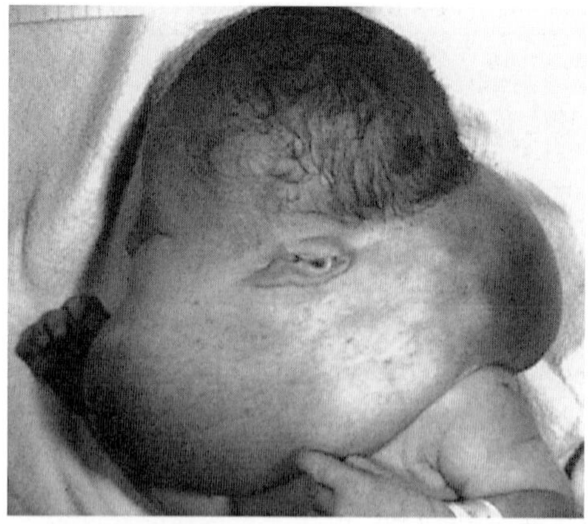

Figure 5-42 Large lymphatic malformation surrounding the neck.

leptomeninges with absence of facial involvement (36). Gyriform calcifications of the cerebral cortex are precipitated by an incompletely understood alteration in the vascular dynamics of the leptomeningeal capillary malformation (Fig. 5-39). Seizures result in about 83% of cases. These can begin during infancy contralateral to the leptomeningeal lesion. Most often seizures are local, but generalized seizures may also occur. Hemiparesis is less frequent and may result in a hypotrophic limb (36).

Capillary malformations of the skin may extend below the head and neck and appear anywhere on the body, including the upper and lower limbs (Fig. 5-40). Limb involvement is dif-

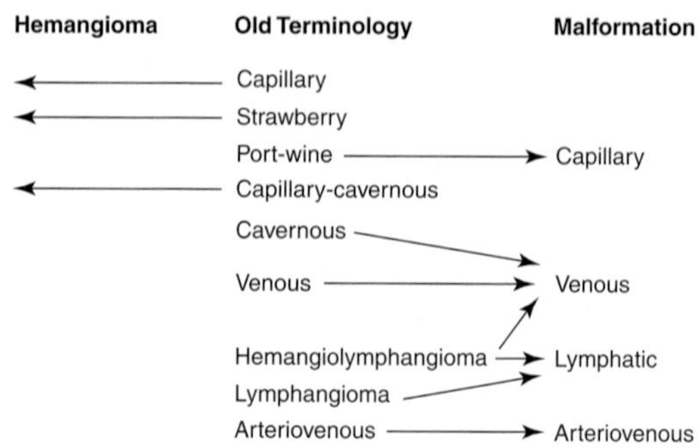

Figure 5-43 Confusion resulting from "standard" terminology of vascular lesions. Adapted from Mulliken, 1993 (80).

Table 5-16 Features of Sturge–Weber, Klippel–Trenaunay, and Parkes Weber Syndromes

Syndromes	Vascular Malformations	Clinical Considerations
Sturge–Weber syndrome	Slow-flow, capillary malformation	Involves leptomeninges with or without choroid and V_1 or V_1–V_2. Capillary malformation can occur elsewhere on body
Klippel–Trenaunay syndrome	Slow-flow, combined vascular malformation (capillary, lymphatic, and venous)	Involves limb(s) and/or trunk
Parkes Weber syndrome	Fast-flow, combined vascular malformation (capillary, arterial, and venous)	Involves upper/lower limbs

From Cohen et al., 2002 (36).

ferent from that found in Klippel–Trenaunay syndrome. The latter has lymphatic malformations, a lateral venous anomaly, lymphatic vesicles and venous flares within the capillary malformation, limb enlargement, and macrodactyly, which do not occur in Sturge–Weber syndrome. Only capillary malformations are found in Sturge–Weber syndrome (36).

REFERENCES

1. Anderson J, Burns HD, Enriquez–Harris P, Wilkie AOM, Heath JK: Apert syndrome mutations in fibroblast growth factor receptor 2 exhibit increased affinity for FGF ligand. Hum Mol Genet 7:1475–1483, 1998.
2. Ardinger HH, Buetow KH, Bell GL, Bardach J, VanDemark DR, Murray JC: Association of genetic variation of the transforming growth factor alpha gene with cleft lip and palate. Am J Hum Genet 45:348–353, 1989.
3. Arvystas M, Shprintzen RJ: Craniofacial morphology in the velocardiofacial syndrome. J Craniofac Genet Dev Biol 4:39–45, 1984.
4. Aylsworth AS, Hicks RPB, Drake AF: New observations in HPE: support for the hypothesis that congenital nasal pyriform aperture stenosis (CNPAS) is a phenotype at the mild end of the holoprosencephaly spectrum. 15th David W. Smith Workshop on Malformations and Morphogenesis, Tampa, Florida, August 4–9, 1994.
5. Barr M Jr, Hanson JW, Currey K, Sharp S, Toriello H, Schmickel RD, Wilson GN: Holoprosencephaly in infants of diabetic mothers. J Pediatr 102:565–568, 1983.
6. Barrow LL, van Bokhaven H, Daack–Hirsch S, Andersen T, van Beersum SEC, Gorlin R, Murray JC: Analysis of the *p63* gene in classical EEC syndrome, related syndromes, and non-syndromic orofacial clefts. J Med Genet 39:559–566, 2002.
7. Belloni E, Muenke M, Roessler E: Identification of *Sonic Hedgehog* as a candidate gene responsible for holoprosencephaly. Nat Genet 14:353–356, 1996.
8. Benke PJ, Cohen MM Jr: Recurrence of holoprosencephaly in families with a positive history. Clin Genet 24:324–328, 1983.
9. Berry SA, Pierpoint ME, Gorlin RJ: Single central incisor in familial holoprosencephaly. J Pediatr 104:877–880, 1984.
10. Breg WR, Steele MW, Miller OJ, Warburton D, de Capos A, Allderdice PW: The cri du chat syndrome in adolescents and adults: Clinical findings in 13 older patients with partial deletion of the short arm of chromosome no 5 (5 "p minus"). J Pediatr 77:782–791, 1970.
11. Brewer CM, Leek JP, Green AJ, Holloway S, Bonthron DT, Markham AF, FitzPatrick DR: A locus for isolated cleft palate, located on human chromosome 2q32. Am J Hum Genet 65:387–396, 1999.
12. Brewster R, Lee J, Ruiz i Altaba A: Gli/Zic factors pattern the neural plate by defining domains of cell differentiation. Nature 393:579–583, 1998.
13. Brown LY, Odent S, David V, Blayau M, Dubourg C, Apacik C, Delgado MA, Hall BD, Reynolds JF, Sommer A, Wieczorek D, Brown SA, Muenke M: Holoprosencephaly due to mutations in *ZIC2*: Alanine tract expansion mutations may be caused by parental somatic recombination. Hum Mol Genet 10:791–796, 2001.
14. Brown SA, Warburton D, Brown LY, Yu C-Y, Roeder ER, Stengel–Rutkowski S, Hennekam REM, Muenke M: Holoprosencephaly due to mutations in *ZIC2*, a homologue of *Drosophila odd-paired*. Nat Genet 20:180–183, 1998.
15. Celli J, Duijf P, Hamel BCJ, Bamshad M, Kramer B, Smits APT, Newbury–Ecob R, Hennekam RCM, Van Buggenhout G, van Haeringen A, Woods CG, van Essen AJ, de Waal R, Vriend G, Habar DA, Yang A, McKeon F, Brunner HG, van Bokhoven H: Heterozygous germline mutations in the *p53* homolog *p63* are the cause of EEC syndrome. Cell 99:143–153, 1999.
16. Cohen ME, Paznekas WA, Francis MK, Jabs EW: Effect of human *TWIST* mutations in Saethre–Chotzen syndrome on TWIST protein

dimerization. American Society of Human Genetics, 49th Annual Meeting, San Francisco, California, October 19–23, 1999.
17. Cohen MM Jr: Syndromes with cleft lip and cleft palate. Cleft Palate J 15:306–328, 1978.
18. Cohen MM Jr: Perspectives on holoprosencephaly. Part I. Epidemiology, genetics, and syndromology. Teratology 40:211–235, 1989.
19. Cohen MM Jr: Perspectives on holoprosencephaly. Part III. Spectra, distinctions, continuities, and discontinuities. Am J Med Genet 34:271–288, 1989.
20. Cohen MM Jr, Kreiberg S: An updated pediatric perspective on the Apert syndrome. Am J Dis Child 147:989–993, 1993.
21. Cohen MM Jr: Perspectives on craniofacial asymmetry. III. Common and/or well-known causes of asymmetry. Int J Oral Maxillofac Surg 24:127–133, 1995.
22. Cohen MM Jr: The Child with Multiple Birth Defects. Second Edition. New York: Oxford University Press, 1997.
23. Cohen MM Jr: Robin sequence and Robin complexes: Causal heterogeneity and pathogenetic/phenotypic variability. Am J Med Genet 84:311–315, 1999.
24. Cohen MM Jr: Etiology and pathogenesis of orofacial clefting. Oral Maxillofac Surg Clin North Am 12:379–397, 2000.
25. Cohen MM Jr: Merging the old skeletal biology with the new. II. Molecular aspects of bone formation and bone growth. J Craniofac Genet Dev Biol 20:94–106, 2000.
26. Cohen MM Jr: On the definition of holoprosencephaly. Am J Med Genet 103:183–187, 2001.
27. Cohen MM Jr: *RUNX* genes, neoplasia, and cleidocranial dysplasia. Am J Med Genet 104:185–188, 2001.
28. Cohen MM Jr: Asymmetry: Molecular, biologic, embryopathic, and clinical perspectives. Am J Med Genet 101:292–314, 2001.
29. Cohen MM Jr: Interface between Robin sequence and ordinary cleft palate. Am J Med Genet 101:288, 2001.
29a. Cohen MM Jr: Vasculogenesis, angiogenesis, hemangiomas, and vascular malformations. Am J Med Genet 108:265–274, 2002.
30. Cohen MM Jr: Hedgehog signaling network. Am J Med Genet 123A:5–28, 2003.
30a. Cohen MM Jr: FGFs/FGFRs and associated disorders. In: Inborn Errors of Development. C.J. Epstein, R.P. Erickson, and A. Wynshaw–Boris, Editors. New York: Oxford University Press. Chapter 33, pp. 380–409, 2004.
31. Cohen MM Jr, Bankier A: Syndrome delineation involving orofacial clefting. Cleft Palate-Craniofac J 28:119–120, 1991.
32. Cohen MM Jr, Gorlin RJ: Genetic considerations in a sibship of cyclopia and clefts. Birth Defects 5(2):113–118, 1969.
33. Cohen MM Jr, Jirásek JE, Guzman RT, Gorlin RJ, Peterson MQ: Holoprosencephaly and facial dysmorphia: Nosology, etiology and pathogenesis. Birth Defects 7(7):125–135, 1971.
34. Cohen MM Jr, Kreiborg S: Suture formation, premature sutural fusion, and suture default zones in Apert syndrome. Am J Med Genet 62: 339–344, 1996.
35. Cohen MM Jr, MacLean RE: Craniosynostosis: Diagnosis, Evaluation and Management, Second Edition, New York: Oxford University Press, 2000.
36. Cohen MM Jr, Neri G, Weksberg R: Overgrowth Syndromes. New York: Oxford University Press, 2002.
37. Cohen MM Jr, Rollnick BR, Kaye CL: Oculoauriculovertebral spectrum: An updated critique. Cleft Palate J 26:276–286, 1989.
38. Cohen MM Jr, Shiota K: Teratogenesis of holoprosencephaly. Am J Med Genet 109:1–15, 2002.
39. Cohen MM Jr, Sulik KK: Perspectives on holoprosencephaly. Part II. Central nervous system, craniofacial anatomy, syndrome commentary, diagnostic approach, and experimental studies. J Craniofac Genet Dev Biol 12:196–244, 1992.
40. de la Cruz JM, Bamford RN, Burdine RD, Roessler E, Barkovich AJ, Donnai D, Schier AF, Muenke M: A loss-of-function mutation in the CPC domain of *TDGF1* is associated with human forebrain defects. Hum Genet 110:422–428, 2002.
41. DeMyer WE: Holoprosencephaly (cyclopia–arhinencephaly). In Handbook of Clinical Neurology. P.J. Vinken, and G.W. Bruyn, editors. Amsterdam: North Holland. pp. 431–478, 1977.
42. DeMyer WE, Zeman W, Palmer CG: The face predicts the brain: Diagnostic significance of median facial anomalies for holoprosencephaly (arhinencephaly). Pediatrics 34:256–263, 1964.
43. Diehl SR, Erickson RP: Genome scan for teratogen-induced clefting susceptibility loci in the mouse: Evidence of both allelic and locus heterogeneity distinguishing cleft lip and cleft palate. Proc Natl Acad Sci USA 94:5231–5236, 1997.
44. Edwards SJ, Gladwin AJ, Dixon MJ: The mutational spectrum in Treacher Collins syndrome reveals a predominance of mutations that create a premature termination codon. Am J Hum Genet 60:515–524, 1997.
45. Fraser FC: Liability, thresholds, malformations, and syndromes. Am J Med Genet 66:75–76, 1996.
46. Gladwin AJ, Dixon J, Loftus SK, Edwards S, Wasmuth JJ, Hennekam RCM, Dixon MJ: Treacher Collins syndrome may result from insertion, deletion, or splicing mutations, which introduce a termination codon into the gene. Hum Mol Genet 5:1533–1538, 1996.
47. Gorlin RJ, Cohen MM Jr: The orofacial region.

In Textbook of Fetal and Perinatal Pathology. Second Edition. J.S. Wigglesworth and D.B. Singer, Editors. Malden: Blackwell Sciences. Chapter 22, pp. 732–778, 1998.
48. Gorlin RJ, Cohen MM Jr, Hennekam RCM: Syndromes of the Head and Neck. Fourth Edition. New York: Oxford University Press, 2001.
49. Gorlin RJ, Meskin LH: Congenital hemihypertrophy. J Pediatr 61:870–878, 1962.
50. Gripp KW, Wotton D, Edwards MC, Roessler E, Adès L, Meinecke P, Richieri–Costa A, Zackai EH, Massagué J, Muenke M: Mutations in *TGIF* cause holoprosencephaly and link NODAL signalling to human neural axis determination. Nat Genet 25:205–208, 2000.
51. Gurrieri F, Trask BJ, van den Engh G, Krauss CM, Schinzel A, Pettenati MJ, Schindler D: Physical mapping of the holoprosencephaly critical region on chromosome 7q36. Nat Genet 3: 247–251, 1993.
52. Harding CO, Green CG, Perloff WH: Respiratory complications in children with spondyloepiphyseal dysplasia congenita. Pediatr Pulmonol 9:49–54, 1990.
53. Hennekam RCM, Van Noort G, de la Fuente FA, Norbruis OF: Agenesis of the nasal septal cartilage: Another sign in autosomal dominant holoprosencephaly. Am J Med Genet 39:121–122, 1991.
54. Hockey A, Crowhurst J, Cullity G: Microcephaly, holoprosencephaly, hypokinesia: Second report of a new syndrome. Prenatal Diagn 8:683–686, 1988.
55. Huang C-C, Emanuel I, Huang S-W, Chen T-Y: Two cases of the de Lange syndrome in Chinese infants. J Pediatr 71:251–254, 1967.
56. Jabs EW, Müller U, Li X, Ma l, Luo W, Haworth IS, Kisak I, Sparkes R, Warman ML, Mulliken JB, Snead ML, Maxson R: A mutation in the homeodomain of the human *MSX2* gene in a family affected with autosomal dominant craniosynostosis. Cell 75:443–450, 1993.
57. Kelley RI, Roessler E, Hennekam RCM, Feldman GL, Kosaki K, Jones MC, Palumbos JC, Muenke M: Holoprosencephaly in RSH/Smith–Lemli–Opitz syndrome: Does abnormal cholesterol metabolism affect the function of *Sonic Hedgehog*? Am J Med Genet 66:478–484, 1996.
58. Kondo S, Schutte BC, Richardson RJ, Bjork BC, Knight AS, Watanabe Y, Howrd E, Ferreira de Lima RLL, Daack–Hirsch S, Sander A, McDonald–McGinn DM, Zackai EH, Lammer EJ, Aylsworth AS, Ardinger HH, Lidral AC, Pober BR, Moreno L, Arcos–Burgos M, Valencia C, Houdayer C, Bahuau M, Moretti–Ferreira D, Richieri–Costa A, Dixon MJ, Murray JC: Mutations in *IRF6* cause van der Woude and popliteal pterygium syndromes. Nat Genet 32:285–289, 2002.
59. Koskinen–Moffett LK, Moffett BC Jr, Graham JM Jr: Cranial synostosis and intrauterine compression: A developmental study of human sutures. In: Factors and Mechanisms Influencing Bone Growth. A.D. Dixon and B.G. Sarnat, Editors. New York: Alan R. Liss. pp. 365–378, 1982.
60. Koyabu Y, Nakata K, Mizugishi K, Aruga J, Mikoshiba K: Physical and functional interactions between Zic and Gli proteins. J Biol Chem 276: 6889–6892, 2001.
61. Kreiborg S, Cohen MM Jr: Syndrome delineation and growth in orofacial clefting and craniosynostosis. In Facial Clefts and Craniosynostosis: Principles and Management. T.A. Turvey, V.K.L. Vig, and R.J. Fonseca, Editors. Philadelphia: WB Saunders. Chapter 3, pp. 57–75, 1996.
62. Kreiborg S, Marsh JL, Cohen MM Jr, Liversage M, Pedersen H, Skovby F, Børgesen SE, Vannier MW: The calvaria and cranial base in Apert and Crouzon syndromes: A three-dimensional CT analysis. J Craniomaxillofac Surg 21:181–188, 1993.
63. Lajeunie E, Le Merrer M, Bonaïti–Pellie, Marchac D, Renier D: Genetic study of nonsyndromic coronal craniosynostosis. Am J Med Genet 55:500–504, 1995.
64. Lajeunie E, Le Merrer M, Bonaïti–Pellie C, Marchac D, Renier D: Genetic study of scaphocephaly. Am J Med Genet 62:282–285, 1996.
65. Lajeunie E, Arnaud E, Le Merrer M, Cinalli G, Marchac D, Renier D: Syndromal and nonsyndromal primary trigonocephaly: Analysis of a series of 237 patients. Am J Med Genet 75:211–215, 1998.
66. Langman J: Medical Embryology. Fourth Edition. Baltimore: Wilkins & Wilkins, 1983.
67. Lemire RJ, Beckwith JB, Warkany J: Anencephaly. New York: Raven Press, 1978.
68. Lewkonia RM, Lowry RB: Progressive hemifacial atrophy (Parry–Romberg syndrome): Report with review of genetics and nosology). Am J Med Genet 14:385–390, 1983.
69. Lidral AC, Romitti PA, Basart AM, Doetschman T, Leysens NJ, Daack–Hirsch S, Semina EV, Johnson LR, Machida J, Burds A, Parnell TJ, Rubinstein JL, Murray JC: Association of *MSX1* and *TGFB3* with nonsyndromic clefting in humans. Am J Hum Genet 63:557–568, 1998.
69a. Marçano ACB, Doudney K, Braybrook C, Squires R, Patton MA, Lees MM, Richieri–Costa A, Lidral AC, Murray JC, Moore GE, Stanier P: *TBX22* mutations are a frequent cause of cleft palate. J Med Genet 41:68–74, 2004.
70. Marden PM, Smith DW, McDonald MJ: Congenital anomalies in the newborn infant, including minor variations. J Pediatr 64:358–371, 1964.
71. Martin RA, Jones KL: Absence of the superior labial frenulum in holoprosencephaly: A new diagnostic sign. J Pediatr 133:150–153, 1998.
72. Martin RA, Jones KL, Benirshke K: Absence of the lateral philtral ridges: a clue to the structural basis of the philtrum. Am J Med Genet 65: 117–123, 1996.

73. McGrath JA, Duijf PHG, Doetsch V, Irvine AD, de Waal R, Vanmolkot KRJ, Wesagowit V, Kelly A, Atherton DJ, Griffiths WAD, Orlow SJ, van Haeringen A, Ausems MGEM, Yang A, McKeon F, Bamshad MA, Brunner HG, Hamel BCJ, van Bokhoven H: Hay–Wells syndrome is caused by heterozygous missense mutations in the SAM domain of p63. Hum Mol Genet 10:221–229, 2001.
74. Ming JE, Kaupas ME, Roessler E, Brunner HG, Golabi M, Stratton RF, Sujansky E, Bale SJ, Muenke M: Mutations in *Patched-1*, the receptor for Sonic Hedgehog, are associated with holoprosencephaly. Hum Genet 110:297–301, 2002.
75. Mitchell LE, Healey SC, Chenevix–Trench G: Evidence for an association between nonsyndromic cleft lip with or without cleft palate and a gene located on the long arm of chromosome 4. Am J Hum Genet 57:1130–1136, 1995.
76. Moore MH, David DJ, Cooter RD: Hairline indicators of craniofacial clefts. Plast Reconstr Surg 82:589–593, 1988.
77. Morse RP, Rawnsley E, Sargent SK, Graham JM Jr: Prenatal diagnosis of a new syndrome: Holoprosencephaly with hypokinesia. Prenatal Diagn 7:631–638, 1987.
78. Muenke M, Beachy PA: Holoprosencephaly. In The Metabolic and Molecular Bases of Inherited Disease. Eighth Edition. C.R. Scriver, A.L. Beaudet, W.S. Sly, D. Valle, B. Childs, K.W. Kinzler, and B. Vogelstein, Editors. New York: McGraw–Hill. pp. 6203–6230, 2001.
79. Müller U, Warman ML, Mulliken JB, Weber JL: Assignment of a gene locus involved in craniosynostosis to chromosome 5qter. Hum Mol Genet 2:119–122, 1993.
80. Mulliken JB: Cutaneous vascular anomalies. Semin Vasc Surg 6: 204–218, 1993.
81. Mulliken JB: Vascular anomalies. In: Grabb and Smith's Plastic Surgery. Fifth Edition. S.J. Aston, R.W. Beasley, and C.H.M. Thorne, Editors. Philadelphia: Lippincott-Raven. pp. 191–203, 1997.
82. Mulliken JB, Glowacki J: Hemangiomas and vascular malformations in infants and children: A classification based on endothelial characteristics. Plast Reconstr Surg 69:412–420, 1982.
83. Mundlos S: Cleidocranial dysplasia: Clinical and molecular genetics. J Med Genet 36:177–182, 1999.
84. Mundlos A, Otto F, Mundlos C, Mulliken JB, Aylsworth AS, Albright S, Lindhout D, Cole WG, Henn W, Knoll JHM, Owen MJ, Mertelsmann R, Zabel BU, Olsen BR: Mutations involving the transcription factor *CBFA1* cause cleidocranial dysplasia. Cell 89:773–779, 1997.
85. Murray JC: Face facts: Genes, environment, and clefts. Am J Hum Genet 57:227–232, 1995.
86. Murray JC: Candidate gene resequencing identifies gene variants in isolated cleft lip and palate. 25th Annual David W. Smith Workshop on Malformations and Morphogenesis, Snowbird Ski and Summer Resort, Utah, August 18–21, 2004.
87. Nanni L, Ming JE, Bocian M, Steinhaus K, Bianchi DW, Die–Smulders C, Gianotti A, Imaizumi K, Jones KL, Campo MD, Martin RA, Meinecke P, Pierpont MEM, Robin NH, Young ID, Roessler E, Muenke M: The mutational spectrum of the *Sonic Hedgehog* gene in holoprosencephaly: *SHH* mutations cause a significant proportion of autosomal dominant holoprosencephaly. Hum Mol Genet 8:2479–2488, 1999.
88. Neilson KM, Friesel RE: Ligand-independent activation of fibroblast growth factor receptors by point mutations in the extracellular, transmembrane, and kinase domains. J Biol Chem 271:25049–25057, 1996.
89. Oldridge M, Lunt PW, Zackai EH, McDonald–McGinn DM, Muenke M, Moloney DM, Twigg SRF, Heath JK, Howard TD, Hoganson G, Gagnon DM, Jabs EW, Wilkie AOM: Genotype–phenotype correlations for nucleotide substitutions in the IgII-IgIII linker of FGFR2. Hum Mol Genet 6:137–143, 1997.
90. OMIM (Online Mendelian Inheritance in Man), 2004.
91. Ouspenskaia MV, Karkera JD, Roessler E, Shen MM, Goldmunts E, Bowers P, Towbin J, Belmont J, Muenke M: Role of *FAST1* gene in the development of holoprosencephaly (HPE) and congenital cardiac malformations in humans. 52nd Annual Meeting of the American Society of Human Genetics, Baltimore, Maryland, October 15–19, Abstract 822, p. 313, 2002.
92. Passarge E, Mecke S, Altrogge HC: Cornelia de Lange syndrome: Evolution of the phenotype. Pediatrics 48:833–836, 1971.
93. Pruzansky S: Not all dwarfed mandibles are alike. Birth Defects 5(2):120–129, 1969.
94. Qiu M, Bulfone A, Ghattas I, Meneses JJ, Christensen L, Sharpe PE, Presley PT, Presley R: Role of the Dlx homeobox genes in proximodistal patterning of the branchial arches: Mutations of *Dlx-1* and *Dlx-2* alter morphogenesis of proximal skeletal and soft tissue structures derived from the first and second arches. Dev Biol 185:165–184, 1997.
95. Qiu M, Bulfone A, Martinez S, Meneses JJ, Shimamura K, Pedersen RA, Rubinstein JL: Null mutation of *Dlx-2* results in abnormal morphogenesis of proximal first and second branchial arch derivatives and abnormal differentiation in the forebrain. Genes Dev 9:2523–2538, 1995.
96. Rintala AE: Freeman–Sheldon's syndrome: Craniocarpo-tarsal dystrophy. Acta Paediatr Scand 57: 553–556, 1968.
97. Roessler E, Belloni E, Gaudenz K, Jay P, Berta P, Scherer SW, Tsui L-C, Muenke M: Mutations in the human *Sonic Hedgehog* gene cause holoprosencephaly. Nat Genet 14:357–360, 1996.
98. Roessler E, Belloni E, Gaudenz K, Vargus F, Scherer SW, Tsui L-C, Muenke M: Mutations in

the C-terminal domain of *Sonic Hedgehog* cause holoprosencephaly. Hum Mol Genet 6:1847–1853, 1997.

99. Roessler E, Du Y, Mullor JL, Casas E, Allen WP, Ellis I, Gillessen–Kaesbach G, Roeder E, Ming JE, Ruiz i Altaba A: Loss-of-function mutations in the human *GLI2* gene cause holoprosencephaly and familial pan-hyopituitarism. 52nd Annual Meeting of the American Society of Human Genetics, October Baltimore, Maryland, 15–19, Abstract 132, p. 190, 2002.
100. Scapoli L, Pezzetti F, Carinci F, Martinelli M, Carinci P, Tognon M: Evidence of linkage to 6p23 and genetic heterogeneity in nonsyndromic cleft lip with or without cleft palate. Genomics 43:216–220, 1997.
101. Sertié AL, Sousa AV, Steman S, Pavanello RC, Passos–Bueno MR: Linkage analysis in a large Brazilian family with van der Woude syndrome suggests the existence of a susceptibility locus for cleft palate at 17p11.2–11.1. Am J Hum Genet 65:433–440, 1999.
102. Shaw D, Ray A, Marazita M: Further evidence of a relationship between the retinoic acid receptor alpha locus and nonsyndromic cleft lip with or without cleft palate (CL ± P). Am J Hum Genet 53:1156–1157, 1993.
103. Shillito J Jr, Matson DD: Craniosynostosis. A review of 519 surgical patients. Pediatrics 41:829–853, 1968.
104. Shprintzen RJ: Pierre Robin, micrognathia, and airway obstruction: The dependency of treatment on accurate diagnosis. Int Anesthesiol Clin 26:64–71, 1988.
105. Shprintzen RJ: The implications of the diagnosis of Robin sequence. Cleft Palate-Craniofac J 29: 205–209, 1992.
106. Smith DW, Bostian KE: Congenital anomalies associated with idiopathic mental retardation. J Pediatr 65:189–196, 1964.
107. Stein J, Mulliken JB, Stal S, Gasser DL, Malcolm S, Winter R, Blanton SH, Amos C, Seemanová E, Hecht JT: Nonsyndromic cleft lip with or without cleft palate: Evidence of linkage to *BCL3* in 17 multigeneration families. Am J Hum Genet 57:257–272, 1995.
108. Suzuki K, Hu D, Bustos T, Zlotogora J, Richieri–Costa A, Helms JA, Spritz RA: Mutations of *PVRL1*, encoding a cell–cell adhesion molecule/herpesvirus receptor, in cleft lip/palate–ectodermal dysplasia. Nat Genet 25:427–430, 2000.
109. Tessier P: Anatomical classifications of facial, cranio-facial and laterofacial clefts. J Max-Fac Surg 4:69–92, 1976.
110. Treacher Collins Syndrome Collaborative Group: Positional cloning of a gene involved in the pathogenesis of Treacher Collins syndrome. Nat Genet 12:130–136, 1996.
111. Twigg SRF, Kan R, Babbs C, Bochukova EG, Robertson SP, Wall SA, Morriss–Kay GM, Wilkie AOM: Mutations of ephrin-B1 (*EFNB1*), a marker of tissue boundary formation, cause craniofrontonasal syndrome. Proc Natl Acad Sci USA 101:8652–8657, 2004.
112. van den Boogaard M-JH, Dorland M, Beemer FA, Ploos van Amstel HK: *MSX1* mutation is associated with orofacial clefting and tooth agenesis in humans. Nat Genet 24:342–343, 2000.
113. Vannier MW: Radiological evaluation of craniosynostosis. In Craniosynostosis: Diagnosis, Evaluation, and Management. M.M. Cohen, Jr, and R.E. MacLean, Editors. New York: Oxford University Press. Chapter 12, pp. 147–157, 2000.
114. Wallis DE, Roessler E, Hehr U, Nanni L, Wiltshire T, Richieri–Costa A, Gillessen–Kaesbach G, Zackai EH, Romens J, Muenke M: Mutations in the homeodomain of the human *SIX3* gene cause holoprosencephaly. Nat Genet 22:196–198, 1999.
115. Warman ML, Mulliken JB, Hayward PG, Müller U: Newly recognized autosomal dominant disorder with craniosynostosis. Am J Med Genet 46: 444–449, 1993.
116. Wassif CA, Maslen C, Kachilele–Linjewile S, Lin D, Linck TM, Connor WE, Steiner RD, Porter FD: Mutations in the human *sterol* Δ^7-*reductase* gene at 11q12–13 cause Smith–Lemli–Opitz syndrome. Am J Hum Genet 63:53–62, 1998.
116a. White KE, Cabral JM, Davis SI, Fishburn T, Evans WE, Ichikawa S, Fields J, Yu X, Shaw NJ, McLellan NJ, McKeown C, FitzPatrick D, Yu K, Ornitz DM, Econs MJ: Mutations that cause osteoglophonic dysplasia define novel roles for *FGFR1* in bone elongation. Am J Hum Genet 76:361–367, 2005.
117. Wilkie AOM, Slaney SF, Oldridge M, Poole MD, Ashworth GJ, Hockley AD, Hayward RD, David DJ, Pulleyn LJ, Rutland P, Malcolm S, Winter RM, Reardon W: Apert syndrome results from localized mutations of *FGFR2* and is allelic with Crouzon syndrome. Nat Genet 9:165–172, 1995.
118. Wilkie AOM, Wall SA: Craniosynostosis: Novel insights into pathogenesis and treatment. Curr Opin Neurol 9:146–152, 1996.
119. Williams HB: Hemangiomas of the parotid gland in children. Plast Reconstr Surg 56:29–34, 1975.
120. Witsch–Baumgartner M, Fitzky BU, Ogorelkova M, Kraft HG, Moebius FF, Glossmann H, Seedorf U, Gillessen–Kaesbach G, Hoffmann GF, Clayon P: Mutational spectrum in the Δ^7-*sterol reductase* gene and genotype–phenotype correlation in 84 patients with Smith–Lemli–Opitz syndrome. Am J Hum Genet 66:402–412, 2000.
121. Wyszynski DF, Beaty TH, Maestri NE: Genetics of nonsyndromic oral clefts revisited. Cleft Palate-Craniofac J 33:406–417, 1996.
122. Yu H, Lee M-H, Starck L, Elias ER, Irons M, Salen G, Patel SB, Tint GS: Spectrum of Δ^7-*dehydrocholesterol reductase* mutations in patients with the Smith–Lemli–Opitz (RSH) syndrome. Hum Mol Genet 9:1385–1391, 2000.

6

Surgical Perspective

Operative correction of the facial skeleton can be divided into two technical subspecial fields: maxillofacial surgery and craniofacial surgery. The arbitrary division is an imaginary line at the level of the cranial base. Although the distinction between the two is useful, some procedures involve both areas simultaneously. The field of maxillofacial surgery developed before craniofacial surgery. The history of surgery involving the facial skeleton is complex and has been particularly well discussed by Wolfe (112).

Surgical and important related topics include craniofacial surgery (4, 10, 11, 14, 42, 54, 55, 57, 58–62, 68, 71, 72, 74, 81, 85–101, 110, 113), neurosurgery (29, 42, 69), surgery for cleft lip-palate repair (15, 22, 46, 52–53b, 56, 57a, 73, 76, 82, 114), orthognathic surgery (2, 16, 30, 64–67, 72, 77, 84, 102, 108, 109), trauma (72), tumors (72), osteodistraction (84b), craniofacial surgical research (43), aesthetic balance and anthropometry (17, 18, 38, 72, 74, 75), cephalometry (27, 34, 78), three-dimensional computerized tomography (3D-CT) (26, 35, 44, 45, 63, 103–105, 115–118), psychosocial studies (33, 39, 40, 49, 70, 79, 80), and craniofacial centers (31, 47, 51).

SOME HISTORICAL VIGNETTES

René Le Fort

At the turn of the century, the French surgeon René Le Fort performed some remarkable studies on maxillary fractures (36, 37, 90, 91). Of thirty-five experiments, two are summarized here:

> Experiment VIII involved a man of about 50 years. The cadaver was supine with the head hanging over the edge of a table. A violent blow was struck with a club on the right upper jaw. The force was directed obliquely from the bottom to the top, as if the subject were standing. When the specimen was dissected, a transverse fracture of the maxilla was observed.
>
> Experiment XXXI involved a man of about 60 years. The cadaver was in decubitus dorsal position. The head was supported by a table. A violent blow with a wooden club was delivered to the face in an anteroposterior direction. At dissection, a fracture separated the bony face from the cranium.

From his experiments, Le Fort (36, 37) concluded that there are three weak lines in the facial skeleton where fractures tend to occur. These transverse fractures are named after him (Fig. 6-1). In the Le Fort I fracture, the weak line is just above the level of the teeth. The Le Fort II fracture is above the nasal aperture, but does not involve the zygomas. In the Le Fort III fracture, the bony face is detached from the cranium.[a] Surgical osteotomies for the correction of midfacial deficiences in Apert syndrome, Crou-

[a]In Le Fort's original papers (36, 37), he used the term Guerin's fracture for what we call Le Fort I; it was his third weak line (III). His first weak line (I), we call Le Fort III.

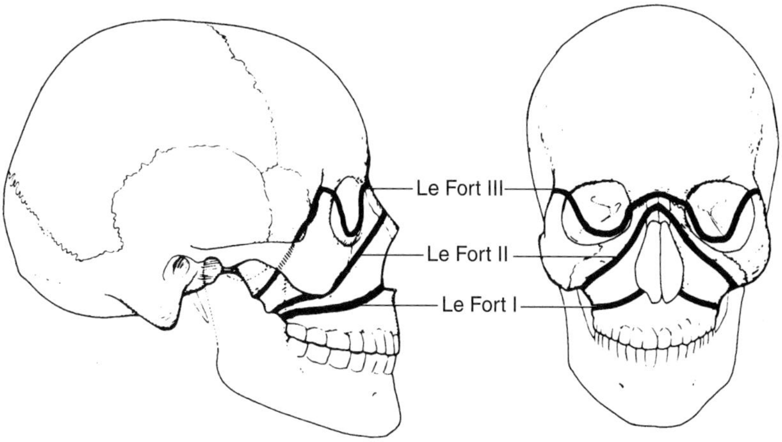

Figure 6-1 Le Fort I, Fracture of the maxilla just above the level of the teeth. Le Fort II, Fracture of the maxilla above the nasal aperture. Le Fort III, Fracture and detachment of the bony face from the cranium.

zon syndrome, and facial defects from old fractures are derived from Le Fort's experiments (91).

Varaztad Kazanjian

After arriving in the United States from Turkey at the turn of the century, Varaztad Kazanjian took a job in a wire mill in Worcester, Massachusetts. Later he received a dental degree from Harvard University. He became an assistant in prosthetic dentistry first and then a professor. In those days, jaw fractures were treated in the Dental School Clinic, rather than in the hospital. It was a time before serious automobile accidents, and most jaw fractures resulted from drunken altercations on weekends and holidays. During the ensuing years, Kazanjian treated more than 400 such patients. Instead of securing intermaxillary fixation by interdental splints made of uncomfortable vulcanite rubber, Kazanjian used wire, derived from his earlier experience at the mill (13).

During the First World War, Kazanjian was selected as the Chief Dental Officer for patients in France. He treated extensive facial war wounds with intermaxillary fixation and also wired both maxillary and mandibular bone fragments together. Facial soft tissue tears were sutured, and innovative prosthetic appliances were fabricated for missing facial parts (13).

Kazanjian had departed the Harvard Dental School as Professor of Prosthetic Dentistry and returned after the war as Professor of Military Oral Surgery. Over the ensuing years, he published many articles on jaw fractures and facial surgery. When Kazanjian was 40 years old, his wife died suddenly, and since he had no children, he attended Boston University for two years to obtain his medical degree. Mostly for political reasons in Boston, he was not permitted to practice the extensive procedures that he had developed during World War I. After receiving his medical degree, a hospital appointment became possible and he became Professor of Clinical Oral Surgery at the Harvard Dental School in 1922 and continued in this capacity until 1941, when he was appointed as the first Professor of Plastic Surgery at the Harvard Medical School (13).

Dr. Paul Tessier

Dr. Paul Tessier, a French plastic surgeon, had a background in ophthalmology and orthopedic surgery as well as an interest in congenital malformations. He is the acknowledged father of craniofacial surgery, but always preferred the term *orbitocranial surgery* to describe his work. Following World War II, Dr. Tessier frequently visited England where he spent time with Archibald McIndoe and also had some exposure to Sir Harold Gillies, the first surgeon to perform a Le Fort III osteotomy (112).

Dr. Tessier performed his first Le Fort III osteotomies on patients with severe Crouzon syndrome as subcranial procedures in the late 1950s and early 1960s. He developed precise operative plans that differed from the approach of Gillies in several ways. He taped multiple pages to the wall in the operating room at the Hôpital Foch, outlining every step of each operation to be performed (112).

Dr. Tessier and Gerard Guiot, a neurosurgeon, developed initially a two-stage transcranial approach for the surgical correction of orbital hypertelorism. This involved reinforcing the dura with a dermal graft over the anterior cranial fossa as a preliminary operation, and then later, this was followed by mobilization of the orbits, resection of the excess interorbital bone, and correction of the hypertelorism. These procedures represented the first true craniofacial operations. Dr. Tessier referred to hypertelorism as *telorbitism* and used the term *useful orbit* to describe the circumferential orbital dissection, which was carried out to within 1 cm in front of each optic foramen. When the orbits were moved medially, the eyeglobes would also move along with them (112).

Dr. Tessier developed a basic approach to the facial skeleton by using wide periosteal dissection through coronal and intraoral incisions, which left no visible scars. He used autogenous bone grafts from the cranium, hip, tibia, and rib. He also developed surgical procedures for Treacher Collins syndrome, hemifacial microsomia, orbital dystopias, orbito-ocular clefts, and many other facial malformations (112).

Based on his extensive experience, Dr. Tessier developed a craniofacial clefting classification (94) that is used universally today (see Figure 5-13). He divided the craniofacial framework into five levels for surgical purposes (97) (Fig. 6-2, Table 6-1), Level A, the cranial vault, is impor-

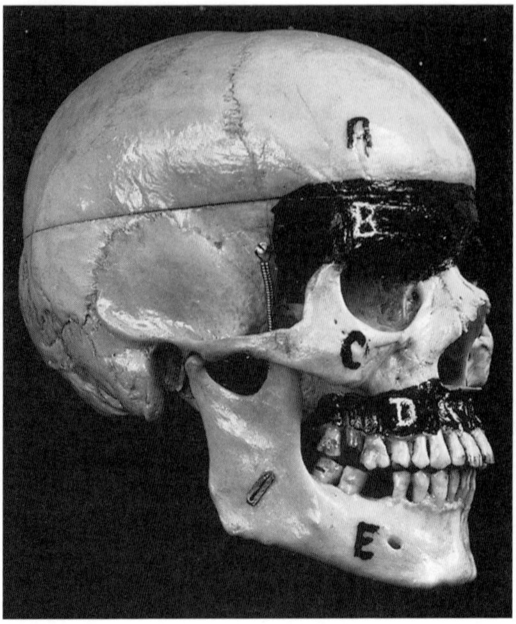

Figure 6-2 Dr. Tessier's craniofacial framework consisting of five different levels (A–E). See Table 6-1. From Tessier, 2000 (97).

Table 6-1 Dr. Tessier's Craniofacial Framework Consisting of Five Different Levels

Level	Craniofacial Component
A	Cranial vault
B	Frontonasoethmoidal complex, upper part of the orbital cavities, orbital roofs, and supraorbital ridges
C	Lower orbits with the body of maxilla and the zygomas
D	Upper jaw
E	Mandible

From Tessier, 2000 (97).

Table 6-2 Dr. Tessier's Classification of the Craniosynostoses

Description	Group Number
Isolated cranial vault dysmorphism	1
Symmetric orbitocranial dysmorphism (with or without telorbitism)	2
Trigonocephaly	2-1
Acro-oxycephaly	2-2
Brachycephaly without telorbitism	2-3
Brachycephaly with euryprosopia and telorbitism	2-4
Asymmetric orbitocranial dysmorphism (plagiocephaly)	3
Pure vertical discrepancy of orbital cavities	3-1
Plagiocephaly without telorbitism	3-2
Plagiocephaly with telorbitism	3-3
Saethre-Chotzen group	4
Crouzon group	5
Regular Crouzon	5-1
Top Crouzon	5-2
Bottom Crouzon	5-3
Trilobular Crouzon	5-4
Apert group	6
Hyperacrocephalic Apert	6-1
Hyperbrachycephalic Apert	6-2
Pfeiffer	6-3
Trilobular Apert	6-4
Carpenter	6-5

From Tessier, 2000 (97).

tant for the treatment of premature synostosis of various sutures (see Chapter 5). It is also important for the intracranial approach necessary for the correction of hypertelorism or for fronto-orbital (monobloc) advancement. Level B is important for treating symmetrical orbitocranial dysmorphism with or without hypertelorism. Levels C, D, and E are involved in maxillofacial procedures and levels D and E are important for orthognathic correction. Several levels may be used simultaneously in some surgical procedures. For instance, levels A–D are important for patients in Dr. Tessier's *Crouzon Group* (Table 6-2).

Hugo Obwegeser

Although he was not the first to describe procedures on the jaws, Hugo Obwegeser, a dual qualified dentist and physician, was more than anyone else responsible for the explosive development of orthognathic surgery. At the Zahnhärtzliches Institut in Zürich, he stressed that almost all orthognathic procedures could and should be performed intraorally, thus avoiding facial scars. He also stressed proper planning and the importance of orthodontic participation to achieve a harmonious result. Obwegeser described the sagittal split osteotomy of the mandible and the sliding genioplasty. He was the first surgeon to describe simultaneous maxillary and mandibular correction. In the 1960s and 1970s, he taught a number of U.S. Army oral and maxillofacial surgeons, who were instrumental in bringing orthognathic surgery to the United States. Obwegeser also presented a three day course at the Walter Reed Army Hospital in Washington, D.C., which had a significant impact on American oral and maxillofacial surgeons (64–67, 112).

EVALUATION FOR SURGICAL CORRECTION

Aesthetic Balance

From a clinical perspective, the craniofacial region should be evaluated systemically and meticulously. Both normal and abnormal anatomical features should be assessed, keeping in mind the surgical goal of normal proportions, symmetry, and balance. Although specific dysmorphic features characterize specific syndromes, each patient is unique (72, 74).

Type of Surgical Intervention and Timing

Three important biological factors must be taken into account: (a) the natural history of the disorder (e.g., does the disorder progressively worsen with time or is it nonprogressive?); (b) growth restriction tends to occur if an operated bone has not yet reached maturity (e.g., cleft palate repair too early in life results in maxillary hypoplasia); and (c) an uncertain relationship between the congenitally affected and surgically altered skeleton on the one hand and the un-

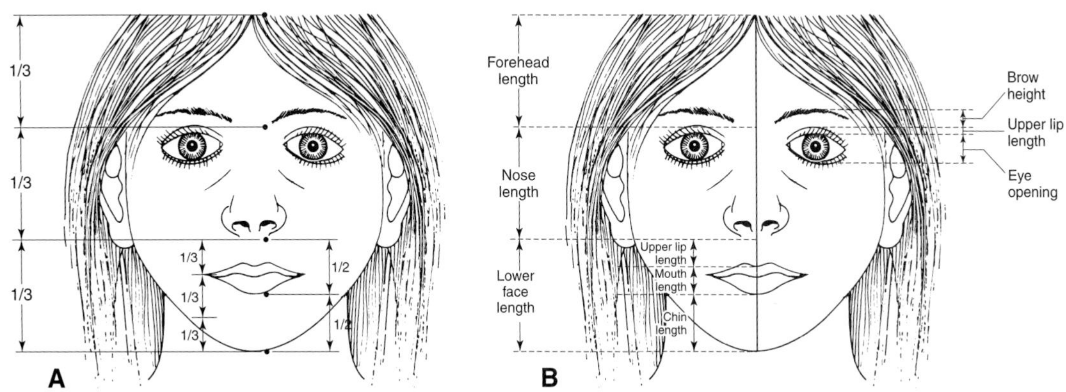

Figure 6-3 A, Vertical facial proportions. B, Vertical measurements of the face. From Lucker, 1981 (38).

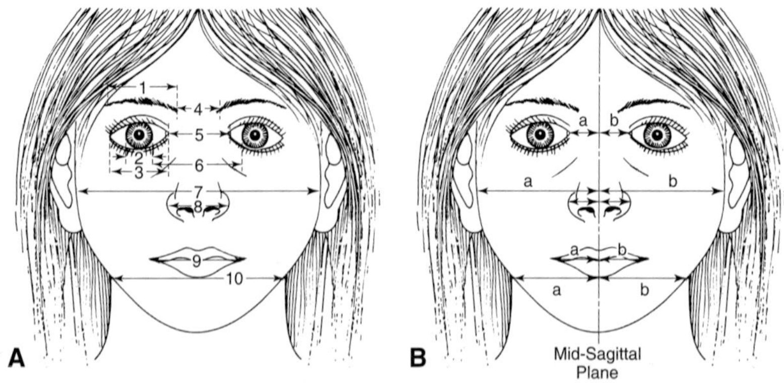

Figure 6-4 A, Horizontal facial measurements. B, Left-right facial symmetry. From Lucker, 1981 (38).

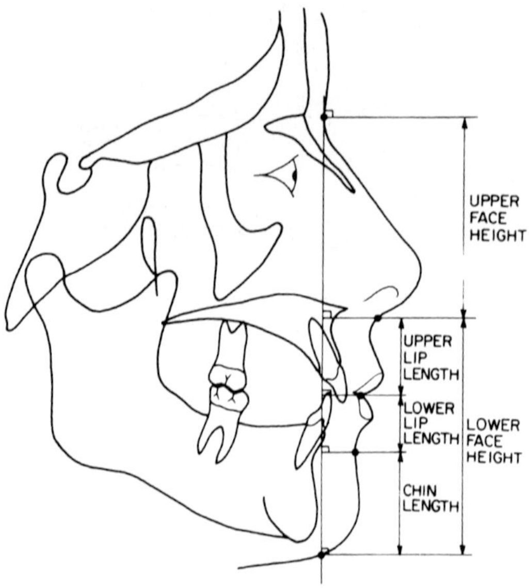

Figure 6-5 Measurements of facial height. From Lucker, 1981 (38).

derlying growing viscera on the other (e.g., the brain and the eyes) (72, 74).

Quantitative Assessment

Quantitative assessment of the craniofacial region either confirms or refutes clinical impressions. Various types of assessment are important for surgical planning: anthropometry, conventional radiography, film-screen tomography, cephalometry, computed tomographic (CT) scans, 3D-CT, magnetic resonance imaging (MRI), and analysis of dental models (72, 74, 103).

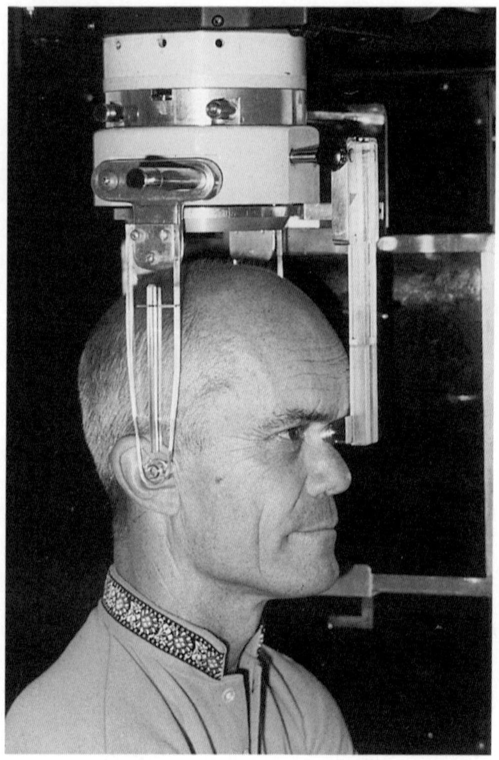

Figure 6-6 Head oriented in the Frankfurt plane by a cephalostat. From Cohen, 1985 (9).

Anthropometric Measurements

Anthropometric measurements on soft tissues can be compared to normal values. Types of measurement include linear distances, angles, arcs, and various ratios that indicate useful craniofacial proportions (17, 18, 38, 75) (Figs. 6-3 to 6-5).

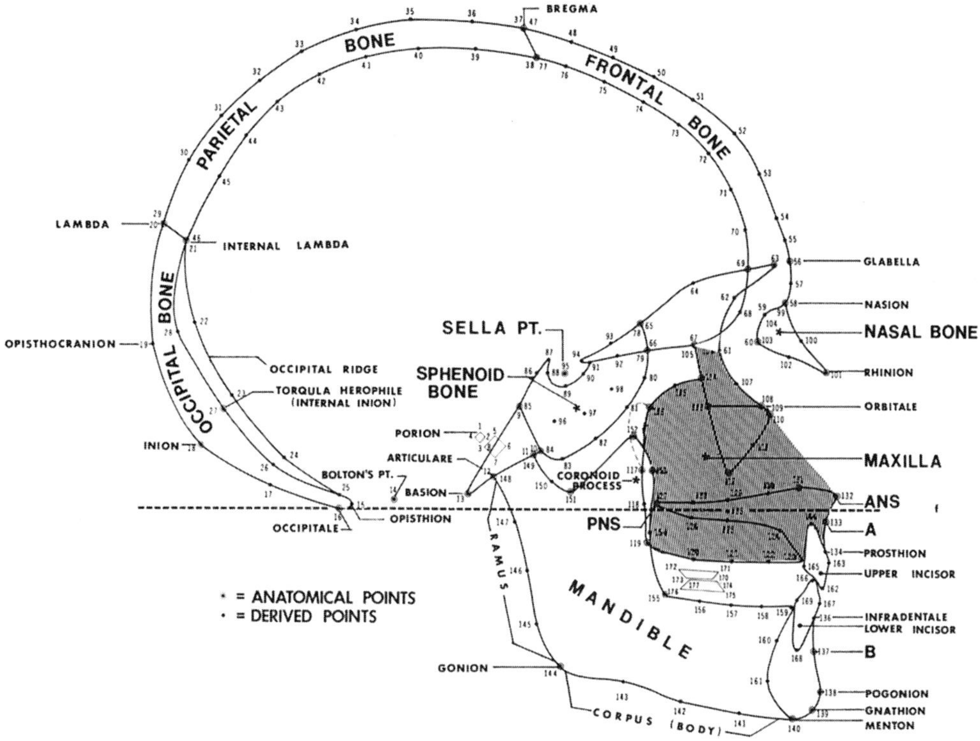

Figure 6-7 Two dimensional model with 177 coordinate points to describe craniofacial morphology. From Walker and Kowalski, 1971 (106).

Figure 6-8 Morphologic structures disarticulated to show measurements of bony areas. With a series of adjacent triangles from sella point to the calvarial bones, an endocranial area may also be determined. From Walker and Kowalski, 1972 (107).

Cephalometric Measurements

Cephalograms are obtained with the head oriented in the Frankfurt plane by a cephalostat (9) (Fig. 6-6). A fixed X-ray source is used with the central ray perpendicular to the film. Anatomic and derived points may be located on cephalometric tracings (106, 107) (Figs. 6-7, 6-8), and various linear and angular measurements can be taken and used for analysis and comparison (74, 75) (Fig. 6-9). A photocephalometric technique permits simultaneous visualization of facial soft tissues and the underlying facial skeleton (Fig. 6-10) and also permits evaluation of craniofacial osteotomies by superimposed images (27) (Fig. 6-11).

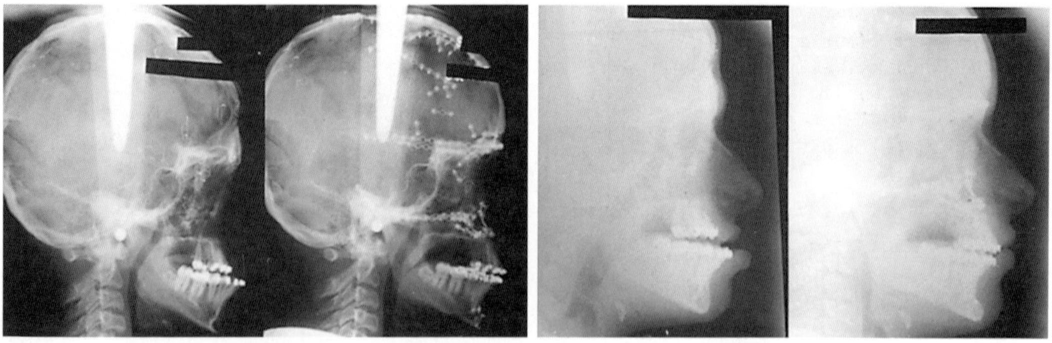

A B

Figure 6-9 A, Lateral cephalometric radiograph before and one year after a Le Fort III midface and forehead advancement in a case of Crouzon syndrome. B, Lateral cephalometric radiographs before and one year after osteotomies. From Posnick, 2000 (74).

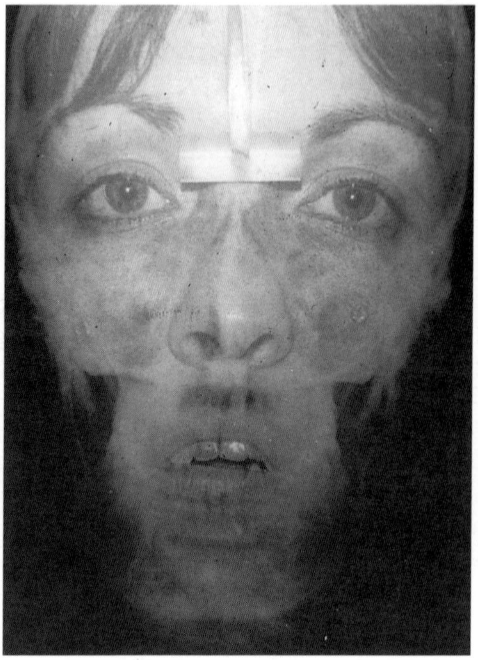

Figure 6-10 Photocephalogram with superimposed facial soft tissues and the facial skeleton in frontal view. From Hohl et al., 1978 (27).

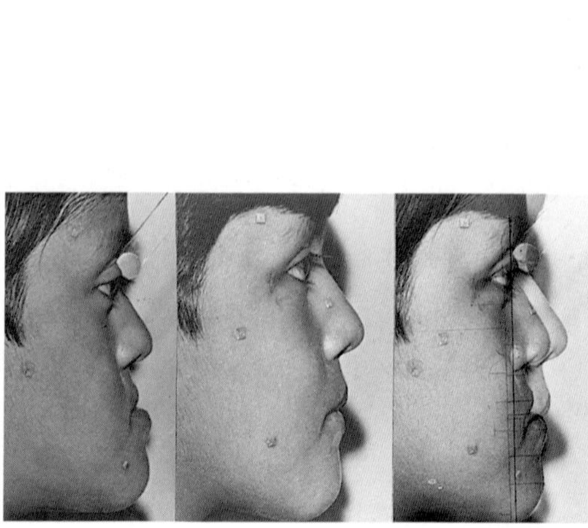

Figure 6-11 Photocephalogram showing superior superimposed images before and after a Le Fort III osteotomy. From Hohl et al., 1978 (27).

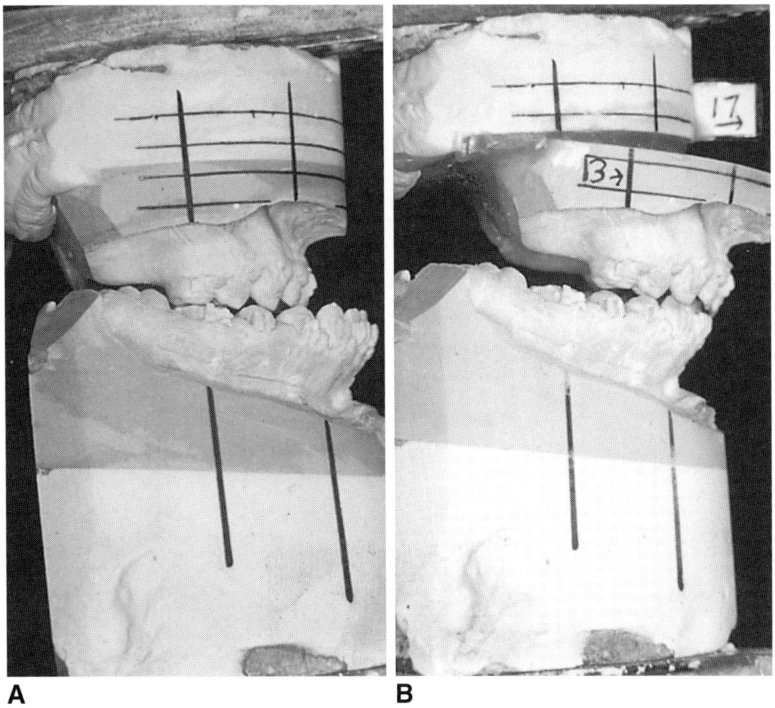

Figure 6-12 A, Articulated dental casts of a Crouzon syndrome patient. B, Model surgery showing predicted results of a Le Fort III osteotomy. From Posnick, 2000 (74).

Analysis of Dental Models

Articulated dental casts may be used to predict results of osteotomies by comparing presurgical models with presurgical *model surgery* (74) (Fig. 6-12).

Three-Dimensional Computed Tomography (3D-CT)

Analysis of 3D-CT scans permits improved anatomical understanding of craniofacial malformations, improves diagnostic accuracy, assists in the staging of surgical correction, and allows preoperative measurements that are used intraoperatively (74, 103) (Figs. 6-13 to 6-16).

FOUR TYPES OF FACES: SURGICAL IMPLICATIONS

Four types of faces are shown in Figure 6-17. Craniofacial surgery is always contraindicated for a normal face or for a grossly dysmorphic face associated with (a) incompatibility with life, (b) extraordinary means of life support, or (c) severe mental deficiency. Minor facial anomalies (see Chapter 5, Figs. 5-3, 5-4 and Table 5-4) can be altered by elective plastic surgical procedures. In Figure 6-18, a girl with the typical facial features of Down syndrome looks perfectly normal following an operation to eliminate her minor

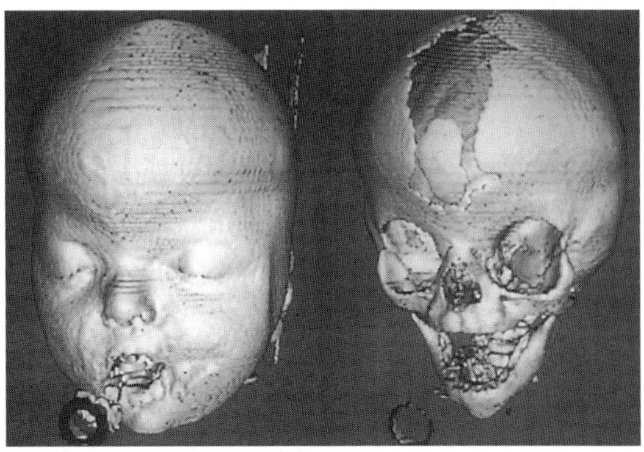

Figure 6-13 Three-dimensional CT reconstruction of an infant with Apert syndrome. Left, Soft tissue appearance. Right, Underlying skull of the same infant. These two views from the same set of data are made possible by changing the threshold in the marching cubes algorithm. Courtesy of M.W. Vannier, Chicago.

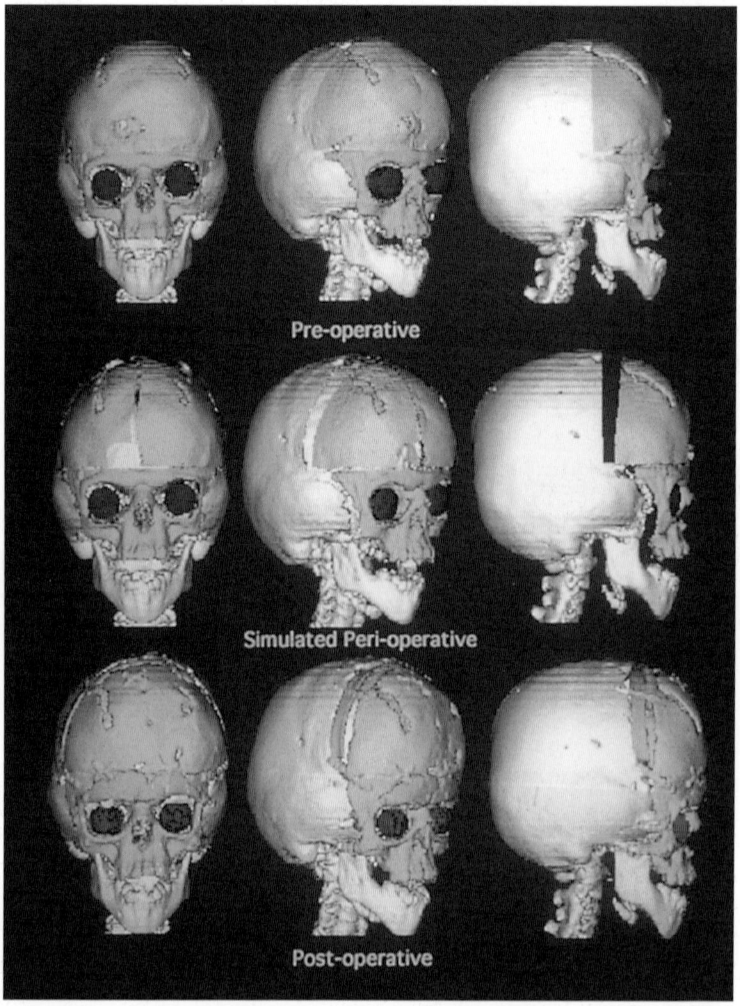

Figure 6-14 Three-dimensional CT reconstruction from a patient with Crouzon syndrome who underwent a midface and frontal bone advancement. Top, Preoperative view. Center, Computer simulation of the surgical procedure. Bottom, Postoperative view. Courtesy of M.W. Vannier, Chicago.

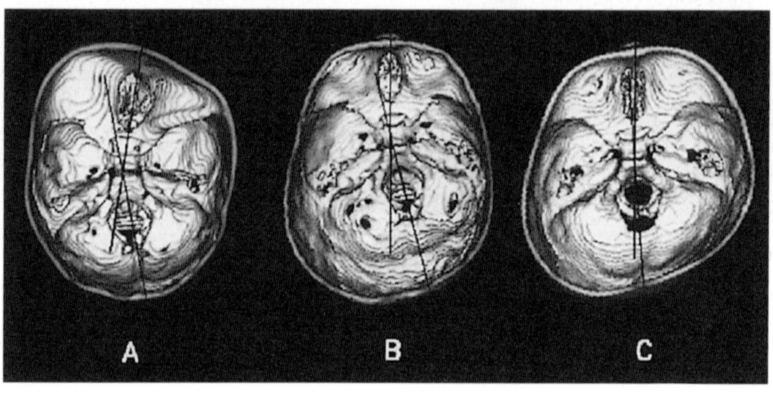

Figure 6-15 Surface 3D-CT views of the cranial base in plagiocephaly. A, Unilateral coronal synostosis. B, Unilateral lambdoid synostosis. C, Deformational plagiocephaly without synostosis. From Vannier, 2000 (103).

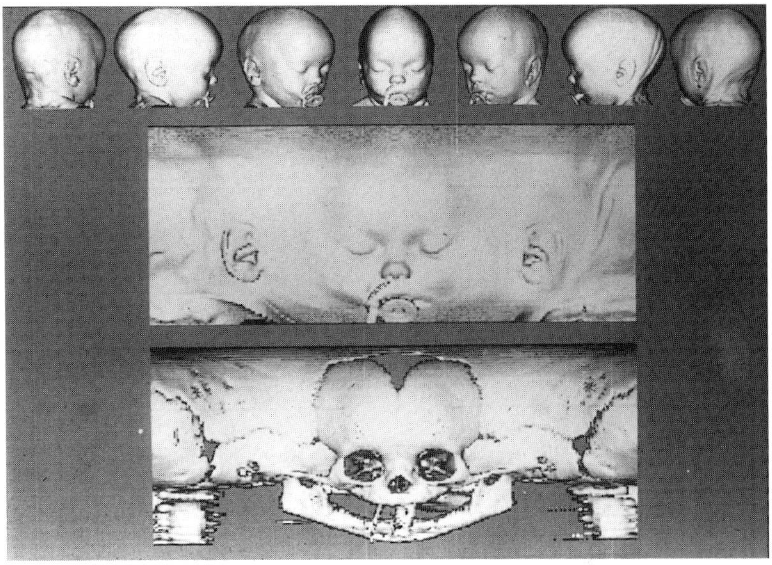

Figure 6-16 Evaluation of the head and skull by panoramic 3D-CT reconstruction. At the top are seven orthographic views of the head in a child with sagittal synostosis from different rotations. To summarize this information in a single view, a cylindrical mapping transformation was used that unwraps the head and skull surface and allows complete 360° coverage in a single view. From Vannier, 2000 (103).

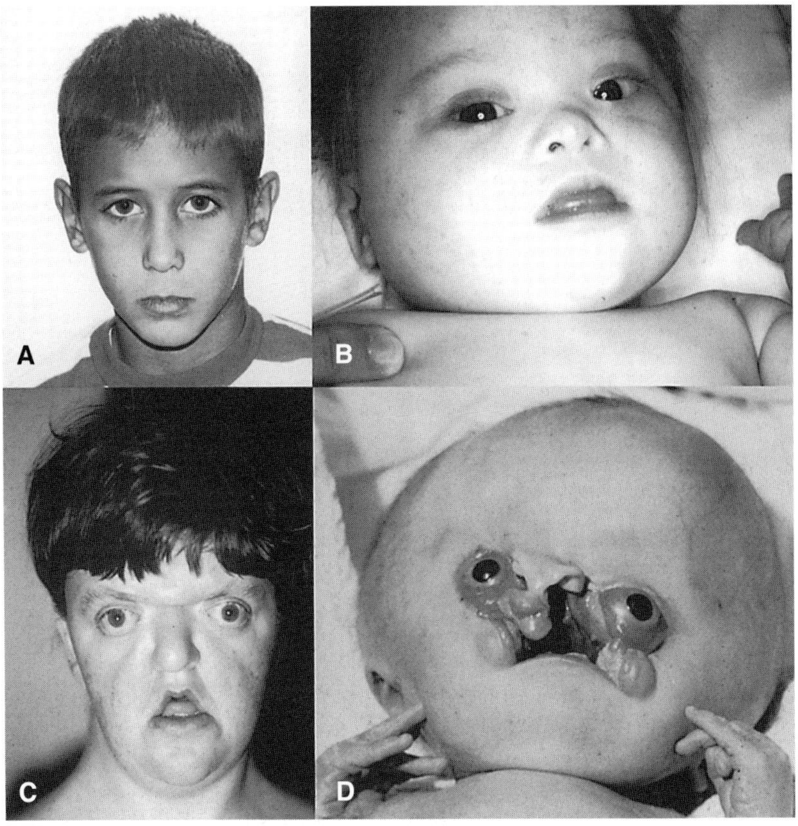

Figure 6-17 Four types of faces. A, Normal face. B, Down syndrome face with minor anomalies. C, Apert syndrome with surgically correctable craniofacial dysmorphism. D, Grossly dysmorphic face incompatible with life. Note large midfacial defect, unprotected eyes, and hydrocephalus.

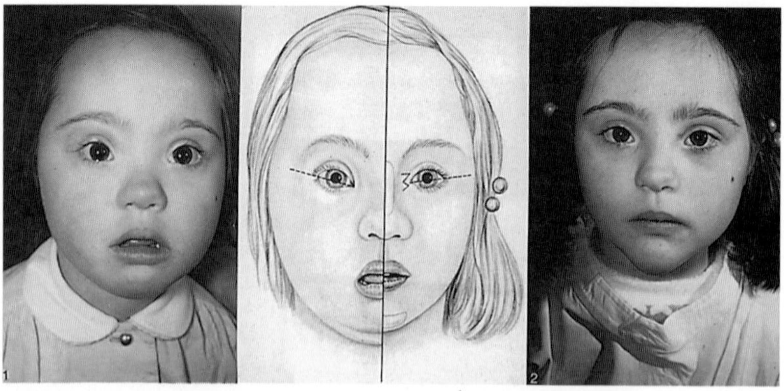

Figure 6-18 Left, 5-year-old girl with typical facial features of Down syndrome. Center, Elective plastic surgical procedures included elimination of epicanthic folds, correction of upslanting palpebral fissures, elevation of nasal dorsum and glabella, wedge excision for macroglossia, advancement genioplasty, and removal of fatty tissue under the chin. All procedures were carried out in a single operation. Right, Same girl postsurgically at 9 years. Note the absence of Down syndrome facial features. From Höhler, 1977 (28). Even though mental deficiency is characteristic of Down syndrome, when the child lives at home, some parents elect such surgical procedures, in part, to reduce the frequency of staring by others.

anomalies (28). A teenage girl with Apert syndrome (Fig. 6-17) and reasonable intelligence can benefit from craniofacial correction. A dramatic result can be achieved for some patients with a very severe abnormality, such as a massive vascular malformation of the face (see Chapter 1, Fig. 1-17).

CLEFT LIP AND CLEFT PALATE

The etiology and pathogenesis of orofacial clefting are complex (see Chapter 5). A simplified surgical summary of procedures for cleft lip–palate is shown in Table 6-3.[b] Clefts vary not only in type but also in degree and severity of distortion. Cleft lip repair is usually carried out at 3–5 months. Cleft palate closure should be done before 1 year of age. Surgeons usually agree on principles but often argue about techniques. Many are known for unilateral and bilateral cleft lip repair, for cleft palate repair, and for secondary procedures. Contributions include, among others, those of Veau, LeMesurier, Tennison, Skoog, Randall, Millard, Delaire, McComb, and Mulliken (52–53b, 56, 57a, 72, 73, 76).

Infants with unilateral and bilateral complete cleft lip undergo preoperative dentofacial orthopedics; advantages include (a) minimizing tension on labial repair, (b) facilitating primary nasal correction, (c) allowing proper proportions in philtral design, and (d) permitting closure of the alveolar gap (53, 53a, 56).

Most surgeons follow Millard's principle for rotation advancement in closing unilateral com-

Table 6-3 A Simplified Summary for Repair of Cleft Lip and Palate

- Primary procedures
  - Repair of unilateral or bilateral cleft lip and nasal deformity
  - Repair of cleft palate
- Secondary procedures
  - Pharyngeal flap
  - Sphinter pharyngoplasty
  - Bone graft to alveolar cleft(s)
  - Nasal revision
  - Labial revision
  - Le Fort I maxillary advancement

[b]Many different classifications are known.

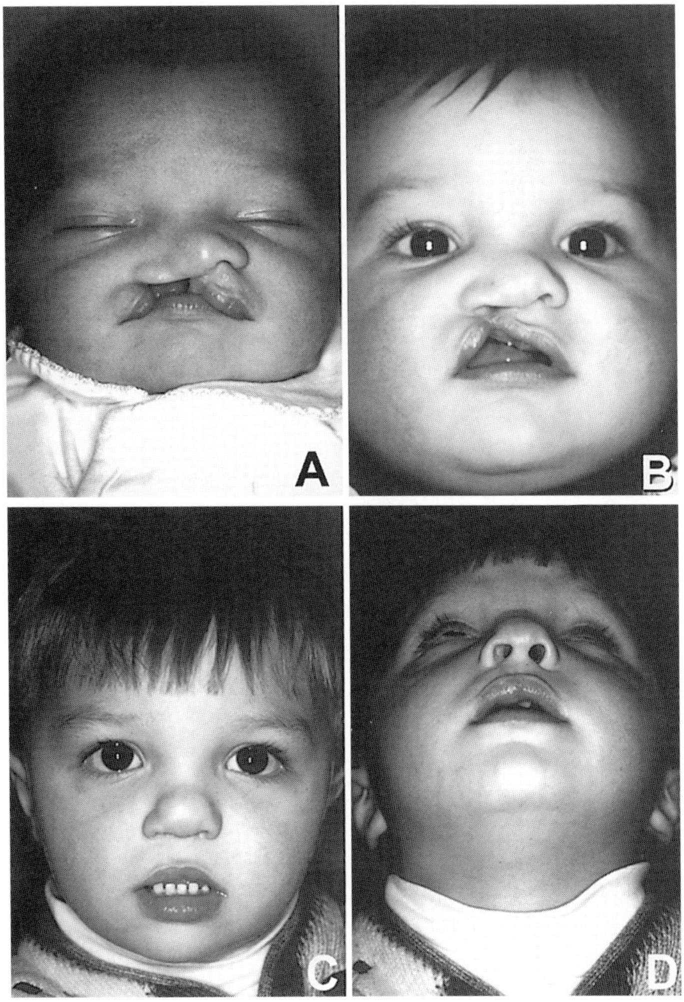

Figure 6-19 Unilateral complete cleft lip–palate. A, Right unilateral cleft lip–palate. B, Following active premaxillary dentofacial orthopedics, lip adhesion, and closure of alveolar cleft (gingivoperiosteoplasty). C, Age 1.5 years. Following rotation-advancement repair and primary nasal correction. D, Age 1.5 years. Submental view demonstrates nostril symmetry. Courtesy of J.B. Mulliken, Boston.

plete cleft lip. Every surgeon has his or her own technical interpretation of Millard's concept (56). The old zigzag labial scars have been replaced by a gentle curvilinear scar along the philtral column. Surgeons do agree on the operative goals: symmetrical Cupid's bow and labial height muscular apposition, and primary correction of the nasal deformity. The alar cartilage must be properly positioned and suspended to the ipsilateral upper lateral cartilage. "Symmetry of the alar base is addressed in three dimensions, including maneuvers to position the deviated anterior–caudal septum, configure the sill, and efface the lateral vestibular web" (56). Nostril splinting has been used postoperatively to maintain the corrected position of the nose. An alternative is to use a resorbable internal splint to shield the repaired alar cartilage from deforming forces caused by scarring (114).

Figure 6-19 shows a patient with unilateral complete cleft lip, cleft palate, and nasal deformity before and after correction by Mulliken. Rotation advancement for repair is diagrammed in Figure 6-20.

Mulliken induced the principles for surgical repair of bilateral complete cleft lip and nasal deformity: (a) symmetry, (b) primary muscular continuity, (c) proper philtral size and shape, (d) formation of the median tubercle from the lateral labial elements, and (e) primary positioning

Figure 6-20 Unilateral complete cleft lip–palate: Two-stage repair. A, First stage: Labionasal adhesion. B, Second stage: Markings for rotation-advancement. C, Dissection of orbicularis oris muscle in lateral lip. D, Hemicolumellar lengthening (a, b, c). E, Closure of orbicularis oris muscle. F, (a) Symmetrical placement of alar base in three dimensions; (b) dissection through nostril rim incision to expose lower lateral cartilage; (c) suspension of lower lateral to upper lateral cartilage; (d) lenticular excision of lateral vestibular web. G, Design of unilimb Z-plasty to lower Cupid's bow handle and second unilimb Z-plasty to augment medial vermilion. H, Closure at philtral column after construction of Cupid's bow. Courtesy of J.B. Mulliken, Boston.

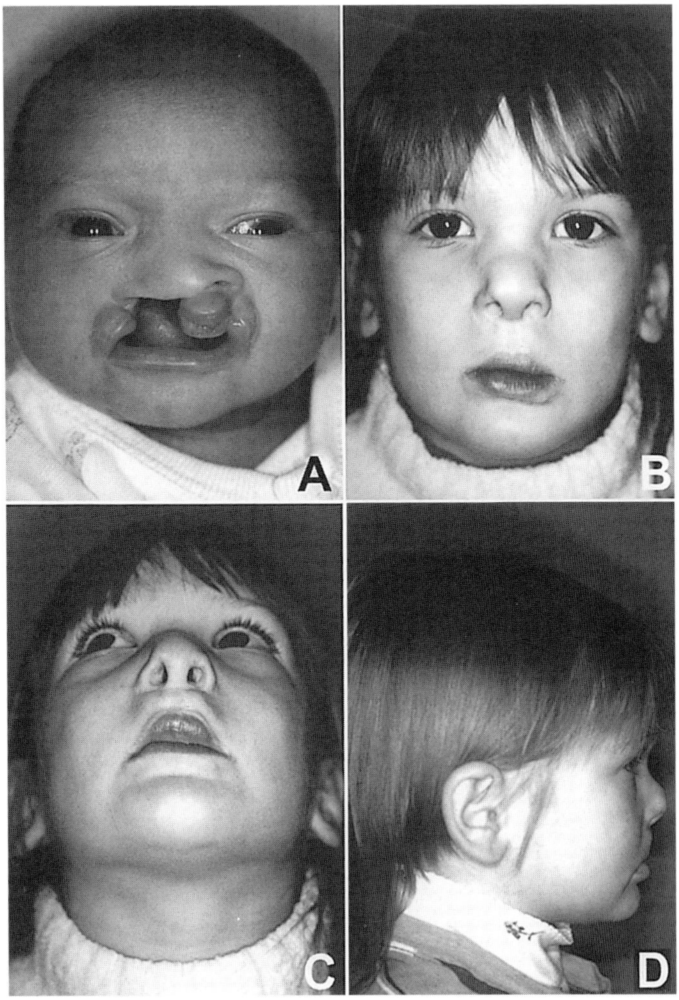

Figure 6-21 Bilateral complete cleft lip–palate. A, Newborn with twisted premaxilla and tiny band at left sill, prior to premaxillary orthopedics. B, Age 5 years, following single stage nasolabioalveolar repair and palatal closure. C, Submental view to demonstrate columellar/nasal tip proportion. D, Lateral view. Columellar-labial angle will diminish with growth. Courtesy of J.B. Mulliken, Boston.

of the alar cartilages to constrict the nasal tip and columella. Management includes preoperative premaxillary orthopedics (53).

It was once commonly accepted that repair of bilateral cleft lip was twice as difficult as for unilateral cleft lip and the results were not as good. With current preoperative and operative methods, the appearance of a child born with bilateral cleft lip can be comparable and, in some cases, surpasses that of a repaired unilateral cleft lip (53a). In Figure 6-21, an infant with bilateral complete cleft lip–palate is shown before and after a single-stage nasolabioalveolar repair by Mulliken (53). The most common revision is reduction of extra mucosa in the lateral labial elements and is often associated with placement of additional tissue to the median tubercle. This is accomplished after the permanent central incisors have erupted to allow for proper dental show (53). Figure 6-22 diagrams the Mulliken method of simple-stage nasolabial repair with a semi-open approach to nasal-tip cartilages through bilateral rim incisions (53).

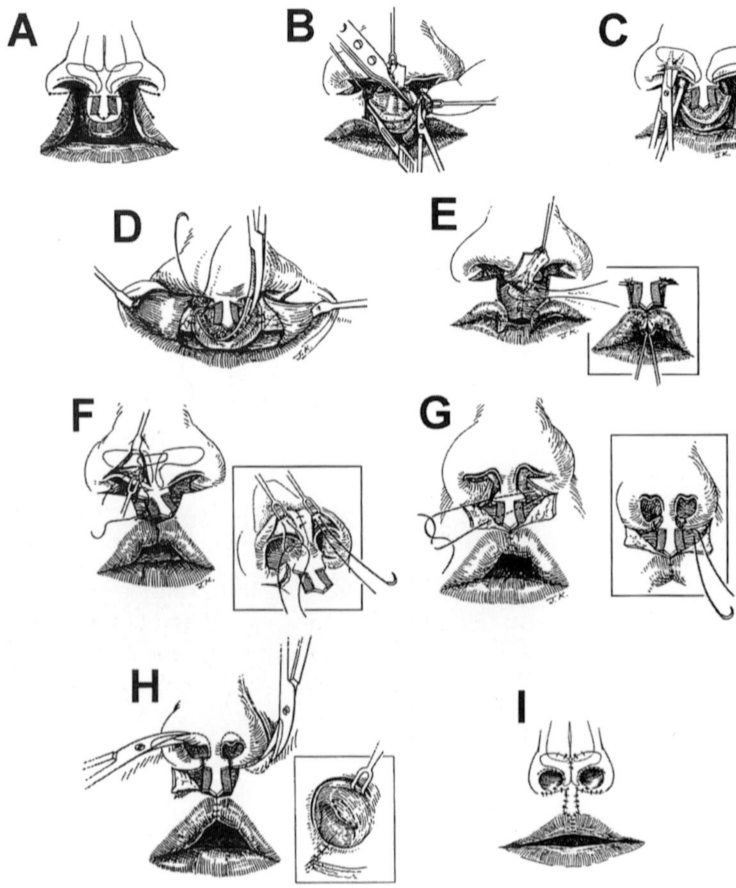

Figure 6-22 The Mulliken method of single-stage repair of bilateral cleft lip and nasal deformity. A, Markings: philtral flap designed with biconcave sides and flanking deepithelialized flaps. B, Dissection of orbicularis oris muscle from lateral labial elements. C, Exposure of lower lateral cartilages through rim incisions. D, Trimming redundant prolabial vermilion-mucosa following gingivoperiosteoplasties. E, Apposition of orbicularis oris. Inset, Construction of median tubercle. F, Interdomal suture and suspension of lower to upper lateral cartilages. G, Inter-alar cinch suture. Inset, Trimming alar base flaps and construction of sills. H, Excision extra nostril skin and trimming upper edge of lateral lip elements. Inset: excision lateral vesicular web. I, After completion of repair. From Mulliken, 2001 (53).

CRANIOFACIAL SURGERY

A classification of craniofacial surgical procedures is shown in Table 6-4.[c] Examples of surgical procedures for Crouzon syndrome, Apert syndrome, hypertelorism, and unilateral coronal synostosis shown here are taken directly from a chapter by Dr. Tessier (97); Wolfe (111) has called it the "most complete discussion of craniofacial surgery ever written," noting that Dr. Tessier wrote it in 1982 for the first edition of a book on craniosynostosis that appeared in 1986. The second edition, published in 2000, has an afterword by Wolfe (111) with updated comments on craniofacial surgery by Dr. Tessier.

Operations are much more complicated, more

[c]Many classifications are possible. I derived this one by modifying the table of contents in the book by Posnick (72). Not included are other subjects covered by Posnick, such as cleft lip, cleft palate, and orthognathic surgical procedures.

Table 6-4 A Classification of Conditions Treated by Craniofacial Surgeons

Craniosynostosis
Scaphocephaly (sagittal synostosis)
Anterior plagiocephaly (unilateral coronal synostosis)
Brachycephaly (bilateral coronal synostosis)
Trigonocephaly (metopic synostosis)
Posterior plagiocephaly (unilateral lambdoid synostosis)
Craniofacial syndromes and anomalies
Crouzon syndrome
Apert syndrome
Pfeiffer syndrome
Cloverleaf skull
Treacher Collins syndrome
Hemifacial microsomia
Binder syndrome
Tessier-type craniofacial clefts
Other syndromes and anomalies
Hypertelorism
Hypertelorism
Frontonasal dysplasia
Craniofrontonasal syndrome
Congenital growths and tumors
Dermoid cysts, gliomas, and encephaloceles
Hemangiomas and vascular malformations
Fibrous dysplasia
Benign and malignant tumors
Craniomaxillofacial trauma

Modified and reordered from the table of contents from the book by Posnick, 2000 (72). Not included in this table are other subjects covered by Posnick, such as cleft lip, cleft palate, and orthognathic surgery.

varied, and individualized than indicated in the brief discussion that follows. Furthermore, progress in craniofacial surgery means altering and improving some procedures and abandoning others.

CROUZON SYNDROME

Crouzon syndrome is discussed in Chapter 5. Figure 6-23 shows a patient at 9 years before craniofacial repair and at age 14 years after correction. The ocular proptosis and midfacial deficiency have been completely transformed by Dr. Tessier, who performed an additional genioplasty (97). A diagram of his procedure is shown in Figure 6-24. A midfacial advancement was carried out by Le Fort III osteotomy. Following advancement, bone grafts were placed for fixation of the zygomatic arches, lateral orbital walls, and between the maxilla and the pterygoid plates, and for reshaping the lower forehead (97).

APERT SYNDROME

Apert syndrome is discussed in Chapter 5. Figure 6-25 shows a patient at 18 years before craniofacial repair and at age 22 years after repair. The ocular proptosis, downslanting palpebral fissures, and midface deficiency have been completely transformed by Dr. Tessier (97). A diagram of his procedure is shown in Figure 6-26. A Le Fort III midfacial advancement was accomplished with a sagittal split of the lateral orbital rims that required an intracranial approach. An additional Le Fort I maxillotomy was combined with the midfacial advancement (Fig. 6-27). Used for adults, the procedure corrects the anterior open bite typical of Apert syndrome and also widens the maxillary arch (97).

HYPERTELORISM

Orbital hypertelorism is covered more extensively than other disorders in this chapter because this subject, unlike the others, has not been discussed previously in Chapter 5, Dysmorphic and Genetic Perspectives. Hypertelorism is etiologically heterogeneous, and Table 6-5 lists some syndromes in which hypertelorism is a feature. Because different causes do not lead to a common pathogenetic pathway, hypertelorism is not a single malformation, sequence, or developmental field. It is pathogenetically heterogeneous as well as etiologically heterogeneous. For example, Apert syndrome, frontonasal dysplasia, and Waardenburg syndrome, type 1 produce "widesetness" of the eyes in different ways (7). However, most cases of severe hypertelorism occur sporadically and no genetic basis has been established to date, although frontonasal dysplasia has been described in two families (7).

Several theories of pathogenesis have been advanced (7). Mann (41) suggested that hypertelorism results from early ossification of the lesser wings of the sphenoid, fixing the orbits in fetal position.

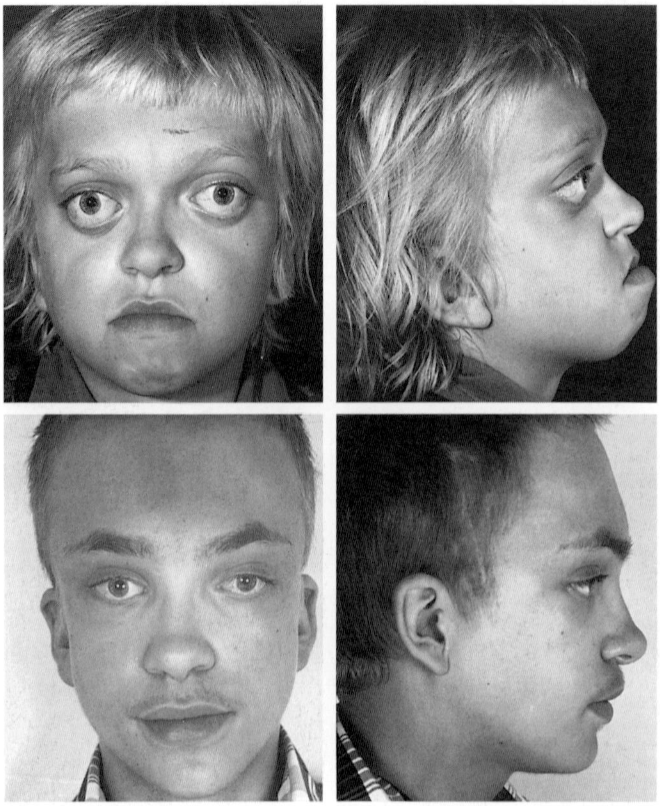

Figure 6-23 Dr. Tessier's procedure for typical, regular Crouzon syndrome (Group 5). Patient at 9 years before surgery (top) and at 14 years following surgery (bottom). Le Fort III midfacial advancement was carried out, using a semi-open skull procedure and a laterally based transverse frontal crescent. A genioplasty was carried out in addition. See Figure 6-24 for details. From Tessier, 2000 (97).

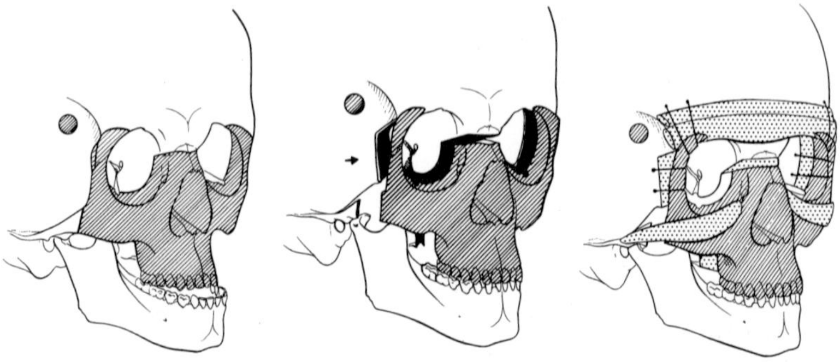

Figure 6-24 Dr. Tessier's procedure for regular Crouzon syndrome (Group 5) or Apert syndrome (Group 6). A Le Fort III midface advancement was carried out, using a semi-open skull procedure and a laterally based transverse frontal crescent. Frontal advancement. Note the bone grafts used for fixation of the zygomatic arches, lateral orbital walls, and between the maxilla and the pterygoids. Note also bone grafts for reshaping the lower forehead and glabellar region. See Figure 6-23. From Tessier, 2000 (97).

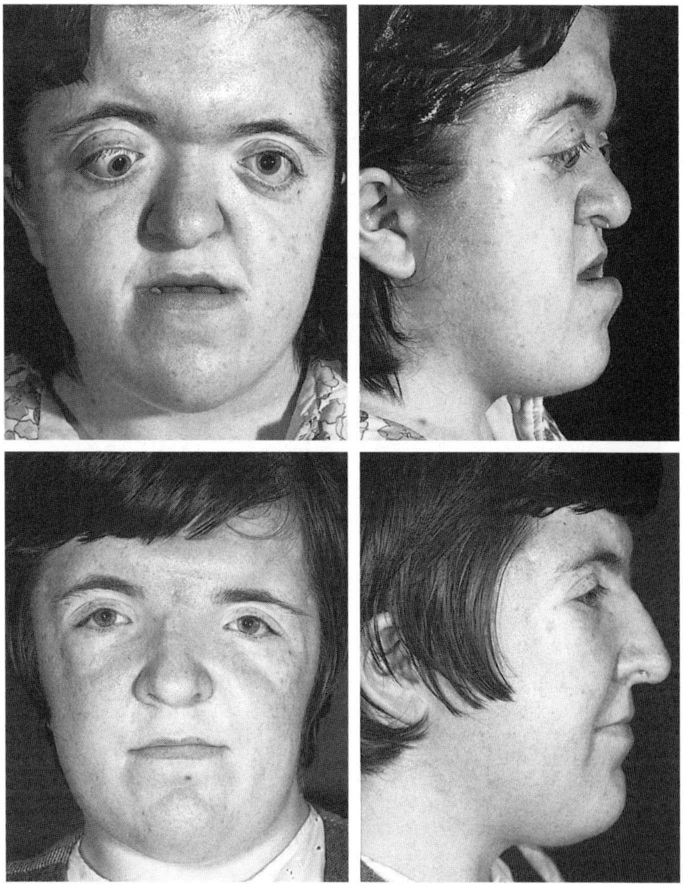

Figure 6-25 Dr. Tessier's procedure for regular Apert syndrome (Group 6-1). Patient at 18 years before surgery (top) and at 22 years following surgery (bottom; procedure now obsolete). A Le Fort III advancement was carried out with a sagittal split at the lateral orbital rims. A Le Fort I maxillotomy was also carried out. See Figures 6-26 and 6-27. From Tessier, 2000 (97).

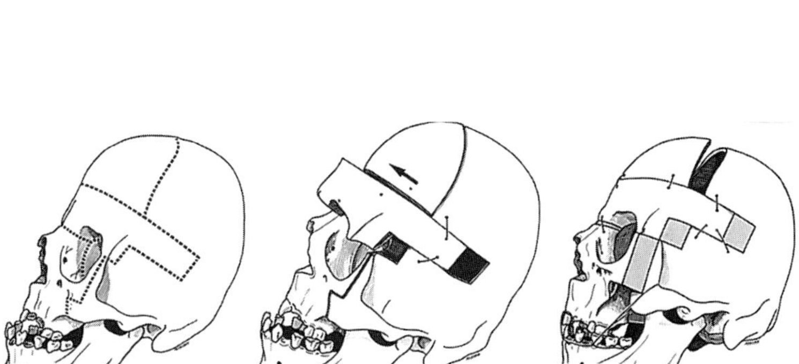

Figure 6-26 Dr. Tessier's procedure for Crouzon and Apert syndromes (Groups 5 and 6), using an intracranial approach. One stage procedure at two levels (C and C+D). Frontofacial advancement and cranial vault remodeling (level A). See Figure 6-25. From Tessier, 2000 (97).

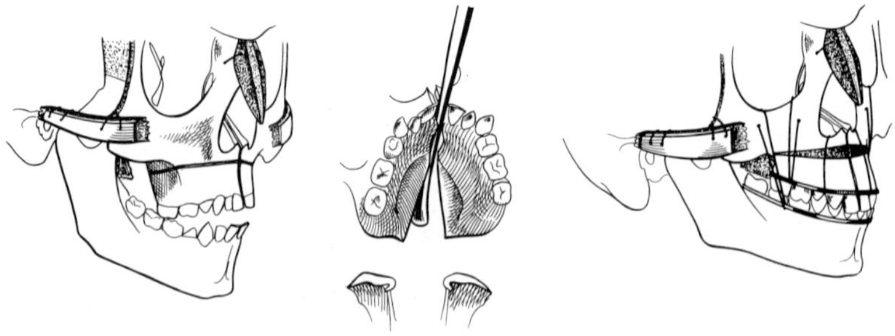

Figure 6-27 Dr. Tessier's Le Fort I maxillotomy for upper jaw displacement, often combined with other types of midface advancement, particularly for Apert syndrome in which there is an anterior open bite. The procedure is used for adults and widens the maxilla. See Figures 6-25 and 6-26. From Tessier, 2000 (97).

A common anatomic finding in many cases of hypertelorism is marked increase in the horizontal width of the ethmoid sinuses together with ethmoidal prolapse (7, 99). Tessier and associates (99) observed increased divergence of the axis of each orbit from the midsagittal plane but a normal or nearly normal distance between the optic foramina (Fig. 6-28).

Median bony clefting of the frontal or frontonasal region is common in severe instances of frontonasal dysplasia.[d] If the nasal capsule fails to develop properly, the primitive brain vesicle fills the space normally occupied by the capsule, producing anterior cranium bifidum, a morphokinetic arrest in the position of the eyes, and lack of elevation of the nasal tip. Frank encephalocele in the midline can produce the same type of morphokinetic arrest found in frontonasal dysplasia (Fig. 6-29). Less commonly, frontal teratoma, frontal lipoma, hamartoma, intracranial cyst, or intrinsic cartilaginous defect affecting the nasal capsule can produce hypertelorism (6–8, 23, 32) (Fig. 6-30).

Hypertelorism can be associated with widow's peak. For example, it is found in about 45% of cases of frontonasal dysplasia (83). We generated this hypothesis from a patient with a rudimentary ectopic left eye displaced to the temporal region with a circular area of hair-growth suppression instead of normal scalp hair (Fig. 6-31). If normal eye development is associated with a circular area of hair-growth suppression, hypertelorism could result in the two circular areas of hair-growth suppression being pulled apart, allowing widow's peak hair prolongation onto the forehead (83) (Fig. 6-32).

Hypertelorism of lesser degree occurs in Apert syndrome, Pfeiffer syndrome, and Crouzon syndrome. The cribriform plate may be prolapsed. The mechanism causing hypertelorism is related to disturbances in the formation of the cranial base (7).

Moss (50), Aduss and associates (1), and Figalová and colleagues (20) showed an increased

Table 6-5 Some Syndromes with Hypertelorism

Frontonasal dysplasia
Craniofrontonasal syndrome
Brachycephalofrontonasal dysplasia
Acrofrontofacionasal dysostosis
Frontofacionasal dysplasia
Greig cephalopolysyndactyly
Naguib syndrome
Nasopalpebral lipoma–coloboma syndrome
Noonan syndrome
Opitz BBB/G syndrome
Aarskog syndrome
LEOPARD syndrome
Waardenburg syndrome, type I
Blepharonasofacial syndrome
Sotos syndrome
Bannayan–Riley–Ruvalcaba syndrome
Robinow syndrome
Apert syndrome
Pfeiffer syndrome
Cloverleaf skull
Nevoid basal cell carcinoma syndrome
Craniometaphyseal dysplasia
Craniodiaphyseal dysplasia

From Cohen et al., 1995 (7).

[d]Common bony clefts include median and paramedian types (see Fig. 5-13).

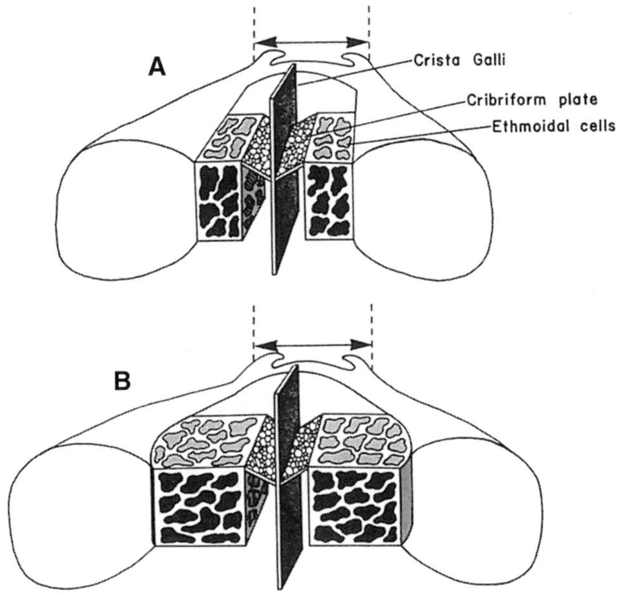

Figure 6-28 A, Normal interorbital distance. B, Hypertelorism with widely spaced orbits. Note enlargement of the ethmoid sinuses, normal distance between the optic foramina (arrows), and laterally curving medial orbital walls. From Cohen et al., 1995 (7). Redrawn from Converse et al., 1970 (11), to better illustrate the curve of the lateral rim of the ethmoid bone and its effect on the medial wall of the orbit.

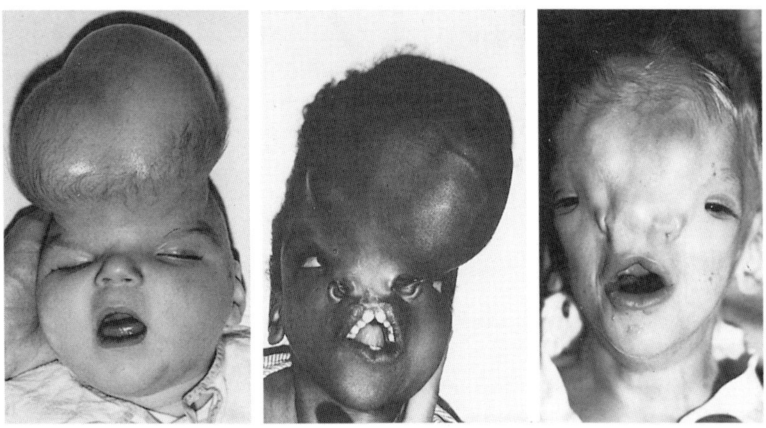

Figure 6-29 From left to right: Frontal encephalocele, frontonasal encephalocele, and frontonasal dysplasia, illustrating hypertelorism. From Cohen and Lemire, 1982 (6).

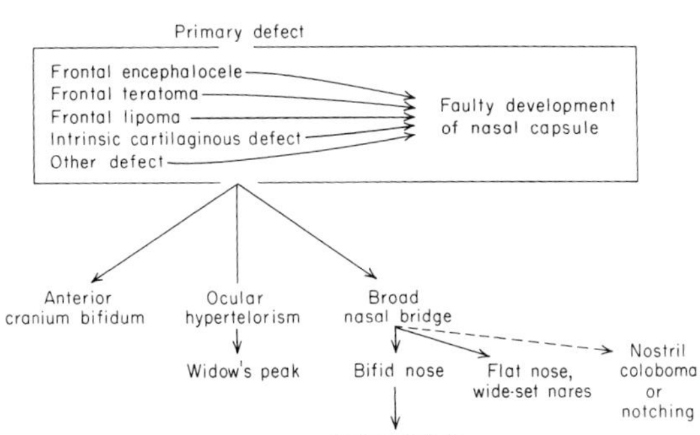

Figure 6-30 Diagram of different types of primary defects resulting in faulty development of the nasal capsule and its sequelae. From Cohen et al., 1995 (7).

interorbital distance in patients with nonsyndromal orofacial clefting. With a large sample size, Aduss and associates (1) showed that (a) the effect was small, the interorbital distance being increased by about one standard deviation, and (b) the increase was primarily in patients with bilateral or unilateral cleft lip with or without cleft palate, but not with isolated cleft palate or Robin sequence only. The mechanism resulting in the association between orofacial clefting and an increased interorbital distance is not known, but it appears to involve a developmental field defect. Moss (50) suggested that a defect in the nasal capsule (unspecified) is related to both or-

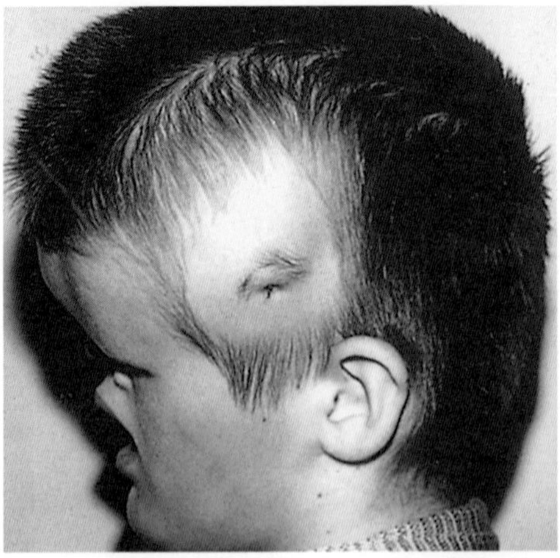

Figure 6-31 Circular area of periocular hair-growth suppression surrounding a rudimentary ectopic left eye displaced to the temporal region. From Smith and Cohen, 1973 (83).

ofacial clefting and increased interorbital distance.

There are a number of abnormalities associated with some cases of hypertelorism, including nonprotruding lipomas of the corpus callosum, posterior encephalocele extending to the ethmoid, calcification of the falx cerebri, duplication of the crista galli, absence of the crista galli, bifid nose, wrinkling of the nose, tissue tags of the nose, and colobomas of the nostrils (7, 8). Most are associated with frontonasal dysplasia, but their significance is unknown at present (23).

Measurements

Various types of measurements are available for different purposes (7). Normal distance between the eyes, primary telecanthus, hypertelorism, and secondary telecanthus are illustrated in Figure 6-33. Bony interorbital measurements from radiographs or CTs are the most accurate (12, 25, 48) and are important for surgical considerations.

The canthal index (24) is useful for consultations by correspondence when facial photographs are provided but no measurements are given in the accompanying clinical summary. Because the canthal index is based on the ratio of inner to outer canthal distance, it can be calculated from clinical photographs alone, regardless of size.

Inner and outer canthal measurements (19), while less accurate than bony interorbital measurements and not necessarily reflective of them (21), are the simplest to use for standard clinical evaluation in clinical genetics.

Operative Correction for Orbital Hypertelorism

Dr. Tessier's surgical classification of interorbital distances and hypertelorism is summarized in Table 6-6. Figure 6-34 shows a patient with pronounced hypertelorism, wide flat nose, and nostril colobomas. Radiographic findings included paramedian encephaloceles with bilateral bony clefts extending from the frontal bone through

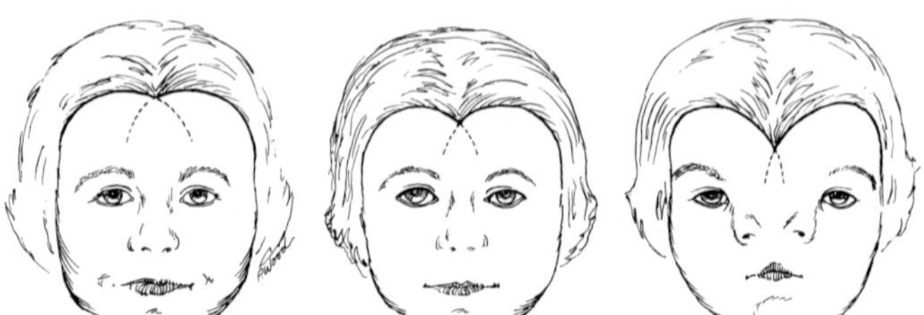

PERIOCULAR FIELDS	NORMAL	DECREASED	NORMAL
INTEROCULAR DISTANCE	NORMAL	NORMAL	INCREASED

Figure 6-32 Left, Girl with normotelorism. Center, Normal girl with reduced size of circular periocular fields of hair-growth suppression and widow's peak. Right, Widow's peak secondary to frontonasal dysplasia. Intersecting lines represent the limits of periocular hair-growth suppression. From Smith and Cohen, 1973 (83).

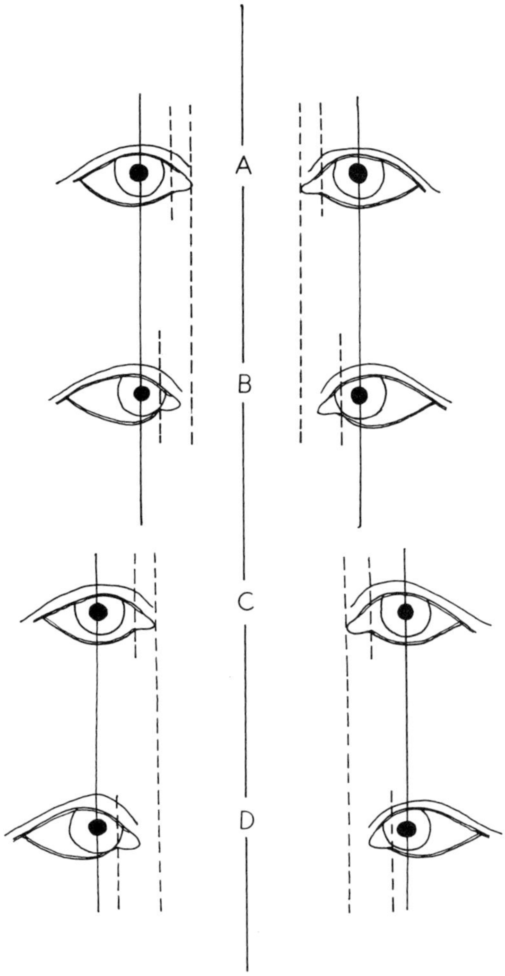

Figure 6-33 Primary telecanthus, secondary telecanthus, and hypertelorism. A, Normal interocular distance. B, Primary telecanthus. Inner canthi are far apart, although outer canthi are normally spaced. Note how a vertical line through the lacrimal punctum cuts the cornea. C, True ocular hypertelorism. Both inner and outer canthi are abnormally far apart. D, True ocular hypertelorism together with secondary telecanthus. Note how a vertical line through the lacrimal punctum cuts the cornea. From Cohen et al., 1995 (7).

the pyriform aperture[e] on both sides. Also found were widening of the olfactory grooves and crista galli, and calcification of the falx cerebri (94). The remarkable surgical results achieved by Dr.

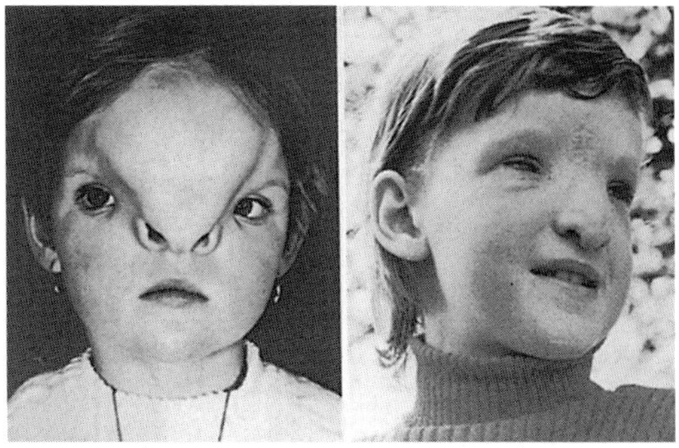

Figure 6-34 Hypertelorism before and after Dr. Tessier's procedure. Left, Pronounced hypertelorism, wide flat nose, and nostril colobomas. Radiographic findings showed paramedian encephaloceles with bony clefting extending from the frontal bone through the quadrangular "pyriform" aperture. Also found were widening of the olfactory grooves and crista galli, and calcification of the falx cerebri. Right, Remarkable surgical results from a nasofrontoethmoidal resection, using an intracranial approach. From Tessier, 1976 (94) and 1977 (101).

Tessier (92, 98-100) are shown in Figure 6-34. He carried out a nasofrontoethmoidal resection using an intracranial approach.

Dr. Tessier's surgical correction of orbital hypertelorism by the intracranial route (98) is summarized in Figures 6-35 and 6-36. Coronal, nasofrontal, and infra-orbital incisions are made. Osteotomies involve the frontal bone, the orbits, the orbital apices, orbital roof, and between the ethmoid and sphenoid. The nasofrontoethmoidal complex is then removed, preserving the cribriform plate and olfactory fibers, followed by resection of the vomer and a transmaxillomalar osteotomy. The orbits are then moved closer together. The nasal mucosa and cartilages, upper and lower, are sutured, and wires are used for medial canthopexies and fixing the orbits together. The base of the maxillary frontal process is resected to enlarge the nasal airway. Bone grafts are placed in the malar, temporal, nasal, and ethmoidal areas. A nasal bone graft is dowelled into the frontal bone. The last stage includes suturing and the placement of drains (98).

Figure 6-37 shows a 5-year-old girl with orbital hypertelorism and a midfacial cleft involving the nose. The procedure was carried out by

[e]Pyriform means pear-shaped. Actually, with pronounced hypertelorism, the nasal aperture is usually quadrangular (92).

Table 6-6 Dr. Tessier's Classification of Interorbital Distances, Hypertelorism, and Surgical Procedures

Types	Degree/Interorbital Distance	Comments	Surgical Procedures
Euryopia or telecanthus	First degree/30–34 mm	Not true hypertelorism	Surgical correction of epicanthic folds, broad nasal root, traumatic telecanthus, and widely spaced eyebrows
Orbital hypertelorism without true ocular malposition or deviation, except in cases of cranio-facial dysostosis in which exotropia and ocular proptosis occur	Second degree/> 34 mm	Face is usually widened but orbits are nearly normal in orientation and shape on a frontal radiograph	Second degree cases are not always amenable to the intracranial approach, but to the extracranial approach if the level of the cribriform plate is reasonably normal
True orbital hypertelorism	Third degree/usually > 40 mm	Face is widened, orbits are lateralized and ovoid in shape on a frontal radiograph, and the cribriform plate is often lateralized	Intracranial approach for removal of the nasofrontoethmoidal complex, preserving the cribriform plate and olfactory fibers, followed by bringing the useful orbits closer together

Adapted from Tessier, 1972 (92).

Posnick (71), who performed a facial bipartition with repositioning (Fig. 6-38). Posnick's remarkable results are shown in Figure 6-37.

UNILATERAL CORONAL SYNOSTOSIS

Figure 6-39 shows a patient with unilateral coronal synostosis and facial asymmetry. Following two early craniectomies for right-sided involvement, further procedures were carried out in four stages. In stage 1, derotation of both frontal orbital segments was carried out from right to left at 11 years. In stage 2, a maxillotomy was utilized to rotate the maxilla from left to right at 12 years. Stage 3 consisted of a monobloc rhinotomy for straightening the nose and an advancement genioplasty at 13 years. In stage 4, the mandible was rotated from left to right at 19 years. The patient is shown at 30 years, 19 years after the stage 1 operation (97). Today, such patients are seen in craniofacial centers and undergo fronto-orbital correction during infancy. The need for revision is low unless the patient is syndromic.

ORTHOGNATHIC SURGERY

Principles and techniques of orthognathic surgery have been discussed by numerous authors (2, 16, 64–67, 72, 77). A surgical classification is shown in Table 6-7.[f] These procedures are sophisticated, flexible, and adaptable. The techniques apply to routine and complex dentofacial deformities and also to a variety of other situations, such as secondary procedures for cleft lip–palate and malunions of facial fractures, particularly those involving the mandibular condyle.

The following sections on evaluation, general approach, and basic orthognathic procedures are based on Posnick (72). Evaluation includes careful clinical examination by the surgeon for skeletal components, occlusion, and particularly for soft tissue components and facial aesthetics. Cephalometric analysis and model surgery are also important, although overreliance on cephalometric norms should never be a substitute for careful clinical analysis of facial aesthetics (72).

The general approach for correcting dentofacial deformities includes (a) presurgical orthodontic treatment, (b) orthognathic surgery, and (c) completion of orthodontic treatment. The

[f]Many classifications of orthognathic surgical procedures are possible. I derived this one by modifying the table of contents in the book by Posnick, 2000 (72). Not included are other subjects covered by Posnick, such as procedures for craniofacial malformations, cleft lip, and cleft palate.

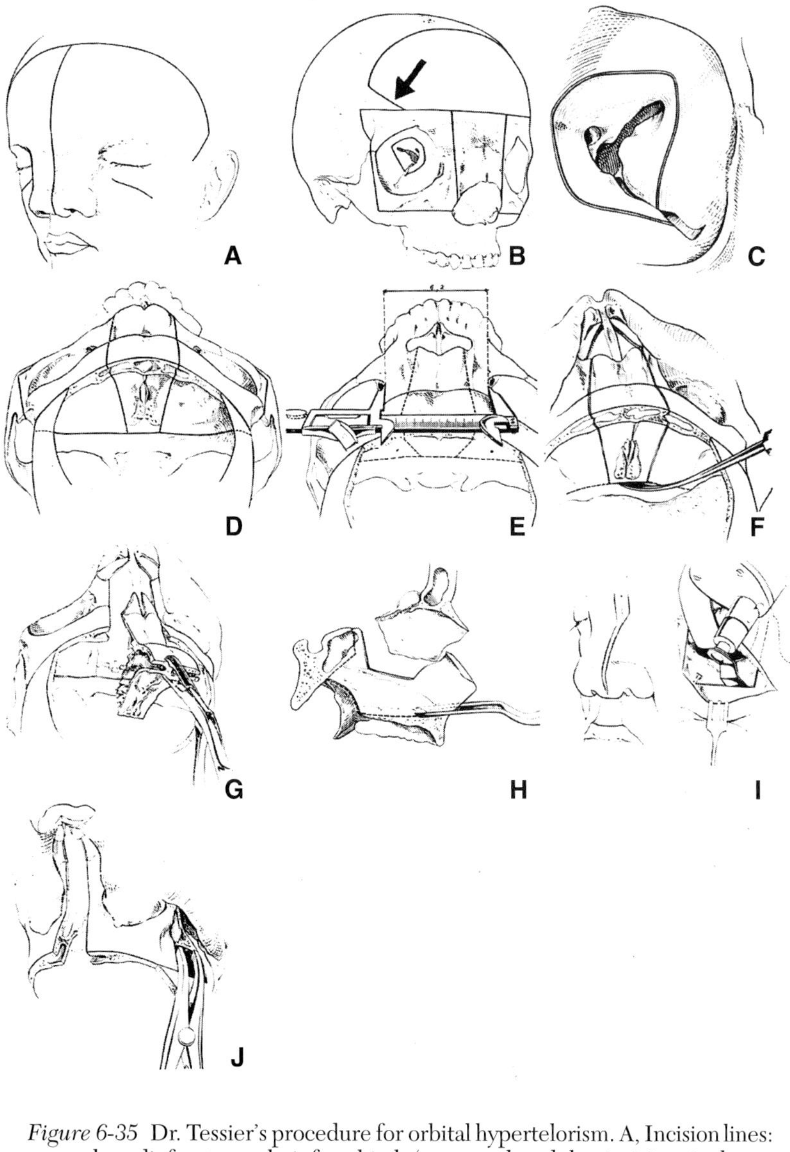

Figure 6-35 Dr. Tessier's procedure for orbital hypertelorism. A, Incision lines: coronal-mediofrontonasal; infraorbital (now replaced by incision in lower fornix). B, Lines of craniotomy: median resection; orbital osteotomy; frontotemporal spur (arrow). C, Osteotomy near the orbital apex. D, Lines of frontoethmoidal resection and osteotomy of the orbital roof. E, Measurement of the interorbital distance. F, Osteotomy between ethmoid and sphenoid. G, Removal of the nasofrontoethmoidal resection. H, Resection of vomer. I, Transmaxillomalar osteotomy; the infra-orbital bundle. J, Transverse displacement of the orbit with the expansion forceps. From Tessier, 1973 (98).

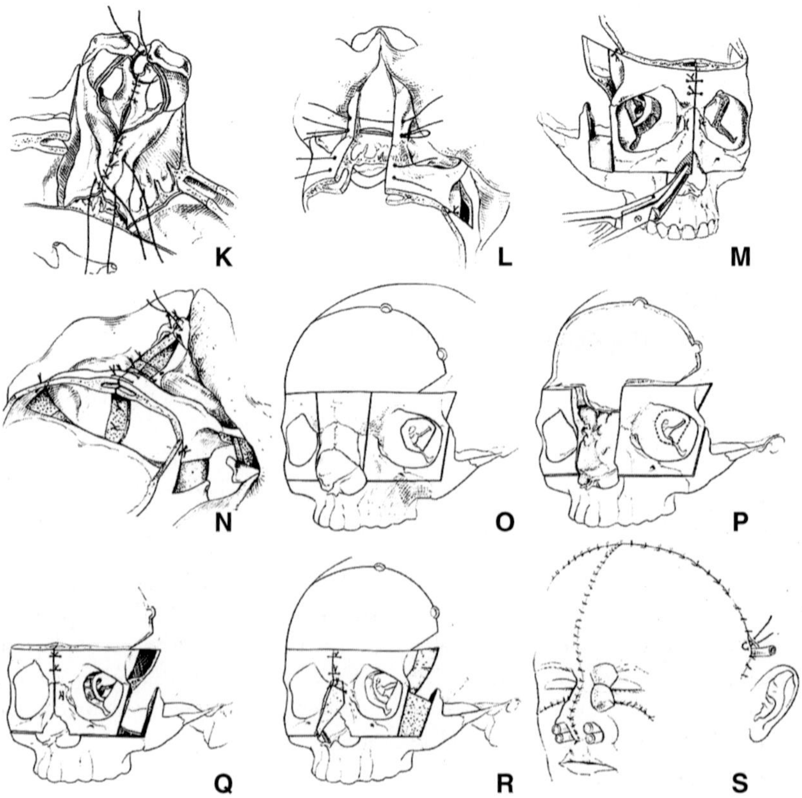

Figure 6-36 Dr. Tessier's procedure for orbital hypertelorism continued. K, Suture of nasal mucosa, upper and lower lateral cartilages. L, Wires for medial canthopexies and fixing the orbits together. M, Resection of the base of the maxillary frontal process for enlarging the nasal airway. N, Bone grafts: malar, temporal, nasal, and ethmoidal. Summary of the procedure: O, Lines of the osteotomies and resection. P, After resection. Q, After bringing the orbits closer together. R, After bone grafting. S, Sutures; drains. From Tessier, 1973 (98).

goal is the improvement of occlusal function, chewing, speech articulation, breathing, and labial closure. This should be achieved in concert with aesthetic refinement of the face (72).

Basic orthognathic surgical procedures include (a) Le Fort I osteotomy, sometimes with segmentation of maxillary components; (b) sagittal splitting of the mandibular ramus; and (c) osteoplastic genioplasty. These procedures are individually designed for each patient and are often combined. Each skeletal unit is repositioned during the operation (Fig. 6-40). Mandibular deficiency occurs most commonly, generally with additional underdevelopment of the chin. A prognathic mandible is frequently associated with maxillary deficiency and an overly long chin. Asymmetric mandibular prognathism probably has heterogeneous causes and often includes asymmetric condylar growth with hyperplasia on one side. In maxillary excess, mouth breathing during early childhood is a frequent common denominator with secondary alterations in the mandible and chin (72).

Figure 6-41 shows a woman before and after correction for mandibular deficiency performed by Stoelinga (84), who used a bilateral sagittal split advancement osteotomy of the mandible. Figure 6-42 shows a 43-year-old woman with excessive maxillary growth, resulting in a long face, open mouth with labial incompetence, and ex-

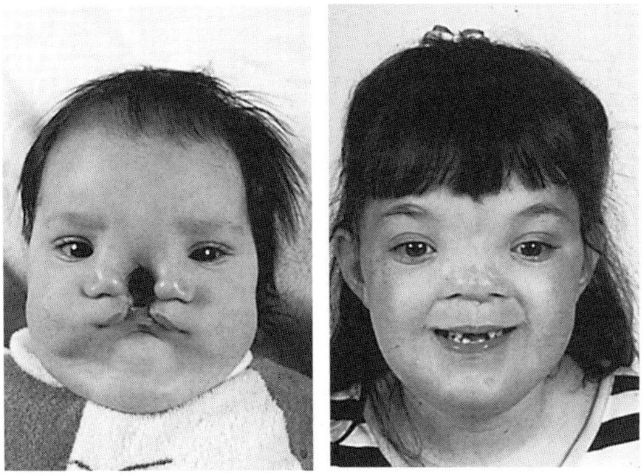

Figure 6-37 A 5-year-old girl with orbital hypertelorism and a midline cleft involving the nose. Preoperative (left) and postoperative (right) views. Posnick performed a monobloc bipartition osteotomy with medial repositioning and remarkable results. From Posnick, 1992 (71).

cessive showing of maxillary central incisors. Surgical correction was performed by West (109), who carried out a Le Fort I procedure with an ostectomy (Fig. 6-43) for intrusion of the maxilla. In the same operation, the mandible was advanced by sagittal splitting of the ramus on each side. Results show shortening of the lower face, improved labial closure, improved nasal form, and a stronger profile (Figs. 6-42, 6-44).

Figure 6-45 shows a young man with midfacial deficiency, short upper lip, and mandibular prognathism. West (108) performed a high Le Fort I osteotomy, repositioning the maxilla anteriorly and inferiorly. The mandible was set back with sagittal ramus osteotomies. Results show vertical lengthening of the lower face, improved labial line, and a proportional mandible (Fig. 6-45).

TRAUMA

Injury to the soft tissues and bones of the face can be caused by various precipitating events, including, among others, traffic accidents, altercations, sports, and other accidents in the home, in the workplace, at school, and outdoors. For psychological considerations of patients with trauma involving the face, see Chapter 7. Two patients who sustained injuries to soft tissue and bone from automobile accidents are shown in Figures 6-46 and 6-47.

TUMORS

Many pediatric and adult tumors are known, including, among others, glioma, hemangioma, rhabdomyosarcoma, melanotic neuroectodermal tumor, ameloblastoma, malignant melanoma, lymphoma, squamous cell carcinoma, and basal cell carcinoma. About 5% of neoplasms involve the oral, pharyngeal, nasal and adjacent sinus, or facial structures. Other malignant tumors invade additional structures above the clavicles such as the larynx, neck, face, and brain. Neoplasms vary in size, location, and histopathology. Treatment may include surgical resection, chemotherapy and radiation in certain cases, surgical reconstruction, and maxillofacial prosthetic rehabilitation (12a, 72). For psychological considerations of patients with tumors involving the face, surgical removal, and reconstruction, see Chapter 7.

Figure 6-48 shows a woman with orbital exenteration for a tumor. A silicone prosthesis was fabricated and inserted. The use of eyeglasses makes the prosthesis less obvious. Because silicone prostheses darken with sunlight, a replacement is required every four years.

OSTEODISTRACTION

Distraction osteogenesis is the biologic process by which new bone forms by applying a tensile force across a gap between two bone segments that have been surgically separated. Although first used for limb lengthening, distraction osteogenesis is now applied to a variety of cranial, midfacial, maxillary, and mandibular procedures (84b).

Bone regeneration is accompanied by expansion of the soft tissue matrix, including blood vessels, nerves, muscles, skin, mucosa, fascia, and ligaments. A distraction rate of 1 mm per day remains the standard for mandibular lengthening, maxillary advancement, and simultaneous maxillary and mandibular procedures. However, faster distraction rates (1.5–3 mm per day) have

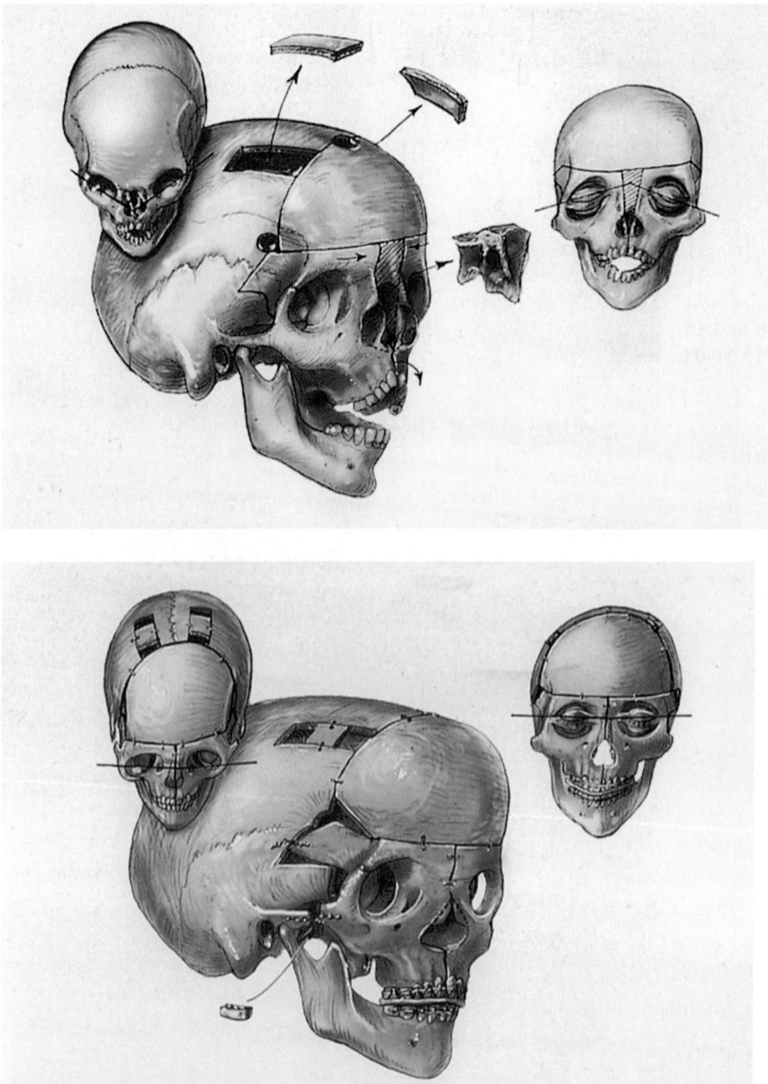

Figure 6-38 Top, Craniofacial skeleton showing proposed facial bipartition osteotomies. Bottom, Craniofacial reconstruction stabilized with miniplates and cranial bone grafts. From Posnick, 1992 (71).

also been reported. A latency period is necessary to allow time for initial callus formation and healing of soft tissues. In craniofacial distraction osteogenesis, a 5–7 day latency period is commonly used, followed by a variable amount of time for consolidation and maturation of the bony callus with the distraction device left in place (6–8 weeks to 3 months or more) (84b).

A variety of distraction devices have been used: monodirectional, bidirectional, multidirectional, interceptive, bone-supported, tooth-supported, implant-supported, and subcutaneous. Clinical parameters that affect treatment outcomes include the patient's age, surgical technique, distraction device, distraction rate, latency period, and consolidation/maturation time (84b).

Robin sequence and Robin complexes are known to be etiologically and pathogenetically heterogeneous (see Chapter 5). If Robin complex is malformational and the mandible is not expected to grow, as in Treacher Collins syndrome, mandibular distraction may avoid tracheostomy or permit earlier decannulation. However, osteodistraction is contraindicated if the infant has deformational Robin sequence, in which mandibular catch-up growth is expected. In such cases, temporary tongue–lip adhesion

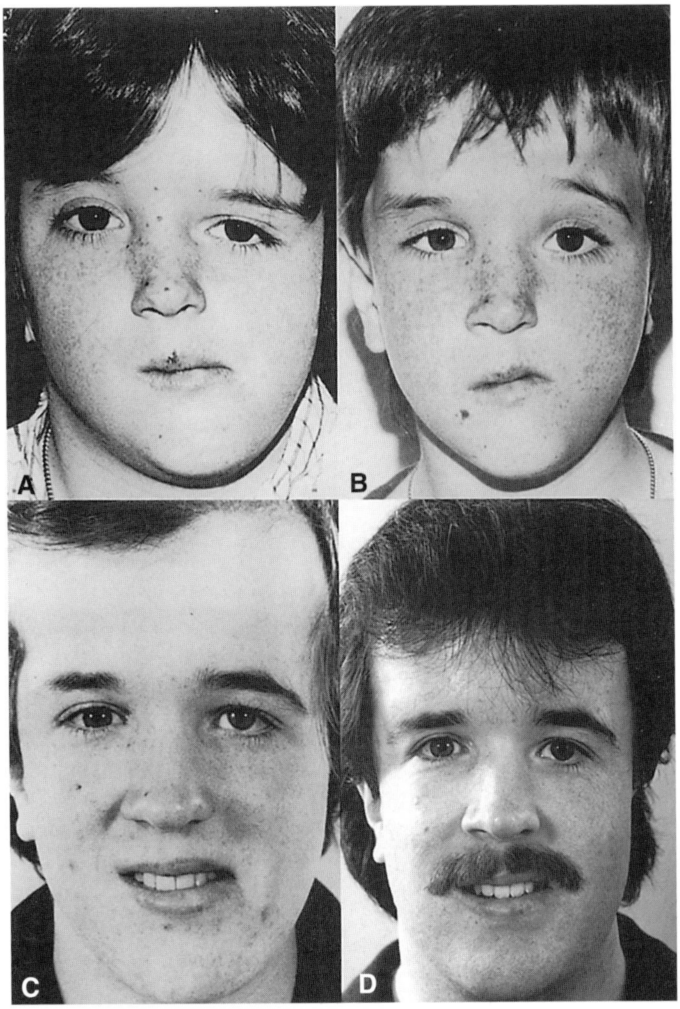

Figure 6-39 Right craniosynostotic plagiocephaly with facial asymmetry, following two early coronal craniectomies. Corrective procedures were carried out in four stages. A, In stage 1, derotation of both frontal orbital segments was carried out from right to left at 11 years (Dr. Tessier, surgeon). B, In stage 2, a maxillotomy was utilized to rotate the maxilla from left to right at 12 years (H. Kawamoto, surgeon). C, Stage 3 consisted of a monobloc rhinotomy for straightening the nose and an advancement genioplasty at 13 years (Dr. Tessier, surgeon). D, In stage 4, the mandible was rotated from left to right at 19 years (H. Kawamoto, surgeon). The patient is shown at 30 years, 19 years after the stage 1 operation. From Tessier, 2000 (97).

Table 6-7 A Classification of Conditions Treated by Orthognathic Surgeons

Condition
Mandibular deficiency
Without maxillary deformity
With maxillary deformity
Mandibular excess and maxillary deficiency
Maxillary excess
Without mandibular deficiency
With mandibular deficiency
Asymmetric mandibular prognathism
Without maxillary deformity
With maxillary deformity
Alterations of the chin

Modified, reordered, and condensed from the outline of the book by Posnick, 2000 (72).

will open the airway and permit normal feeding until the time of palatal closure. Rarely, severe Robin sequence requires tracheostomy.

In children, osteodistraction for Le Fort III advancement in syndromic synostoses, such as Apert, Pfeiffer, and Crouzen syndromes, obviates the need for harvesting bone grafts from the cranium, ileum, or ribs. There are two technical strategies. A halo is used to *pull* the midfacial segment forward more commonly than a semiburied device to *push* it forward. A halo may also be used for Le Fort I maxillary advancement for

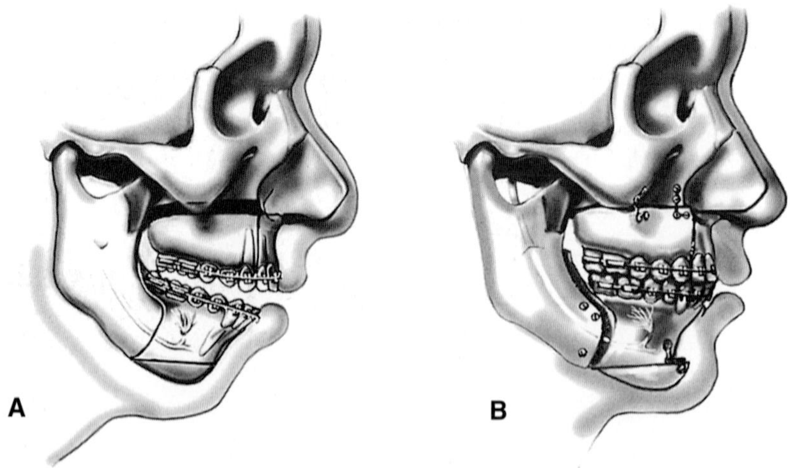

Figure 6-40 A, Lateral preoperative view. Patient has had preoperative orthodontic treatment. Lines show surgical procedures to be carried out: a three piece Le Fort I osteotomy, sagittal splitting of the ramus of the mandible, and osteoplastic genioplasty. B, Lateral postoperative view. Results of surgical procedures with repositioning of skeletal components. Note the titanium plates and screws. From Posnick, 2000 (72).

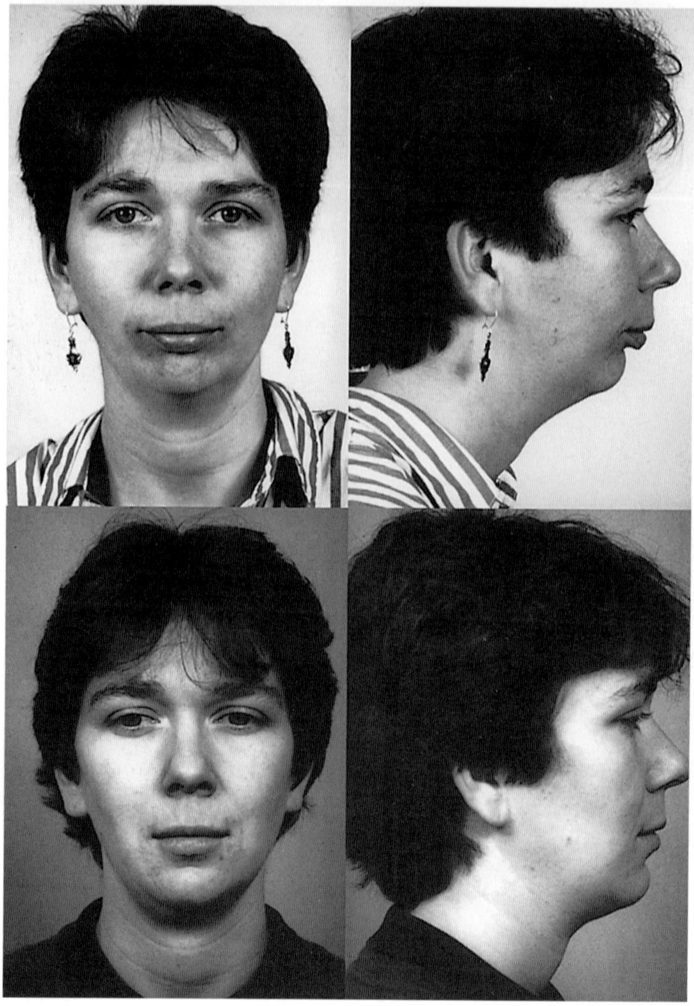

Figure 6-41 Before and after a procedure for mandibular deficiency using a bilateral sagittal split advancement osteotomy. From Stoelinga, 1990 (84).

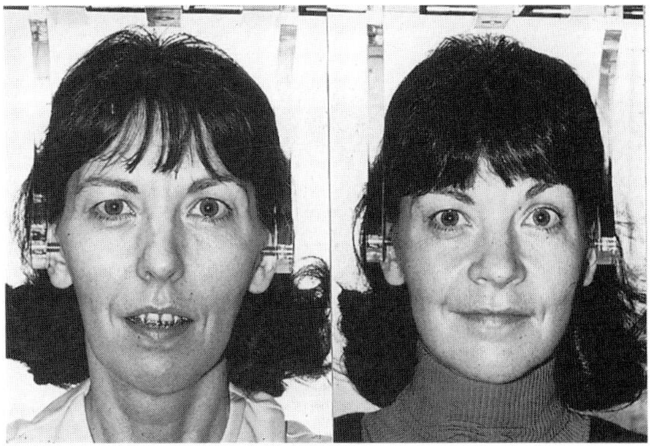

Figure 6-42 Before and after a procedure for maxillary excess. Left, Long face, open mouth with lip incompetent, and excessive showing of maxillary central incisors. Right, Results of a Le Fort I procedure with an ostectomy for intrusion of the maxilla. In the same operation, the mandible was advanced by sagittal splitting of the ramus on both sides. Results show shortening of the lower face, improved lip closure, and an improved shape of the nose. From West, 1975 (109).

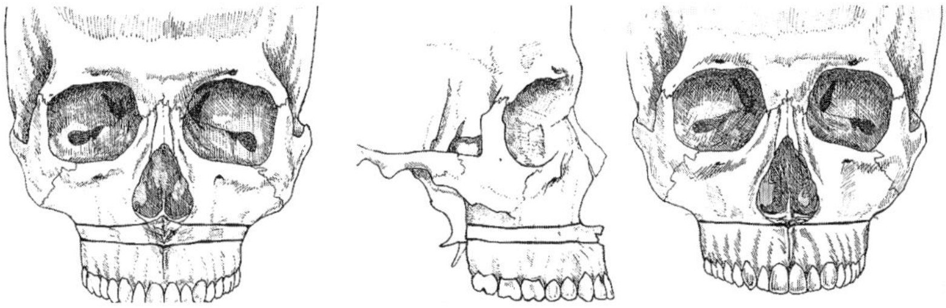

Figure 6-43 Le Fort I procedure with an ostectomy for intrusion of the maxilla. From West, 1975 (109).

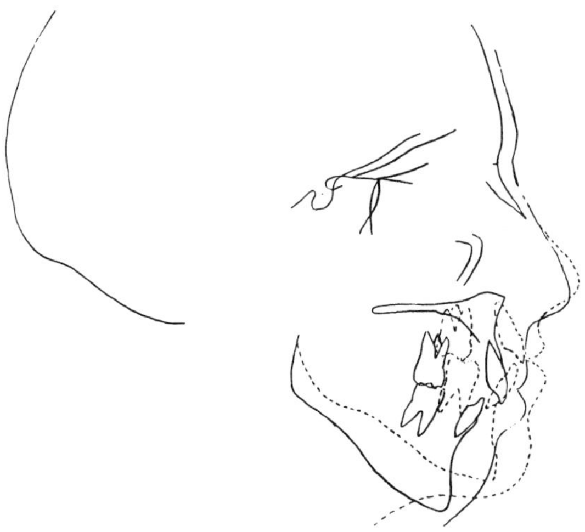

Figure 6-44 Presurgical (solid lines) and postsurgical (dashed lines) results. Note the shortening of the face and alteration in the shape of the nasal profile. From West, 1975 (109).

retrusion in patients with cleft lip and palate and, rarely, for Le Fort II advancement, as in patients with Binder syndrome. Semiburied devices can be used for mandibular distraction in patients with hemifacial microsomia, Treacher Collins syndrome, Nager acrofacial dysostosis, and posttraumatic condylar hypoplasia. Distraction may be used in many other conditions, such as expansion of the mandibular arch by an intraoral device (84b).

Further technical advances will involve computerized programming as well as newer, smaller, and internal devices that may someday make osteodistraction as common as orthodontic treatment.

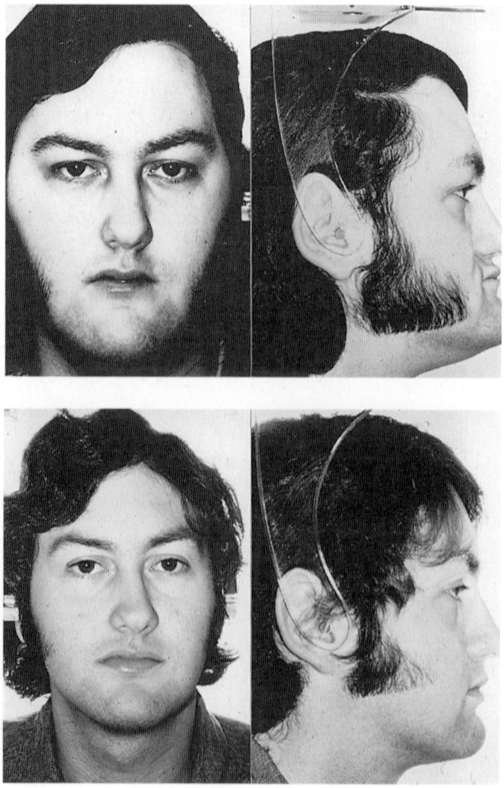

Figure 6-45 Maxillary-mandibular dysharmony. Top, Midface deficiency, short upper lip, and mandibular prognathism. Bottom, Results of a high Le Fort I osteotomy, the maxilla being repositioned anteriorly and inferiorly. The mandible was set back with sagittal ramus osteotomies. Note the lengthening of the lower face, improved lip line, and a proportional mandible. From West, 1978 (108).

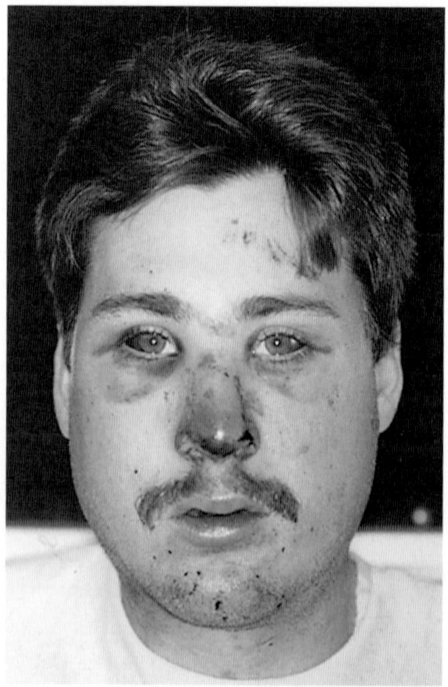

Figure 6-46 Automobile accident: 24 year old man. Le Fort III fracture. Nasal, periorbital, and conjunctival ecchymoses. Courtesy of A. Morrison, Halifax.

THE OUTER LIMITS

Conjoined Twins

Conjoined twins arise from a single zygote that develops into a partially divided inner cell mass during the second and possibly the early part of the third week of ovulation age. With a birth prevalence of 1/100,000 and a female predilection, more than half are stillborn. Cephalothoracopagus twins with partial fusion of the face and thorax represent a lethal condition. Craniopagus twins who are joined at their head tops have a birth prevalence of 1/2,000,000 (3).

Surgical separation of conjoined twins has technical, psychological, and ethical implications. Not all twins can be separated. If a decision is made to separate them, both may live,

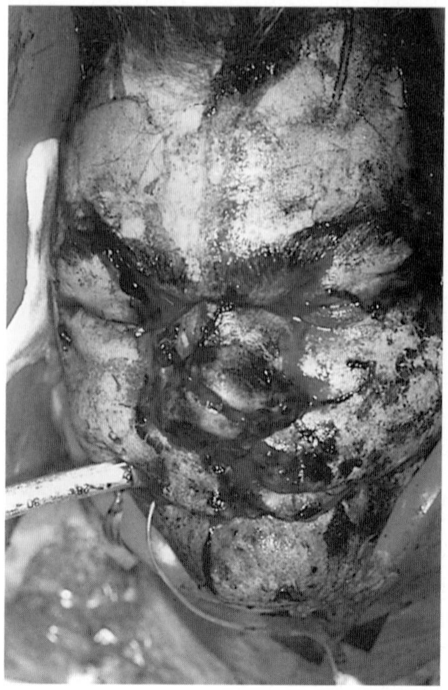

Figure 6-47 Head-on motor vehicle crash: 50 year old man. Extensive facial trauma with multiple fractures and soft tissue injuries, including partial nasal and lower lip avulsion. Courtesy of A. Morrison, Halifax.

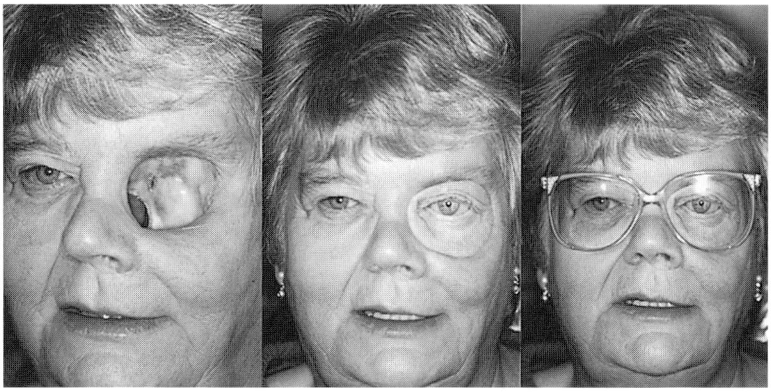

Figure 6-48 Left, Orbital exenteration for a tumor. Center, Orbital silicone prosthesis. Right, Prosthesis in place with eyeglasses. Courtesy of S. Richardson, Halifax.

one or both may be medically compromised, or one or both may die (3).

Lori and Reba Schappel, who have been documented on television, share 30% of their brain tissue and their nonfunctional left eye. They cannot be separated nor do they wish to be. Reba has spina bifida and lies on a stretcher that Lori pushes when they travel anywhere. Lori works with computers and Reba is developing a career as a country singer. Obviously, Lori has to take time off from work when Reba sings.

Craniopagus twins joined at their skull tops share some vascular structures including the sagittal sinus. Anesthesia must be given by two sets of equipment. Surgical complications include formation of thrombi in newly constructed blood vessels, intracranial bleeding, cardiac sequelae, and infections (3).

Facial Transplantation

Facial transplantation, a controversial and untried surgical possibility, involves harvesting recently deceased donor facial skin, including the underlying muscles of facial expression, and transplanting them to a recipient with a severely disfigured face from an automobile accident or fire, for example. The size and extent of donor tissue depend on the size of the recipient's facial defect. In some instances, the perioral and periorbital muscles may remain intact. After dissecting away the scarred and contracted tissues of the face, it has been proposed to transplant the donor skin/muscle complex to the underlying vascular bed. A decision must be made whether to innervate some or all of the donor musculature, allowing some intrinsic musculature to degenerate and resorb. It is estimated that once transplant innervation is complete, only 50% to 75% function can be expected (1a, 5a, 68a).

The proposed procedure is highly experimental and hazardous. The issues are technical, psychological, and ethical. The best success rate for large organ transplantation is about 84%. Thus, about one in five facial transplants could fail. In addition to the possibility of surgical failure and infection, a lifelong requirement for immunosuppression is a serious matter. Patients would also face the risk of chronic rejection, resulting in complete allograft failure. In renal transplant recipients, chronic rejection after 10 years may be as high as 50% (5, 5a).

Although some ethicists have expressed concern that the recipient's face may look like the donor's face, this is not likely because the allograft would be placed on a different facial skeleton. Careful patient selection and attention to serious psychological and ethical issues have been addressed by Osborne et al. (68a).

At this writing, facial transplantation has been rejected by the Royal College of Surgeons of England and by the French National Ethics Advisory Committee, but has been approved by the Cleveland Clinic's Institutional Review Board. The University of Louisville has been instrumental in researching facial transplantation and has suggested that the time has come to enter the clinical phase (68a).

In France, a woman severely disfigured by a dog bite received the world's first partial facial transplant (nose, lips, and chin) in late November, 2005. Although facial transplantation has been rejected by the French National Ethics Advisory Committee, apparently the door was left open for a partial procedure (3a).

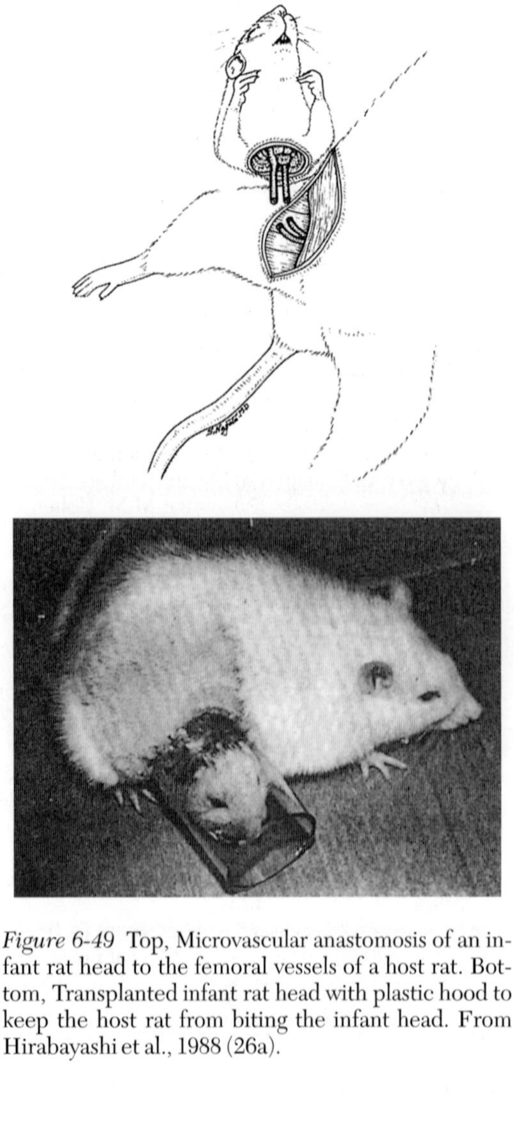

Figure 6-49 Top, Microvascular anastomosis of an infant rat head to the femoral vessels of a host rat. Bottom, Transplanted infant rat head with plastic hood to keep the host rat from biting the infant head. From Hirabayashi et al., 1988 (26a).

Heterotopic Rat Head Transplant

Heads of infant rats have been transplanted to the bodies of isohistogeneic adult rats by microvascular anastomoses. The descending aorta and the caudal vena cava of the donor were anastomosed to the femoral vessels of the host (26a) (Fig. 6-49). Because the transplanted heads have no moving soft tissues, growth of the craniofacial skeleton can be studied independently of soft tissue and muscular attachments (84a).

REFERENCES

1. Aduss H, Pruzansky S, Miller M: Interorbital distance in cleft lip and palate. Teratology 4:171–182, 1971.
1a. Barker JH: Personal communication, 2005.
2. Bell WH, Proffit WR, White RP Jr: Surgical Correction of Dentofacial Deformities. Philadelphia: W.B. Saunders, 1980.
3. Bergsma D, Editor: Conjoined twins. Birth Defects 3(1):1–147, 1967.
3a. Bianchi C: World's first partial face transplant done in France on a woman disfigured by dog bite. Associated Press, December 1, 2005.
4. Bruneteau RJ, Mulliken JB: Frontal plagiocephaly: Synostotic, compensational, or deformational. Plast Reconstr Surg 89:21–31, 1992.
5. Butler P: Division. Plast Reconstr Surg 115:36e–37e, 2005.
5a. Check E: Surgeons seek go-ahead to perform first face transplant. Nature 431:389, 2004.
6. Cohen MM Jr, Lemire RJ: Syndromes with cephaloceles. Teratology 25:161–172, 1982.
7. Cohen MM Jr, Richieri–Costa A, Guion–Almeida ML, Saavedra D: Hypertelorism: Interorbital growth, measurements, and pathogenetic considerations. Int J Oral Maxillofac Surg 24: 387–395, 1995.
8. Cohen MM Jr, Sedano HO, Gorlin RJ, Jirásek JE: Frontonasal dysplasia (median cleft face syndrome): Comments on etiology and pathogenesis. Birth Defects 7(7):117–119, 1971.
9. Cohen MM Jr, Walker GF, Phillips C: A morphometric analysis of the craniofacial configuration in achondroplasia. J Craniofac Biol Dev Biol Suppl 1:139–165, 1985.
10. Converse JM, McCarthy JG, Wood–Smith D, Editors: Symposium on Diagnosis and Treatment of Craniofacial Anomalies. St. Louis: Mosby, 1979.
11. Converse JM, Ransohoff J, Mathews ES, Smith B, Molenaar A: Ocular hypertelorism and pseudohypertelorism: Advances in surgical treatment. Plast Reconstr Surg 45:1–13, 1970.
12. Costaras M, Pruzansky S, Broadbent BH Jr: Bony interorbital distance (BIOD), head size, and level of the cribriform plate relative to orbital height. I. Normal standards for age and sex. J Craniofac Genet Dev Biol 2:5–18, 1982.
12a. Curtis TA, Zlotolow IM: Sexuality and head and neck cancer. Front Radiat Ther Oncol 14:26–33, 1980.
13. Deranian HM: With a passion for humanity: The story of Dr. Varaztad H. Kazanjian. Harvard Dent Alumni Bull 38:42–53, 1978.
14. Edgerton MT, Jane JA, Berry FA, Marshall KA: New surgical concepts resulting from cranio-orbito-facial surgery. Ann Surg 182:228–238, 1975.

15. El Deeb M, Waite DE, Curran J: Oral and maxillofacial surgery and management of cleft lip and palate. In Cleft Palate: Interdisciplinary Issues and Treatment. K.T. Mollers and C.D. Starr, Editors. Austin: Pro-Ed. Chapter 4, pp. 79–120, 1993.
16. Epker BN, Wolford LM: Dentofacial Deformities: Surgical–Orthodontic Correction. St. Louis: C.V. Mosby, 1980.
17. Farkas LG: Anthropometry of the Head and Face in Medicine. New York: Elsevier, 1981.
18. Farkas LG, Munro IR, Editors: Anthropometric Facial Proportions in Medicine. Springfield: C.C. Thomas, 1987.
19. Feingold M, Bossert WH: Normal values for selected physical parameters: An aid to syndrome delineation. Birth Defects 10(13):1–15, 1974.
20. Figalová P, Hajnis K, Smahel Z: The interocular distance in children with cleft before the operation. Acta Chir Plast 16:65–77, 1974.
21. Freihofer HPM: Inner intercanthal and interorbital distances. J Max-Fac Surg 8:324–328, 1980.
22. Grabb WC, Rosenstein SW, Bzoch KR, Editors: Cleft Lip and Palate. Boston: Little, Brown, 1971.
23. Guion–Almeida M-L, Richieri–Costa A, Saavedra D, Cohen MM Jr: Frontonasal dysplasia: Analysis of 21 cases and literature review. Int J Oral Maxillofac Surg 25:91–97, 1996.
24. Günther H: Konstitutionelle Anomalien des Augenabstandes und der Interorbitalbreite. Virchows Arch Pathol Anat Histopathol 290:373–384, 1933.
25. Hansman C: Growth of interorbital distance and skull thickness as observed in roentgenographic measurements. Radiology 86:87–96, 1966.
26. Hemmy DC, Zonneveld FW, Lobregt S, Fukuta K: A decade of clinical three-dimensional imaging: A review. Part I. Historical development. Invest Radiol 29:489–496, 1994.
26a. Hirabayashi S. Harii K, Sakurai A, Takaki EK, Fukada O: An experimental study of craniofacial growth in a heterotopic rat head transplant. Plast Reconstr Surg 82:236–243, 1988.
27. Hohl TH, Wolford LM, Epker BN, Fonseca RJ: Craniofacial osteotomies: A photocephalometric technique for the prediction and evaluation of tissue changes. Angle Orthodont 48:114–125, 1978.
28. Höhler H: Changes in facial expression as a result of plastic surgery in mongoloid children. Aesthetic Plast Surg 1:245–250, 1977.
29. Jane JA, Persing JA: Neurosurgical treatment of craniosynostosis. In Craniosynostosis: Diagnosis, Evaluation, and Management. Second Edition. M.M. Cohen, Jr. and R.E. MacLean, Editors. New York: Oxford University Press. Chapter 18, pp. 209–227, 2000.
30. Kaban LB: Biomedical technology revolution: Opportunities and challenges for oral and maxillofacial surgeons. Int J Oral Maxillofac Surg 31: 1–12, 2002.
31. Kaye CI, Rollnick BR: Samuel Pruzansky and the center for craniofacial anomalies. In Craniofacial Dysmorphology: Studies in Honor of Samuel Pruzansky. M.M. Cohen, Jr. and B.R. Rollnick, Editors. New York: A.R. Liss. pp. 25–30, 1985.
32. Kitlowski EA: Congenital anomaly of the face. Inclusion cyst arising from mucous membrane of pharynx with deformity of nose and cleft lip and palate. Plast Reconstr Surg 23:64–68, 1959.
33. Kiyak HA, Hohl T, Sherrick P, West RA, McNeill RW, Bucher F: Sex differences in motives for and outcomes of orthognathic surgery. J Oral Surg 39:757–764, 1981.
34. Kreiborg S: The application of roentgencephalometry. In Craniofacial Dysmorphology: Studies in Honor of Samuel Pruzansky. M.M. Cohen, Jr. and B.R. Rollnick, Editors. New York: A.R. Liss. pp. 31–41, 1985.
35. Kreiborg S: 3D-Lab, University of Copenhagen.
36. Le Fort R: Experimental study of fractures of the upper jaw. Parts I and II. Plast Reconstr Surg 50:497–506, 1972.
37. Le Fort R: Experimental study of fractures of the upper jaw. Part III. Plast Reconstr Surg 50: 600–605, 1972.
38. Lucker GW: Esthetics and a quantitative analysis of facial appearance. In Psychological Aspects of Facial Form. G.W. Lucker, K.A. Ribbens, and J.A. McNamera, Editors. Ann Arbor: Center for Human Growth and Development. pp. 49–79, 1981.
39. Lucker GW, Ribbens KA, McNamera JA Jr, Editors: Psychological Aspects of Facial Form. Ann Arbor: Center for Human Growth and Development, 1981.
40. MacGregor FC: After Plastic Surgery. New York: Praeger, 1979.
41. Mann I: Developmental Abnormalities of the Eye. London: Cambridge University Press, 1970.
42. Marchac D, Renier D: Craniofacial Surgery for Craniosynostosis. Boston: Little, Brown, 1982.
43. Marsh JL: Surgical research on craniosynostosis. In Craniosynostosis: Diagnosis, Evaluation, and Management. Second Edition. M.M. Cohen, Jr. and R.E. MacLean, Editors. New York: Oxford University Press. Chapter 21, pp. 292–306, 2000.
44. Marsh JL, Vannier MW: Surface imaging from computerized tomographic scans. Surgery 94: 159–165, 1983.
45. Marsh JL, Vannier MW: Three-dimensional surface imaging from CT scans for the study of craniofacial dysmorphology. J Craniofac Genet Dev Biol 9:61–75, 1989.
46. McCarthy JG, Editor: Cleft Lip & Palate and

Craniofacial Anomalies. Plastic Surgery, Volume 4. Philadelphia: W.B. Saunders, 1990.

47. McCarthy JG: The concept of a craniofacial anomalies center. In Symposium on Diagnosis and Treatment of Craniofacial Anomalies, J.M. Converse, J.G. McCarthy, and D. Woodsmith, Editors. St. Louis: C.V. Misty, Chapter 1, pp. 3–10, 1979.
48. McFee MF, Pruzansky S, Corrales MM, Valvassori GE, Dobben DG, Capek V: Computed tomography in the evaluation of the orbit and the bony interorbital distance (B.I.O.D.). Am J Neuroradiol 7:265–269, 1986.
49. Micheli–Pellegrini V, Manfrida GM: Rhinoplasty and its psychological implications: Applied psychology observations in aesthetic surgery. Aesthetic Plast Surg 3:299–319, 1979.
50. Moss ML: Hypertelorism and cleft palate deformity. Acta Anat 61:547–557, 1965.
51. Mouradian WE: Who decides? Patients, parents, or gatekeepers: Pediatric discussions in the craniofacial center. Cleft Palate-Craniofac J 32:510–514, 1995.
52. Mulliken JB: Bilateral complete cleft lip and nasal deformity: An anthropometric analysis of staged to synchronous repair. Plast Reconstr Surg 96:9–23, 1995.
53. Mulliken JB: Primary repair of bilateral cleft lip and nasal deformity. Plast Reconstr Surg 108: 181–194, 2001.

53a. Mulliken JB: Bilateral cleft lip. Clin Plast Surg 31:209–220, 2004.

53b. Mulliken JB: The changing faces of children with cleft lip and palate. New Engl J Med 351: 745–747, 2004.

54. Mulliken JB, Bruneteau RJ: Surgical correction of the craniofacial anomalies in Apert syndrome. Clin Plast Surg 18:277–289, 1991.
55. Mulliken JB, Godwin SL, Pracharktam N, Altobelli DE: The concept of the sagittal orbital-globe relationship in craniofacial surgery. Plast Reconstr Surg 97:700–706, 1996.
56. Mulliken JB, Martinez–Peréz D: The principle of rotation-advancement for repair of unilateral cleft lip and nasal deformity. Technical variations and analysis of results. Plast Reconstr Surg 104: 1247–1260, 1999.
57. Mulliken JB, Vander Woude DL, Hansen M, Goumnerova L, Scott RM: Analysis of posterior plagiocephaly: Deformational vs. synostotic. Plast Reconstr Surg 103:371–380, 1999.

57a. Mulliken JB, Wu JK, Padwa BL: Repair of bilateral cleft lip: Review, revisions, and reflections. J Craniofac Surg 14:609–620, 2003.

58. Murray JE, Mulliken JB, Kaban LB, Belfer M: Twenty year experience in maxillocraniofacial surgery. An evaluation of early surgery on growth, function and body image. Ann Surg 190: 320–331, 1979.
59. Murray JE, Swanson LT: Midface osteotomy and advancement for craniosynostosis. Plast Reconstr Surg 41:299–307, 1968.
60. Murray JE, Swanson LT, Cohen M, Habul MB: Correction of midfacial deformities. Surg Clin North Am 51:341–352, 1971.
61. Murray JE, Swanson LT, Strand RD, Hricko GM: Evaluation of craniofacial surgery in the treatment of facial deformities. Ann Surg 182: 240–265, 1975.
62. Murray JE, Swanson LT, Strand RD, Hricko GM: Reconstructive surgery for major craniofacial deformities. Surg Clin North Am 56:495–551, 1976.
63. Neumann K, Moegelin A, Temminghoff N, Radlanski RJ, Langford A, Unger M, Langer R, Bier J: 3D-computed tomography: A new method for the evaluation of fetal cranial morphology. J Craniofac Genet Dev Biol 17:9–22, 1997.
64. Obwegeser H: The surgical correction of mandibular prognathism and retrognathia with consideration of genioplasty. Oral Surg 10:677–689, 1957.
65. Obwegeser H: Indications for surgical correction of mandibular deformity by sagittal splitting technique. Br J Oral Surg 1:157–168, 1964.
66. Obwegeser HL: Surgical correction of small or retrodisplaced maxillae: The dish-faced deformity. Plast Reconstr Surg 43:351–365, 1969.
67. Obwegeser HL: Correction of the skeletal anomalies of oto-mandibular dysostoses. J Maxillofac Surg 2:73–92, 1974.
68. Obwegeser HL, Weber G, Freihofer HP, Sailer HF: Facial duplication. The unique case of Antonio. J Max-Fac Surg 179–198, 1978.

68a. Osborne W, Barker J, Martinez S, Vossan M, Maldonado C, Grossi F, François C, Cunningham M, Perez–Abadia G, Kon M, Banis J: On the ethics of face transplantation research. Am J Bioethics 4:1–12, 2004.

69. Persing JA, Jane JA, Editors: Craniofacial Disorders. Neurosurg Clin North Am. 2:507–711, 1991.
70. Pertschuk MJ: Reconstructive surgery: Objective change of objective deformity. In Body Images: Development, Deviance, and Change. T.F. Cash and T. Pruzinsky, Editors. New York: Guilford Press. Chapter 11, pp. 237–252, 1990.
71. Posnick JC: The role of plate and screw fixation in the treatment of craniofacial malformations. In Rigid Fixation of the Craniomaxillofacial Skeleton. M.J. Yaremchuk, J.S. Gruss, and P.N. Manson, Editors. Stoneham: Butterworth. pp. 512–516, 1992.
72. Posnick JC: Craniofacial and Maxillofacial Surgery in Children and Young Adults. Philadelphia: W.B. Saunders, 2000.
73. Posnick JC: The staging of cleft lip and palate reconstruction: Infancy through adolescence. In Craniofacial and Maxillofacial Surgery in Chil-

dren and Young Adults. Philadelphia: W.B. Saunders, Chapter 32, pp. 785–826, 2000.
74. Posnick JC: Craniosynostosis and the craniofacial dysostosis syndromes: Current surgical management. In Craniosynostosis: Diagnosis, Evaluation, and Management. M.M. Cohen, Jr. and R.E. MacLean, Editors. New York: Oxford University Press. Chapter 20, pp. 269–291, 2000.
75. Powell N, Humphreys B: Proportions of the Aesthetic Face. New York: Thieme-Stratton, 1984.
76. Precious DS, Editor: Cleft Lip and Palate. Oral Maxillofac Surg Clin North Am. 12: 361–522, 2000.
77. Proffit WR, White RP, Sarver DM: Contemporary Treatment of Dentofacial Deformity. St. Louis: C.V. Mosby, 2003.
78. Pruzansky S: Roentgencephalometry of infants: 1949–1973. Trans Third Int Orthodont Congr. London: Crosby Lockwood Staples. pp. 101–117, 1975.
79. Pruzinsky T, Edgerton MT: Body-image change in cosmetic plastic surgery. In Body Images: Development, Deviance, and Change. T.F. Cash and T. Pruzinsky, Editors. New York: Guilford Press. Chapter 10, pp. 217–236, 1990.
80. Richman LC, Eliason M: Psychological characteristics associated with cleft palate. In Cleft Palate: Interdisciplinary Issues and Treatment. Austin: Pro-Ed. Chapter 12, pp. 357–380, 1993.
81. Rougerie J, Tessier P: Craniostenosis: Craniosynostosis and craniofacial dysmorphia: A study of 335 cases with reference to neurosurgery and craniofacial surgery. Z Kinderchir 34:179–183, 1981.
82. Shons AR: Surgical issues and procedures. In Cleft Palate: Interdisciplinary Issues and Treatment. K.T. Moller and C.D. Starr, Editors. Austin: Pro-Ed. Chapter 3, pp. 49–78, 1993.
83. Smith DW, Cohen MM Jr: Widow's peak scalp-hair anomaly and its relation to ocular hypertelorism. Lancet ii:1127–1128, 1973.
84. Stoelinga PJW: Personal communication, 1990.
84a. Sugawara Y, Hirabayashi S, Harii K: Craniofacial growth in a whole rat head transplant: How does a non-functional head grow? J Craniofac Genet Dev Biol 19:102–108, 1999.
84b. Swennen G, Schliephake H, Dempf R, Schierle H, Malevez C: Craniofacial distraction osteogenesis: A review of the literature. Part 1: Clinical studies. Int J Oral Maxillofac Surg 30:89–103, 2001.
85. Tessier P: Colobomas: Vertical and oblique complete facial clefts. Panminerva Med 11:95–101, 1969.
86. Tessier P: Relationship of craniostenoses to craniofacial dysostoses and to faciostenoses: A study with therapeutic implications. Plast Reconstr Surg 48:224–237, 1971.
87. Tessier P: The definitive plastic surgical treatment of the severe facial deformities of craniofacial dysostosis: Crouzon's and Apert's diseases. Plast Reconstr Surg 48:419–442, 1971.
88. Tessier P: Total osteotomy of the middle third of the face for faciostenosis or for sequelae of Le Fort III fractures. Plast Reconstr Surg 48:533–541, 1971.
89. Tessier P: Traitement des dysmorphies faciales propres aux dysostoses crânio-faciales (FCF). Maladies de Crouzon et d'Apert. Neuro-Chir 17: 295–322, 1971.
90. Tessier P: Biographical sketch of René Le Fort. Plast Reconstr Surg 50:606–607, 1972.
91. Tessier P: Commentary on Le Fort's papers. Plast Reconstr Surg 50:605–606, 1972.
92. Tessier P: Orbital hypertelorism I. Scand J Plast Reconstr Surg 6:135–155, 1972.
93. Tessier P: The conjunctival approach to the orbital floor and maxilla in congenital malformation and trauma. J Max-Fac Surg 1:3–8, 1973.
94. Tessier P: Anatomical classification of facial, craniofacial and laterofacial clefts. J Max-Fac Surg 4:69–92, 1976.
95. Tessier P: Recent improvements in treatment of facial and cranial deformities of Crouzon's disease and Apert's syndrome. Symposium on Plastic Surgery in the Orbital Region. P. Tessier, A. Callahan, J.C. Mustardé, and K.E. Salyer, Editors. St. Louis: C.V. Mosby. pp. 271–290, 1976.
96. Tessier P: Réflexions sur la chirurgie crâniofaciale d'aujourd'hui et son avenir chez l'enfant. Ann Chir Plast 24:109–119, 1979.
97. Tessier P: Craniofacial surgery in syndromic craniosynostosis. In Craniosynostosis: Diagnosis, Evaluation, and Management. Second Edition. M.M. Cohen, Jr. and R.E. MacLean, Editors. New York: Oxford University Press. Chapter 19, pp. 228–268, 2000.
98. Tessier P, Guiot G, Derome P: Orbital hypertelorism II. Scand J Plast Reconstr Surg 7:39–58, 1973.
99. Tessier P, Guiot G, Rougerie J, Delbet JP, Pastoriza J: Cranio-naso-orbito-facial osteotomies. Hypertelorism. Ann Chir Plast 12:103–118, 1967.
100. Tessier P, Guiot G, Rougerie J, Delbet JP, Pastoriza P: Hypertelorism: Cranio-naso-orbito-facial and subethmoid osteotomy. Panminerva Med 11: 102–116, 1967.
101. Tessier P, Rougerie J, Hervouet F, Woillez M, Lekieffre M, Derome P: Plastic Surgery of the Orbit and Eyelids: Report of the French Society of Ophthalmology, 1977. Translated by S.A. Wolfe. New York: Masson Publishing USA, 1981.
102. Trimble LD, West RA, McNeill RW: Cleidocranial dysplasia: Comprehensive treatment of the dentofacial abnormalities. J Am Dent Assoc 105:661–666, 1982.

103. Vannier MW: Radiologic evaluation of craniosynostosis. In Craniosynostosis: Diagnosis, Evaluation, and Management. M.M. Cohen, Jr. and R.E. MacLean, Editors. New York: Oxford University Press. Chapter 12, pp. 147–157, 2000.
104. Vannier MW, Marsh JL: Three-dimensional imaging, surgical planning, and image-guided therapy. Radiol Clin North Am 34:545–563, 1996.
105. Vannier MW, Marsh JL, Warren JO: Three dimensional CT reconstruction images for craniofacial surgical planning and evaluation. Radiology 150:179–184, 1984.
106. Walker GF, Kowalski CJ: A two-dimensional coordinate model for the qualification, description, analysis prediction, and simulation of craniofacial growth. Growth 35:191–211, 1971.
107. Walker GF, Kowalski CJ: Computer morphometries in craniofacial biology. Comput Biol Med 2:235–249, 1972.
108. West RA: Personal communication, 1978.
109. West RA, McNeill RW: Maxillary alveolar hyperplasia, diagnosis and treatment planning. J Max-Fac Surg 3:239–250, 1975.
110. Wolfe SA: A rationale for the surgical treatment of exophthalmos and exorbitism. J Max-Fac Surg 5:249–257, 1977.
111. Wolfe SA: Afterword to Dr. Tessier's chapter: Craniofacial surgery in syndromic craniosynostosis. In Craniosynostosis: Diagnosis, Evaluation, and Management. Second Edition. M.M. Cohen, Jr. and R.E. MacLean, Editors. New York: Oxford University Press. Chapter 19, pp. 256–268, 2000.
112. Wolfe SA: Looking back, looking ahead: How we arrived at our current thinking about craniofacial and maxillofacial surgery. In Craniofacial and Maxillofacial Surgery in Children and Young Adults. By J.C. Posnick. Philadelphia: WB Saunders. Chapter 1, pp. 3–21, 2000.
113. Wolfe SA, Berkowitz S: Plastic Surgery of the Facial Skeleton. Little, Brown, 1989.
114. Wong GB, Burvin R, Mulliken JB: Resorbable internal splint: An adjunct to primary correction of unilateral cleft lip–nasal deformity. Plast Reconstr Surg 110:385–391, 2002.
115. Xia J, Wang D, Samman N, Yeung RWK, Tideman H: Computer-assisted three-dimensional surgical planning and simulation: 3D color facial model generation. Int J Oral Maxillofac Surg 29: 2–10, 2000.
116. Zonneveld FW: A decade of clinical three-dimensional imaging: A review. Part III. Image analysis and interaction, display options, and physical models. Invest Radiol 29:716–725, 1994.
117. Zonneveld FW, Fukuta K: A decade of clinical three-dimensional imaging: A review. Part II: Clinical applications. Invest Radiol 29:574–589, 1994.
118. Zonneveld FW, Lobregt S, van der Meulen JCH, Vaandrager JM: Three-dimensional imaging in craniofacial surgery. World J Surg 13:328–342, 1989.

7

Psychological Perspective

INFORMATION DERIVED FROM FACES

Table 7-1 summarizes various types of information that can be derived from faces. These include facial likeness, facial structure, attributed information, specific information, facial expression, and facial speech (19). Functional models for face recognition, which are not discussed here, have been proposed by several authors (19, 60, 83, 136).

FACE PROCESSING AND RECOGNITION

Face Processing and Species Specificity during Early Infancy

Infants of 4–6 months can discriminate phonetic differences in their own language as well as unfamiliar languages. By 10–12 months, infants discriminate phonetic differences only in their own language. A similar phenomenon has been shown for facial recognition. Infants of 6 months of age are equally good at recognizing the facial identity of both human and nonhuman primates. However, infants at 9 months and adults have a marked ability for recognizing only human faces (Fig. 7-1.)[a] What occurs during infancy for phonetic differences in languages and for differences in face processing suggests a perceptual narrowing in the neural networks involved in early cognition (122).

[a]For methodology about how these results were obtained, see Pascalis et al., 2002 (122).

Early Patterned Visual Deprivation and Face Processing

Visual experience during the first few months of life is essential for the development of refined facial processing. Infants who have been deprived of patterned visual input from birth until 2–6 months because of congenital cataracts are impaired at differentiating faces with variable spacing of features, but are normal at differentiating the shape of specific features. Computerized spatial changes in the position of the eyes and the mouth, and replacement of the eyes and the mouth are shown in Figure 7-2. Following surgical treatment, impairment of differentiating faces with variable spacing of features is still evident 9 years later (102).

Faces of Fear and Deception: Thermal Imaging

Imaging in a dark room with a thermal camera can be used to characterize facial thermal signals for specific activities. A sudden loud, frightening noise is accompanied by alterations in facial blood flow with instantaneous periorbital warming and cooling of the cheeks (105) (Fig. 7-3).

High-definition thermal imaging[b] permits quantification of rapid changes in facial bloodflow and can detect attempted deception. Vol-

[b]Accuracy of the technique is similar to that of a polygraph examination by an expert (123).

Table 7-1 Information That Can Be Derived From Faces

Types of Information	Meaning	Comments
Facial likeness	Photograph	Used in some types of research. Not important in everyday life except for passports and various types of identity cards for occupational or security purposes.
Facial structure	Arrangement of features in relation to one another	More abstract than a facial picture *per se*. Facial configuration permits the identity of celebrities, even from blurred pictures or caricatures.
Attributed information	Attributing information about a face from an unfamiliar photograph	Judgement of age and sex from the face of an unfamiliar person is reasonably accurate. Other characteristics, such as honesty, humor, or friendliness are commonly wild guesses. In most instances, the face reminds the viewer of someone that he or she knows.
Specific information	Giving true information about a known person or celebrity from a familiar photograph	Giving specific characteristics, such as name, occupation, and personality characteristics.
Facial expression	Interpretation of happiness, sadness, anger, disgust, etc.	Not involved in facial recognition. Used in some types of research.
Facial speech	Suggested words from movement of the lips and tongue in a picture or movie	Not involved in facial recognition. Used in some types of research.

Extensive modification of Bruce and Young, 1986 (19).

Figure 7-1 Example of paired different human faces and paired different monkey faces. From Pascalis et al., 2002 (122).

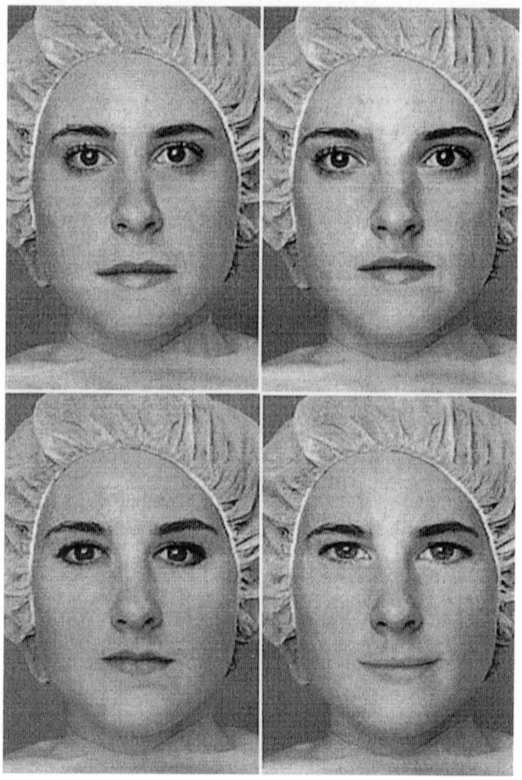

Figure 7-2 Computerized spatial changes in the position of the eyes and the mouth (top) and replacement of the eyes and the mouth (bottom). From Le Grand et al. (2001), (102).

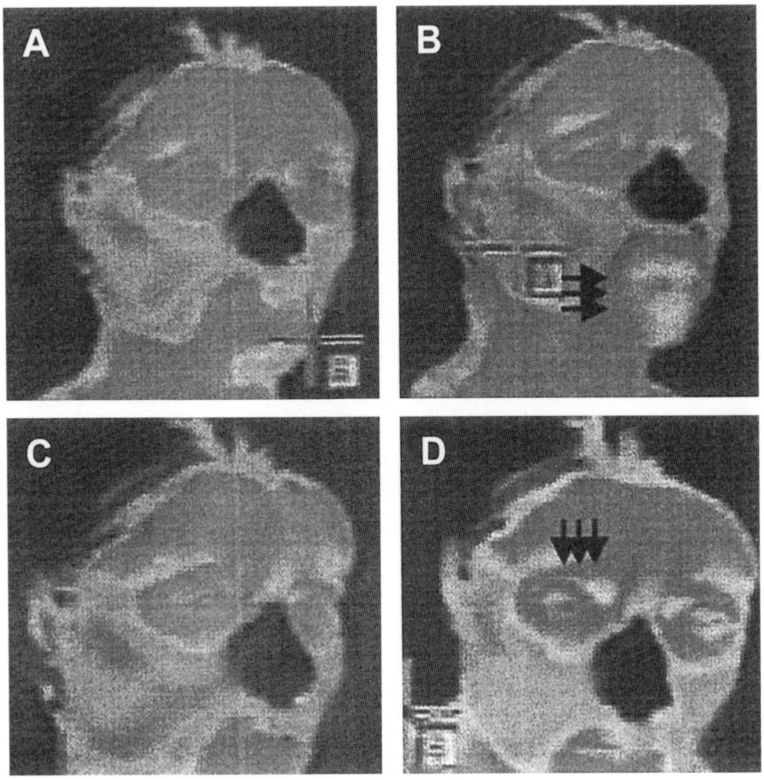

Figure 7-3 High-definition thermal imaging. Characteristic facial thermal signatures indicate alterations in blood flow. Chewing gum is accompanied by warming over the mandibular area (A) before and (B) after chewing; arrows indicate local warming. Instantaneous periorbital warming and cheek cooling following a sudden loud frightening noise (C) before and (D) after the startle; arrows indicate local warming. From Levine et al., 2001 (105).

unteers were asked to commit a mock crime[c] and then assert their innocence. About 83% of "guilty" individuals were correctly identified by thermal warming around the eyes (Fig. 7-4). About 90% of innocent control individuals were detected by lack of thermal warming (123).

Prosopagnosia

An area that governs facial recognition is found at the base of the brain (see Fig. 1-15). Any lesion that destroys this area impairs the ability to identify faces—a condition known as prosopagnosia (67). This is discussed further in Chapters 1 and 4.

[c]They were asked to stab a mannequin, rob it of $20.00, and then assert their innocence of the "crime" (123).

Global and Feature-Based Processing

Perception and recognition of faces result from both global and feature-based information, the former possibly serving as a front end for the latter (16, 18). Faces are more easily recognized than various objects (13). Strong evidence suggests that face recognition involves more global and configurational processing than object recognition.

Faces are normally viewed right side up, which facilitates distinguishing one face from another. When faces are upside down, distinguishing one from another is very difficult. A simple experiment demonstrates the problem of upside down faces. When a face is altered with the mouth in place of the nose and the nose in place of the mouth, the face appears grotesque, but if the face is inverted, it is much more difficult to detect the abnormality (8) (Fig. 7-5).

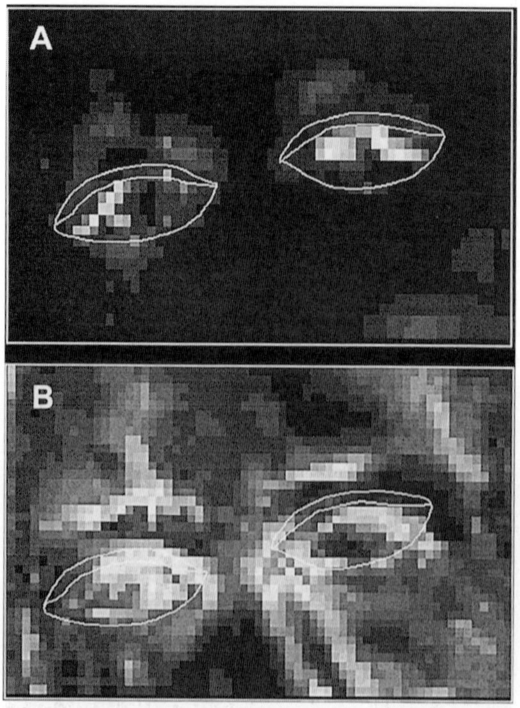

Figure 7-4 Periorbital, high resolution thermal imaging. Images (A) before and (B) after lying about committing a mock crime. White lines indicate eye contours. Note the increased periorbital blood flow that accompanies lying (B). From Pavlidis et al., 2002 (123).

Spatial Frequency Analysis and Facial Identification

In spatial frequency analysis, low-frequency components contribute to global processing, whereas high-frequency components contribute to feature-based processing (69, 79, 105b, 150). Block portraits can be generated from real portraits by reducing the information content by a computer. Selective frequency filtering influences the ease with which block portraits are recognized. Figure 7-6 shows four versions of Abraham Lincoln. The original block portrait is at the top left. Filtering out spatial frequencies greatly enhances facial recognition (top right). Partial filtering produces the two lower pictures (bottom left and right) (79). Viewing the block portraits (top left, bottom left and right) by squinting helps identify them as Abraham Lincoln.

The spatial resolution of the visual system changes with the distance from the center of gaze. Central vision is dominated by much higher spatial frequencies than peripheral vision. Only a few degrees from the center of gaze, vision is much blurrier. Thus, when viewing the Mona Lisa by Leonardo da Vinci (Fig. 7-7), her smile is much more apparent to peripheral vision than to central vision (Fig. 7-8). When staring directly at her mouth, her smile disappears. As Livingstone (105b) noted "Is it warm? Is it real? Or just a low spatial frequency?"

Facial Features and the Process of Recognition

Humans can recognize people they know by their facial features, although voice, hand gestures, body movement, gait, height, body build, and clothing also play a role. Using videos of

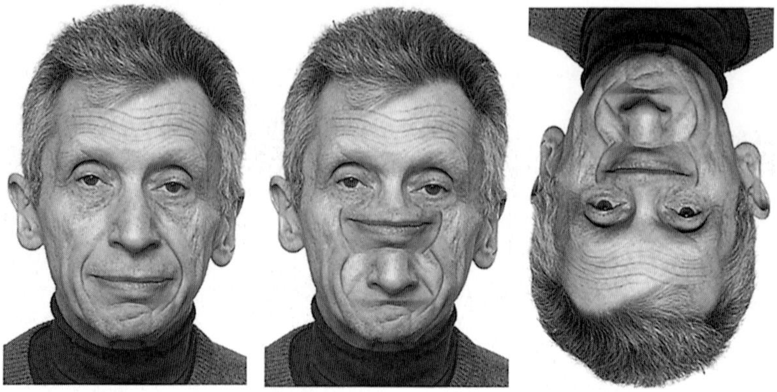

Figure 7-5 When a normal face (left) is altered with the mouth in place of the nose and the nose in place of the mouth, the facial appearance is grotesque (middle), but when the face is inverted, it is much more difficult to recognize the abnormality (right).

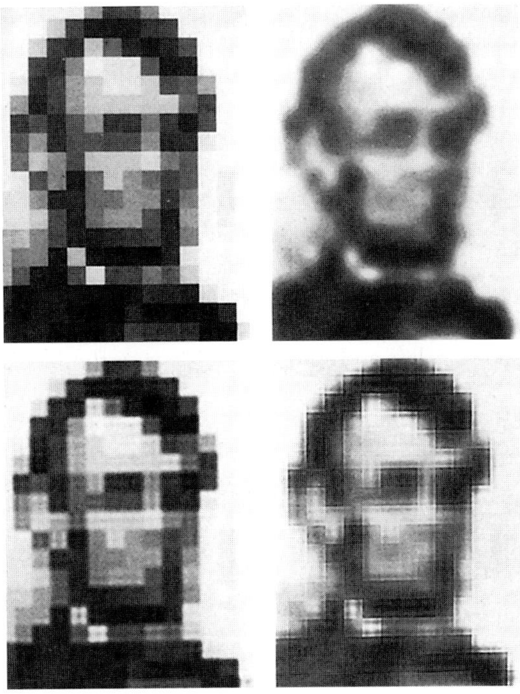

Figure 7-6 Selective frequency filtering influences the ease with which block portraits of Abraham Lincoln are recognized. From Harmon, 1973 (79). See text.

Figure 7-7 Mona Lisa, by Leonardo da Vinci. c. 1502. Oil on wood, 77 × 53 cm, Musée de Louvre, Paris.

sheep's faces and recordings of face-sensitive neurons, it has been shown that sheep can recognize other sheep by their facial features. They can even match the frontal view of a particular sheep to their profile view (86a). Thus, if one sheep knows another, he or she might think "ewe look familiar."

The facial outline, hair, eyes, and mouth are important in perceiving and remembering faces. The upper part of the face is more useful for facial recognition than the lower part. The nose in frontal view usually plays an insignificant role, although a distinctive nose, particularly in profile, may be important (16, 153).

Distinctive faces are more easily recognized than other faces. Attractive faces are more easily

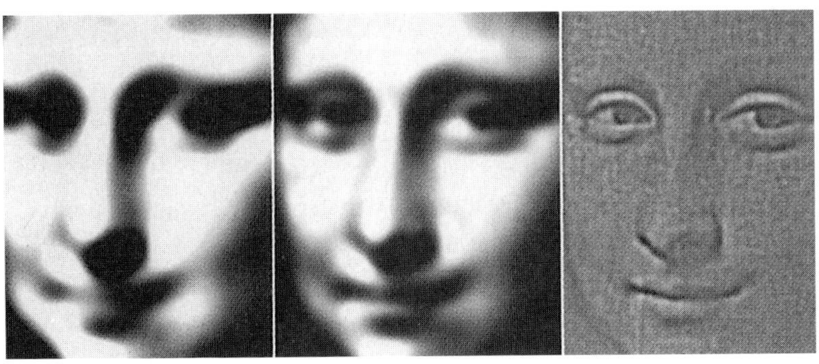

Figure 7-8 Face of Mona Lisa filtered to reveal very low spatial frequencies (left), low spatial frequencies (center), and high spatial frequencies (right). From Livingstone, 2000 (105b). See text.

recognized than the least attractive faces, but average faces are the most difficult to remember. Famous faces are easily remembered and are easier to identify in moving sequences than in still photographs (17, 93, 153).

Memory for faces is viewpoint dependent. Generalization from one profile to another is poor, whereas from one three-quarter view to another is good. In general, top lighting is better for facial recognition than bottom lighting, and recognition from facial negatives is difficult (86, 90).

AUTOMATED SYSTEMS OF FACE RECOGNITION AND ANIMATION

Computer Systems for Face Recognition

Machine recognition of faces can identify or verify one or more persons from a database of faces. Three-dimensional faces are transformed into two-dimensional images. Initial face detection may require extraction from a cluttered scene. Collateral information, including race, gender, age, facial expression, and speech may aid in narrowing the search. Table 7-2 lists some applications of facial recognition systems (178).

Automated machine recognition of faces involves three key steps: face detection, feature extraction, and face recognition (Fig. 7-9). Face detection and feature extraction can be achieved simultaneously. Depending on the type of application (Table 7-2), some of the subtasks can be very challenging, for example, dealing with the variability of cluttered backgrounds (178).

Computer systems for face recognition constitute an ongoing area of research. Face recognition techniques include FERET, FRVT, XM2VTS, and many commercial systems. These can be broadly classified into two groups: static images and videos (178).

A database of faces may have hundreds or thousands of faces from which an average face can be generated (Fig. 7-10). Because some facial characteristics are correlated with each other, it is possible to create mathematical variants known as *eigenfaces*[d] (Fig. 7-11) by

[d]*Eigenface*, from the German prefix meaning *own* or *individual*.

Table 7-2 Applications of Facial Recognition Systems

Categories	Applications
Entertainment	Video games
	Virtual reality
	Training programs
	Human–robot interaction
	Human–computer interaction
Smart cards	Drivers' licenses
	Entitlement programs
	Immigration
	National identification cards
	Passports
	Voter registration
	Welfare fraud
Information security	Parental control of television
	Personal device logon
	Desktop logon
	Application security
	Database security
	File encryption
	Internet security
	Internet access
	Medical records
	Secure trading terminals
Law enforcement and surveillance	Advanced video surveillance
	Closed-caption television control
	Portal control
	Post-event analysis
	Shoplifting
	Suspect tracking and investigation

From Zhao et al., 2003 (178).

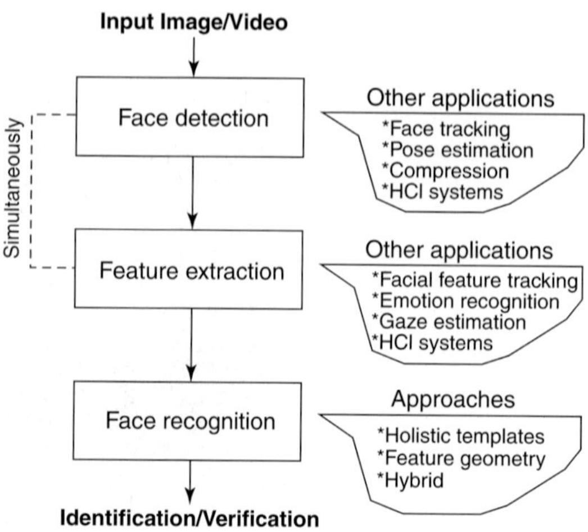

Figure 7-9 Automated machine recognition of faces involves three key steps: face detection, feature extraction, and face recognition. From Zhao et al., 2003 (178).

Figure 7-10 Average face generated from a database of faces. Courtesy of R. Chellappa and G. Aggarwal, Baltimore.

subtracting the characteristics of the average face from each face in the database (148, 152, 178).

The computer can analyze a target photograph fed into the system for identification (Fig. 7-12). The target face is analyzed by a subset of the strongest eigenfaces, which allows the computer to reconstruct and identify the target face (Fig. 7-13) (148, 152, 178).

Facial Modeling and Animation

Two general groups of facial modeling and animation are those based on geometric manipulation and those based on image manipulation. The former accounts for the shape and alterations unique to human facial expression. The latter accounts for the reflective properties of facial skin and hair to permit small-scale detail that is difficult to model by geometric manipulations alone. Both groups have several subcategories (Fig. 7-14), resulting in a diversity of approaches to facial animation. Exact classifications are complicated by overlapping boundaries between various methods, and more recent approaches frequently integrate several methods to produce better results (119).

Parametric models are often inspired by anatomic knowledge, such as facial alterations of the skin caused by facial musculature (171) (Fig. 7-15). Physics-based approaches simulate the influence of muscle contraction on the skin surface by approximating the biomechanical properties of the skin. By matching a three-dimensional model to one or more facial photographs, photo-realistic images can be produced. However, animation is still limited because each facial expression is captured in

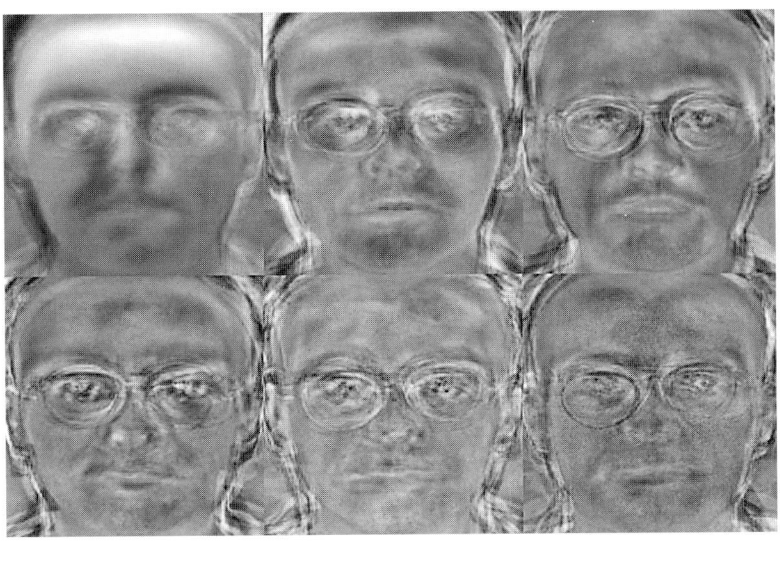

Figure 7-11 Several eigenfaces selected from a much larger series for illustration. Courtesy of R. Chellappa and G. Aggarwal, Baltimore.

Figure 7-12 Target photograph for identification. Courtesy of G. Aggarwal, Baltimore.

advance, and blending of expressions is achieved by linear blends, neglecting the dynamics of facial gestures. In performance-based methods, real faces are tracked to produce facial emotion on the model. Several facial animation systems have included speech synchronization. Movements of the eyes, tongue, and lower teeth on jaw opening have been attempted for realistic facial animation (77) (Fig. 7-16). Furthermore, a dynamic model for simulating facial wrinkling and skin aging has been proposed (174, 175). Finally, skin color changes or "vascular expressions" have been simulated in a few facial animation models (119).

The facial action coding system describes movements of the facial muscles, jaw, and tongue based on an analysis of facial anatomy (55). Table 7-3 lists some single facial action units and Table 7-4 lists examples of sets of action units for basic facial expressions. For example, combining the inner brow raisers, cheek raisers, lid corner pullers, and dimplers creates a happy expression. The facial action coding system and the action units developed by Ekman and Friesen (55) can be simulated by physical muscle modeling, which mathematically describes the properties and behavior of facial skin, muscle, and bone. They can also be simulated by pseudomuscle modeling, which mimics the dynamics of human tissues with heuristic geometric alterations (119).

A variety of other computer morphing techniques can be used, such as mixing faces from two or more individuals, mixing human and animal faces for fun, and growing a Pinocchio-like nose on an individual for fun.

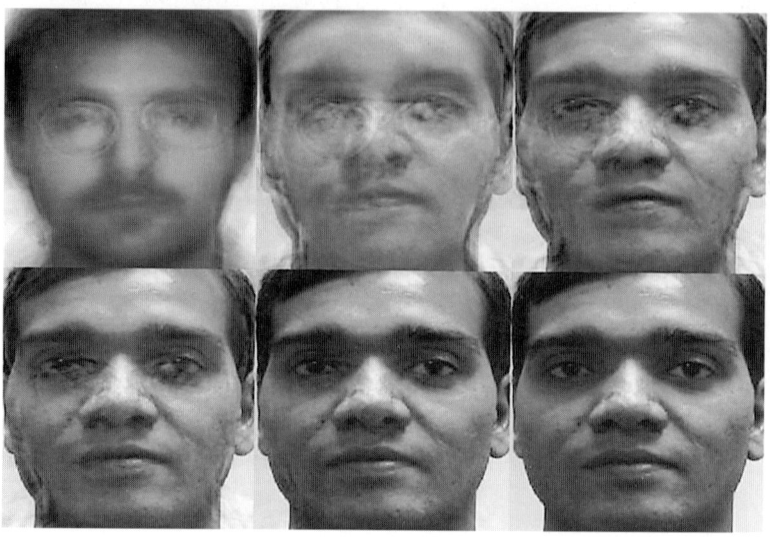

Figure 7-13 Target face shown in Figure 7-12 is analyzed by the strongest eigenfaces, which allow computer reconstruction and identification. Selected from a much larger series for illustration. Courtesy of G. Aggarwal, Baltimore.

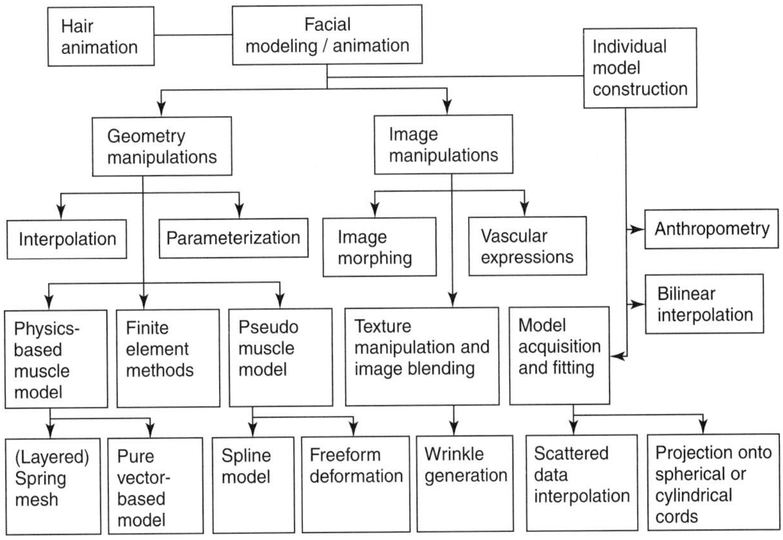

Figure 7-14 Facial modeling and animation. Two general groups of facial modeling include those based on geometric manipulation and those based on image manipulation. See text. From Noh, 2004 (119).

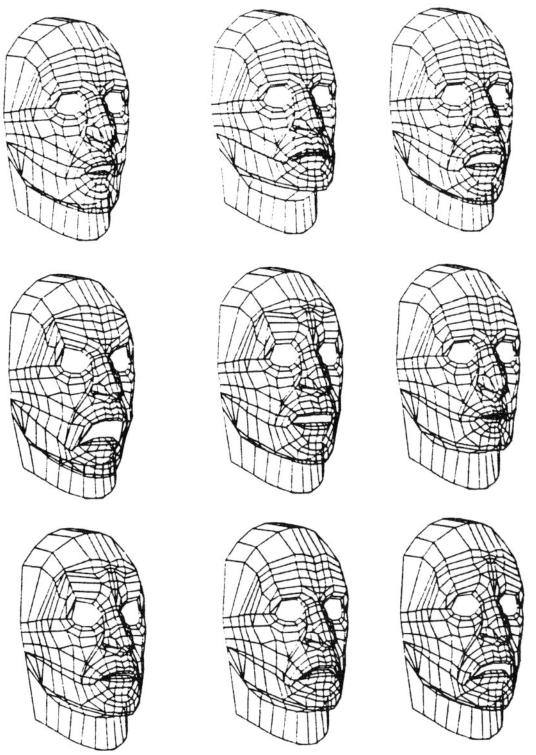

Figure 7-15 Muscle model for animating three-dimensional facial expressions. Nine different expressions are shown in both upper and lower parts of the face. From Waters, 1987 (171).

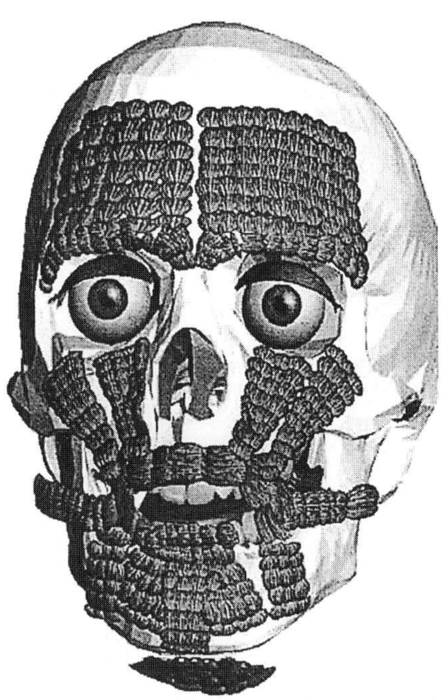

Figure 7-16 Facial modeling. Reference skull with eyes, teeth, tongue, and muscles. From Haber, 2001 (77).

Table 7-3 Facial Action Coding System: Sample of Single Facial Action Units

AU[a]	FACS Name[b]	AU[a]	FACS Name[b]
1	Inner Brow Raiser	12	Lid Corner Puller
2	Outer Brow Raiser	14	Dimpler
4	Brow Lower	15	Lip Corner Depressor
5	Upper Lid Raiser	16	Lower Lip Depressor
6	Cheek Raiser	17	Chin Raiser
7	Lid Tightener	20	Lip Stretcher
9	Nose Wrinkler	23	Lip Tightener
10	Upper Lid Raiser	26	Jaw Drop

[a]AU, action units.

[b]FACS, Facial Action Coding System.

From Ekman and Friesen, 1978 (55).

Table 7-4 Examples of Sets of Action Units for Basic Expressions

Basic Expressions	Involved Action Units (AU)[a]
Surprise	1, 2, 5, 15, 16, 20, 26
Fear	1, 2, 4, 5, 15, 20, 26
Disgust	2, 4, 9, 15, 17
Anger	2, 4, 7, 9, 10, 20, 26
Happiness	1, 6, 12, 14
Sadness	1, 4, 15, 23

[a]For meaning of numbers, see Table 7-3.

From Ekman and Friesen, 1978 (55).

Table 7-5 Factors Causing High Error Rates in Facial Recognition Software

Alterations in hair style
Hair dyeing, wigs, or hair transplantation
Beards, mustaches, or lack thereof
Simple disguises, including eyeglasses, sunglasses, open mouth, moustaches, and beards
Three-quarter facial view, profile view, or face lowering
Weight gain or weight loss
Aging (can be accentuated by alcohol, drugs, or excessive sun exposure)
Disease states
Plastic surgery, orthognathic surgery, or adult orthodontics
Facial trauma

For further discussion, see Taister et al. (2000) (161), Shenk (2003) (152), and the American Civil Liberties Union Websites (2).

Error Rates in Facial Recognition Software

Many factors cause high error rates (Table 7-5) with both false positives and false negatives (152, 161). The Federal Bureau of Investigation and the American Civil Liberties Union do not recommend this technology and find it ineffective.[e] Nevertheless, at this writing, officials at Logan Airport in Boston and T.E. Green Airport in Providence have announced that they will be installing facial recognition technology (2, 161). Surprisingly, there are three million surveillance cameras throughout Britain (162).

[e]Several government agencies have abandoned facial recognition systems. One example is the Immigration and Naturalization Service, which had experimented with this technology to identify people in cars at the Mexican–US border (2).

Fingerprint identification has been very effective in criminal investigations, particularly since the FBI began using a computerized network system. Iris-recognition technology leaves less room for error than fingerprint-recognition technology because iris patterns are more complex than fingerprint patterns. At Schiphol Airport in Amsterdam, frequent flyers can save time with a program that stores individual iris patterns on cards that they can swipe (152).

Projecting Facial Aging

Two main methods of modeling skin with wrinkle formation and aging are used: an image-based method and a model-based method (113). Nancy Burson, a New York artist who has explored the link between art and computer technology, conceived and helped develop with MIT

Figure 7-17 The Method and Apparatus for Producing an Image of a Person at a Different Age, a piece by Artist Nancy Burson (1976). Courtesy of Nancy Burson, New York, 2004 (21, 23). The image on the right projects how the person on the left will probably look in 25 or 30 years (121).

an *Age Machine*[f] that can alter facial images to reflect aging (24). Figure 7-17 shows a Nancy Burson piece titled *The Method and Apparatus for Producing an Image of a Person's Face at a Different Age* (21, 23). The image on the right projects how the person on the left will probably look in 25 or 30 years (121). The Age Machine has enabled the FBI and the National Center for Missing and Exploited Children to help locate victims of kidnapping. It would also be useful to law enforcement officials for locating fugitives who have been at large for an extended period of time.

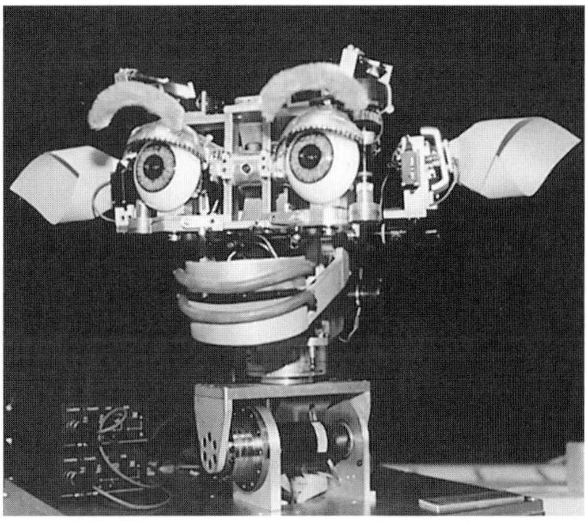

Figure 7-18 Kismet, an expressive robotic head capable of interacting face-to-face with humans. Courtesy of Donna Coveney/MIT.

Sociable Robots

Many robots in science fiction are well-known, such as Data from *Star Trek*, R2D2 from *Star Wars*, and Number 5 from *Short Circuit*. These have been an inspiration for designing real robots (Table 7-6), which have been called *Robo sapiens* (110). Many articles and books about robots and robotics have been published (15, 37a, 62c, 93a, 110). In addition, many robotic conferences have been held (20, 38, 164), including a large exhibition at EXPO 2005 in Japan (62a).

Robots can be used to assist in the social therapy of family high-functioning autistic children (39), to act as nursemaids—"nursebots"—to help the elderly (38), to design robotics for medical applications (37a, 93a), to be tour guides in museums, which has great appeal for children (20, 164), to develop a robotic doll capable of facial expressions (110), and to design a sociable, interacting robot like Kismet[g] (15) (Fig. 7-18).

Kismet is an expressive robotic head capable of interacting face-to-face with humans. The robot can interpret social cues from visual and auditory channels, and can respond with gaze

[f]Nancy Burson has also created a *Human Race Machine*, which permits viewers to see themselves with the characteristics of six different races, a *Couples Machine*, which morphs photographs of men and women with varying percentages or permits couples to view what their offspring might look like or envision different percentages of male and female in one face, and an *Anomaly Machine*, in which participants can view themselves with simulated facial anomalies (22).

[g]Kismet is a Turkish word that means *destiny* or *fate* (15).

Table 7-6 Science Fiction and Real Robots

Science Fiction Robots	Comments
Lt. Commander Data	Android from *Star Trek: The Next Generation*
The Borg	Organic beings with cybernetically enhanced implants. From *Star Trek: The Next Generation*
R2D2, C3PO	Robots from *Star Wars*
Number 5	From the movie *Short Circuit*. Military robot Number 5 gets zapped by lightning and becomes alive and sociable. Three memorable lines are as follows. "Number 5 is alive." When a young woman named Stephanie kisses Number 5 on his face, he exclaims "Right on the sensor!" Finally, when trying to ward off one of several military robots trying to capture him, Number 5 yells "Hey, laser lips. Your mama was a snow blower!"
Do Androids Dream of Electric Sheep?	Title from a science fiction book by Philip K. Dick. Movie titled *Blade Runner* is based on the book.

Real Robots	Comments
P3	Honda's bipedal walking robot. P3 can walk to a door, open it, go through, close the door, and then climb a flight of stairs.
ASIMO	New version of Honda P3. Besides doing everything that P3 can do, ASIMO can run twice as fast, push a cart, serve tea, wave, avoid obstacles, and carry on simple conversations.
Qrio robot	Sony is developing a small humanoid robot with a wireless connection to a grid of 300 high-powered computers, giving it the ability to "think."
Nursebot	Robot that can act as a nursemaid for the elderly
da Vinci Robotic Surgical System	Robotic small surgical arms that can perform minimally invasive laparoscopic surgical procedures. Controlled by a surgeon sitting at a console
iRobot's Bit	Prototype robotic doll with various facial expressions
Sweet Lips	Museum tour guide robot
Waseda-Okino Jaw No. 1	Human skull robotically controlled to study mastication
Human-like robotic head	Developed at Hara-Kobayashi Laboratory, Science University of Tokyo
Kismet	Sociable robotic head developed at MIT

Based on the TV series *Star Trek: The Next Generation* and the movies *Star Wars* and *Short Circuit*. Also based on the writings of Dick, 1990 (44), Menzel and D'Aluisio, 2000 (110), Breazeal, 2002 (15), Lewis, 2004 (105a), and Associated Press, 2005 (7a).

direction, posture, facial expressions, and vocalizations. Kismet is run by 15 networked computers, including 9 for vision, 1 for incoming speech signals, 1 for voice synthesizing, and 4 for emotional processing and behavior. Twenty-one motors in the head provide 3 degrees of freedom for the neck, 3 for the eyes, 4 for the lips, 2 for the ears, 2 for the eyebrows, and 1 for the eyelids (15) (Fig. 7-18).

FACIAL EXPRESSION

Nonhuman Primate Expressions, Human Homologies, and Phylogenetic Origins

Facial expressions of nonhuman primates involve movements of the jaws, lips, tongue, eyes, eyelids, eyebrows, ears, and upper head skin. Because so many components are coordinated, facial expressions are compound in nature. Facial changes are often associated with vocalizations and movements of the body or body parts. Macaques recognize the correspondence between the auditory and visual components of their own calls. Some nonhuman primate facial expressions are homologous to human expressions and have the same phylogenetic origin. However, categories of emotion that correspond to human facial expressions may cross-cut the categories found in nonhuman primates (68, 131, 166).

Primate facial expressions involve behavior such as attacking, fleeing, approaching to associate with others, or interacting socially in various ways. The main types of compound facial expressions in macaques are listed in Table 7-7 and illustrated in Figure 7-19.

Although yelling and screaming can be associated with human anger, no studies of human vocalization that accompany this facial expression have been carried out to date. However, yelling and screaming seem to correspond to the scream faces of all nonhuman primates. Compressing the lips, sometimes associated with the human anger face, resembles the compressed-lips face of chimpanzees when charging and bluffing. It is assumed that the compressed-lips face of chimpanzees and humans has its precursor in the tense-mouth face found in several species of macaques and baboons (131).

Table 7-7 Main Compound Facial Expressions in Macaques

Compound Facial Expressions	Meanings
Tense-mouth face	Tendency to attack
Open-mouth face	Tendency to attack or sometimes to flee
Staring bared-teeth scream face	Tendency to attack or sometimes to flee
Frowning bared-teeth scream face	Strong tendency to flee
Silent bared-teeth face	Weak tendency to flee
Lip-smacking face	Tendency to approach or weak tendency to flee
Protruding-lips face	Tendency to approach and often to mate
Pout face	Baby expression indicating a "wish" to be with the mother
Relaxed open-mouth face	Occurs during social play or leads to social play

Adapted from van Hooff, 1969 (166).

Comparative evidence suggests that human laughing and smiling are derived from two independent facial expressions—the play face and the bared-teeth face—that converged at a later time. Furthermore, comparison with nonhuman primates suggests that both the human smile and the human fear expression are derived from the silent bared-teeth expression found in most species of Old World monkeys and apes and in several species of New World monkeys. Species with dominance-oriented societies tend to separate laughing and smiling, whereas species with rather egalitarian societies tend to use laughter and smiling interchangeably (131).

Crying is uniquely human and is not found in any nonhuman primate. However, chimpanzees can vocalize in a manner reminiscent of weeping. This is associated with a stretched pout–whimper, in which the teeth are partially bared with laterally retracted corners of the lips, slight opening of the jaws, and puckering, particularly of the upper lip. This expression is not seen in humans and no precursors are found in monkeys. Sadness in monkeys is expressed by slow movements and weak muscle tone, not by facial expression (131).

The open-mouth threat face of Old World monkeys does not correspond to any human expression, and the human expression of disgust has no corresponding expression in nonhuman primates (131).

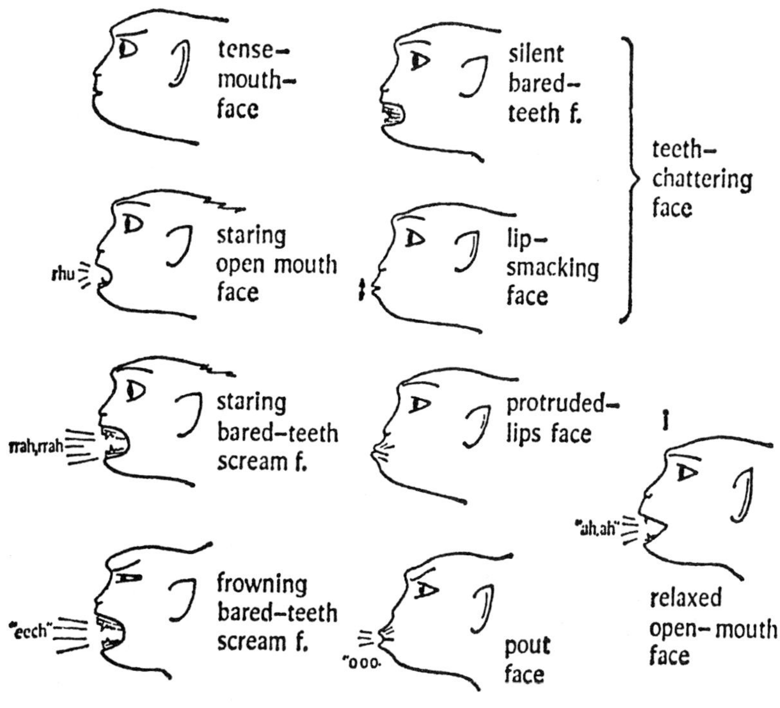

Figure 7-19 Main types of compound facial expressions of macaques. From van Hooff, 1969 (166).

Human Facial Expression

The face has three general indicators: static ones, such as the shape and location of facial features; slow alterations, such as forehead wrinkles and crow's feet; and rapid changes, such as facial expressions. All of these indicators may be modified. For example, the shape of various facial features may be modified by plastic surgery. Forehead wrinkles can be masked by women who have "bangs." Crow's feet can be masked by women who have long hair that covers the sides of the face. Cosmetics and sunglasses may also be used. Facial expressions can be masked by inhibiting or exaggerating a particular expression (54). In human communication, facial expressions work in concert with vocalization, gaze, hand gestures, and body movements (5, 88, 120).

Ekman and Friesen (54) have distinguished six basic facial expressions: happiness, surprise, anger, fear, disgust, and sadness. Blends may occur, such as surprise and happiness, surprise and anger, surprise and fear, and surprise and disgust. Expressions may be spontaneous or voluntary. Fear might be combined with a slight smile to indicate that an individual can "grin and bear it." A child might be taught never to show anger to his or her father and never to show sadness when disappointed. Thus, facial expressions may be modified or controlled on an individual basis, by family upbringing, by social convention, or by cultural rules (54).

Ekman and his coworkers (49, 51, 54) found cross-cultural support for the six basic facial expressions. They proposed a neurocultural theory to account for pancultural expressions and cultural alterations. Three groups of cultural factors interact with the basic expressions: antecedents, modifications, and consequences (Fig. 7-20).

Among the antecedents (Fig. 7-20), some may be universal, such as a sudden loud noise that might produce surprise or fear (see section in this chapter titled Faces of Fear and Deception: Thermal Imaging). However, antecedents occur through learning, which differs in various societies. For example, one society might show upturned lips, nasolabial folds, and closed eyelids at funerals, whereas another society might show downturned lips, stretched lower lip, partially closed eyelids, and dilated nostrils (54, 81).

Modifications under voluntary control are sociocultural and learned early in life. These include intensification, de-intensification, neutralization, and masking (Fig. 7-20). If modest happiness is felt on receiving a gift, it may be interpreted as great happiness. Winning a race may be de-intensified by downplaying great happiness to a modest smile. If fear is felt, it may be neutralized to a "poker face" with no evidence of a fearful face. Smiling may mask anger or fear. It is probably the most common mask because it is easy to do and it counters negative expressions (54, 81).

Consequences are of several types (Fig. 7-20). For facial expression, one briefly felt emotion may give rise to a second felt emotion that supersedes the first emotion or blends with it. One example is a person who is angry and then becomes disgusted by his or her own anger. Other consequences include motor adaptive patterns, such as flight because of fear or attack because of anger. Cultural variations may include various verbal and vocal utterances and physiological responses (54, 81).

Facial expressions are variable in their degree of intensity (54). Ekman and his coworkers (51, 55) have used the term "family of expressions" to characterize this variability. In the anger family of expressions, 60 have been identified and these can be distinguished from the family of

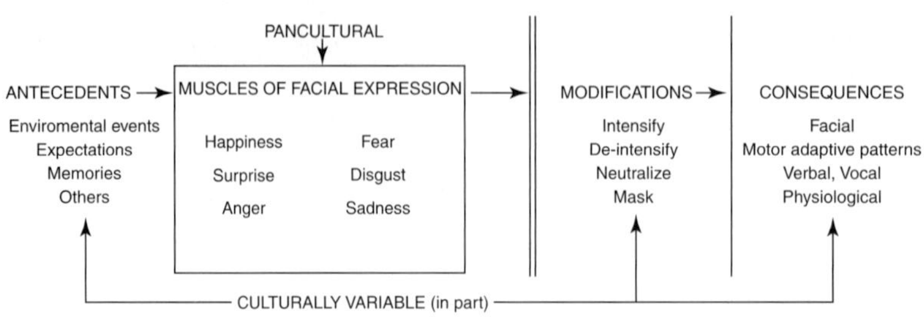

Figure 7-20 Diagrammatic representation of the neurocultural theory of Ekman and his coworkers. See text. Modified after Harrison, 1977 (81).

fear expressions. Although emotions may possibly have distinctive vocal expressions, emotions do not have specific facial expressions. Several emotions may share a particular form of smiling, such as amusement, relief, pride, sensory pleasure, and exhilaration (50, 51).

Although emotions generate facial expressions, surprisingly, the opposite is true also. Levenson et al. (104) studied heart rate, skin conductance, finger temperature, and forearm flexor muscle tension in subjects coached to make facial expressions. Emotion-specific autonomic nervous system activity was found. Autonomic distinctions were found between negative and positive emotions and also among negative emotions.

For testing purposes, facial expressions may be shown in several ways, such as posed photographs, movies, videos, and computer imaging (9). Ekman and his coworkers (55, 57) have provided an objective, anatomically-based method for measuring facial movement known as the *The Facial Action Coding System* (Table 7-3).

Analysis of videos in slow motion can document rapid micro-expressions lasting a fraction of a second that may be informative. For example, a fleeting forehead wrinkle from a film clip might indicate when someone is lying about a subject (47, 48, 53, 59). True smiles can be distinguished from masked smiles that "leak through" to show negative emotions such as a hint of disgust by raising the upper lip slightly or a hint of sadness by lowering the corners of the mouth slightly (56, 58).

In facial happiness, the corners of the lips are retracted sideways and upwards with prominence of the nasolabial folds. The mouth may be closed, as in smiling. The lower jaw may drop open, exposing the teeth, and the eyes may be squinted (Fig. 7-21). A happy face with an open mouth may be accompanied by laughter (54).

A whole volume has been written about smiles (165). Cross-cultural differences have been found. The British tend to smile by pulling their lips back and upwards, exposing their lower teeth. They are most likely to raise their cheeks with crow's feet appearing at the corners of their eyes. In contrast, Americans tend to part their lips and stretch the corners of their mouths (78a). India has about 600 laughter clubs where people congregate just to laugh as a group, suggesting that laughter may be "the best medicine."

In a study of happiness, electrodes were applied to subjects for encephalographic tracings. When the subjects viewed film clips to elicit the experience of happiness, they had left-sided activation in the anterior temporal region (41).

In facial surprise, the eyebrows are raised, the forehead shows wrinkling, the eyes are wide open, and the lips are parted, showing the teeth (54) (Fig. 7-22).

In facial anger, the eyebrows are lowered and

Figure 7-21 Posed facial happiness.

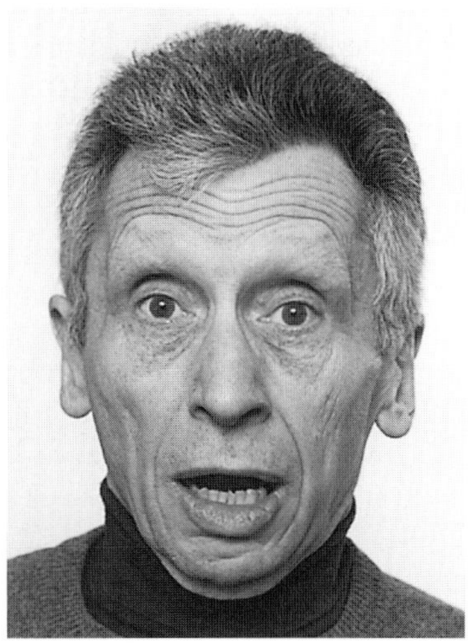

Figure 7-22 Posed facial surprise.

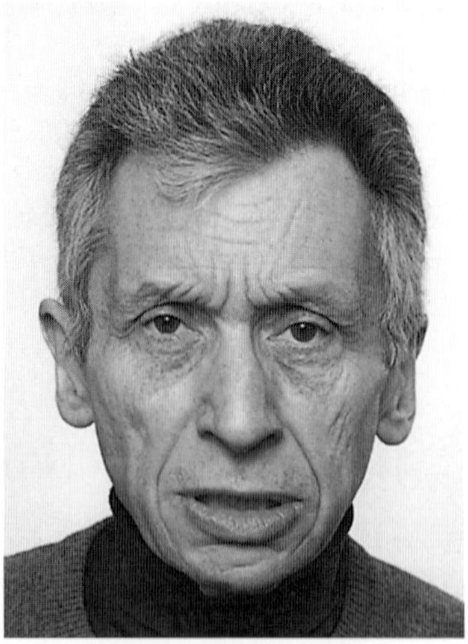

Figure 7-23 Posed facial anger.

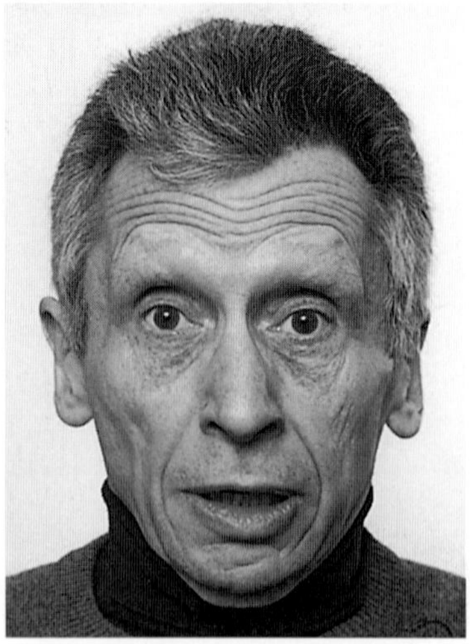

Figure 7-24 Posed facial fear.

drawn together with vertical lines appearing between the brows. The eyes appear to be staring. The lips may be pressed together or opened (Fig. 7-23). Angry words may accompany the expression (54). During a videotaped type A structural interview, men with coronary artery disease were studied for possible transient myocardial ischemia by measuring left ventricular wall motion and the amount of blood pumped from the ventricle. Patients who exhibited ischemia expressed more facial anger[h] (139).

In facial fear, the eyebrows are raised and the forehead is wrinkled. The eyes are widened and the mouth is open with the lips being tense (Fig. 7-24) or stretched laterally (54). Neuroimaging studies have shown that activation of the amygdala occurs when a person looks at a fearful face, specifically at the wide-eyed expression in which the sclerae show superiorly. With bilateral lesions of the amygdala, the patient has selective impairment for recognizing fearful facial expressions because of an inability to look normally at the eye region, which is important in identifying the face of fear. However, when the patient was instructed to look explicitly at the eyes, she could recognize fearful facial expressions (1a, 168a, 171b).

[h]These patients also exhibit more nonenjoyment smiles.

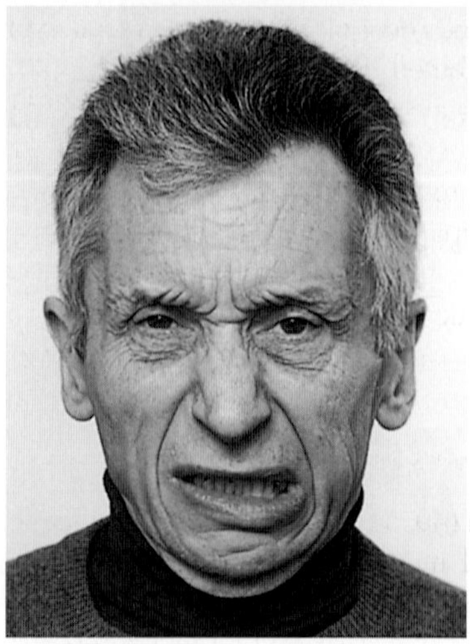

Figure 7-25 Posed facial disgust.

In facial disgust, the eyebrows are lowered, the nose is wrinkled, and the upper lip is raised (54) (Fig. 7-25). In a study of disgust, electrodes were applied to subjects for electroencephalic tracings. When the subjects viewed film clips to elicit the experience of disgust, they had right-sided activation in the frontal and anterior temporal regions (41). However, impaired ability to recognize the disgust face occurs with left-sided

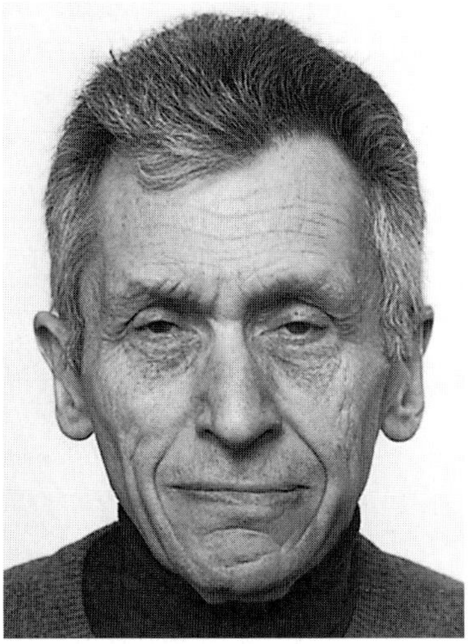

Figure 7-26 Posed facial sadness.

central nervous system damage to the insula and putamen; recognition of other facial expressions is unaffected (28).

In facial sadness, the inner corners of the eyebrows may be drawn up and the lips are closed and stretched sideways (Fig. 7-26) or the corners of the lips may turn down and the lip may tremble (54). In some instances, sadness may be accompanied by vocal expressions of grief or by crying.

Controversies

Ekman (49, 51) reviewed the evidence for cross-cultural universality of basic facial expressions, but this universality has been debated (52, 89, 141–143). In a study of the United States and India, some facial expressions were found to be more recognizable than others, and a gradient of recognition was established (78). Evidence for minimal universality has been reviewed elsewhere (144).

When facial expressions are spontaneous rather than produced by coaching, agreement about these expressions decreases among subject interpreters (115, 176). Also, when senders of facial expressions and receivers of these expressions interact, the receivers often mimic the expressions of the senders (45). The same is true of vocal utterances. Laughter tends to elicit laughter and yawning tends to elicit yawning (82, 143).

One theorist has proposed that facial expressions show nothing about how a person feels, but what his or her intended actions might be. Thus, expressions are viewed as primarily social, implying that the facial expressions of senders have evolved to influence receivers in ways that are beneficial to senders (64).

Pathologic Conditions Affecting Facial Expression

Moebius syndrome is characterized by abnormalities of the sixth and seventh cranial nerves together with limb reduction defects in some cases. The face is mask-like without the ability to make facial expressions or to gaze laterally. Labial sounds like "m" and "b" are difficult to make. Drooling and drinking difficulties also occur. Feelings of isolation, loneliness, and low self-esteem are common (35, 74).

Muscle transplantation can permit smiling in individuals with Moebius syndrome. It also produces improvement in speech, drinking, and drooling. A segment of the gracilis muscle from the thigh is transplanted through a preauricular incision and is anchored at the oral commissure and in front of the ear. Facial vessels are used for revascularization, and the motor nerve to the masseter muscle is used for reinervation. Procedures are carried out on both sides of the face (179).

Bell's palsy is acquired and results in limitation of facial expression on one side, although bilateral involvement has been documented on occasion. Parkinson's disease is characterized by reduced or absent facial animation together with tremor, slow movement, and difficulty in initiating movement (32, 35, 74).

GAZING, KISSING, AND TONGUE SHOWING

Visual Gazing and Social Interaction

Physiological arousal—measured by galvanic skin responses, electroencephalograms, and other central nervous system testing—underlies gazing.

All primates—monkeys, apes, and humans—use gazing and mutual gazing to attract, to affiliate, to seduce, and among nonhuman primates, to repel and to attack as well. It has been shown that continued human staring at a caged rhesus monkey without gaze eversion tends to produce a rapid foreword attack to the front of the cage directed at the human experimenter outside the cage (62). Affiliative gazing emerges early in life, and primate infants, including humans, have eye-to-eye contact with the mother during breast feeding. They can exchange mutual gazes by the fifth week and respond positively to eyes and to eye patterns with dilated pupils. During the first year of life, babies can play the well-known intermittent mutual gazing game—peek-a-boo (5, 6).

Generally, a proper amount of gazing is considered acceptable. If too much gazing occurs, a person may be regarded as disrespectful, insulting, supercilious, or even threatening. If too little gazing occurs, a person may be regarded as inattentive, impolite, submissive, or even dishonest. Extroverts tend to gaze more than introverts, women more than men, and adults more than adolescents. People with schizophrenia, depression, or neurosis tend to avert their gaze, particularly when interviewed about their problems. Children with autism gaze the least (5, 61).

Different functions of gazing in social interaction include information seeking, communication of interpersonal attitudes, synchronization of speech, gazing proximity behavior, and gazing distraction behavior. These functions and their characteristic features (7, 10, 61, 72, 92) are summarized in Table 7-8. Probably two or more of these functions can operate simultaneously.

Table 7-8 Functions of Gazing

Functions	Characteristic Features
Information seeking	Looking may supplement auditory information by observing eye movements and facial expressions.
Communication of interpersonal attitudes	People tend to look more at people they like. Mutual gazing tends to be particularly high in affectionate couples and may also act as a cue for intimacy. People who are high in dominance tend to look more during competitive situations. Those high in affiliation tend to look more during cooperative situations.
Synchronization of speech	Shifts in gazing help to coordinate the timing of speech between two people having a conversation.
Gazing proximity behavior	Close proximity tends to be associated with increased gazing in people with a high need to affiliate or who are affectionate or who belong to contact cultures in which men tend to stand close together and frequently touch each other.
Gazing distraction behavior	Speakers tend to look at listeners more during fluent speech than during hesitant speech. The hesitant speaker tends to look away at the beginning of the utterance while he or she is planning what to say. The overall amount of gazing is inversely related to the cognitive difficulty in the topic of conversation.

Derived from data in Argyle et al., 1973 (7), Beattie, 1981 (10), Exline, 1972 (61), Goldman–Eisler, 1968 (72), and Kendon, 1967 (92).

Shifts in gazing help coordinate the timing of speech between two people in conversation. However, eye contact is only one modality in social interaction; it is not used during telephone conversations, which can be coordinated by pauses in speech between the two speakers.

Each society tends to have its own rules that govern gazing. Table 7-9 summarizes cultural variations in gazing among the Navaho, Wituto, Bororo, Luo, Mende, Tuareg, and Japanese. There are also contact and noncontact cultures, which have different implications for gazing. In contact cultures, like Latin Americans and Southern Europeans, men tend to stand close together and frequently touch each other. They are more likely to show an increase in mutual gazing. In noncontact cultures, like Northern Europeans and Pakistanis, men typically do not stand close together and tend to have less frequent mutual gazing (5).

Kissing

The word *kiss* may have been derived from the German word *kussen*. The Romans practiced a passionate lip kiss known as the *basium*, which may have led to the English word *buss*, rather than kiss, in the writings of Chaucer and Shakespeare. Kissing is by no means universal (103). Nose rubbing is found among the Inuit.

Besides "necking" and passionate kissing, many other types of kisses are used. Politicians and military leaders in France, Spain, and

Table 7-9 Cultural Variations in Gazing

Society	Geographical Area	Cultural Rules of Gazing
Navaho	Southwest area of the United States	Navaho do not gaze directly at another person during conversation.
Wituto Bororo	South America	Speaker and listener look at irrelevant objects during conversation. Story teller turns away from his audience, facing the back of the hut.
Luo	Kenya	Men may not look at their mother-in-law.
Mende	Sierra Leone	Mende believe that the dead are recognizable because they never look a live person in the face. Therefore, they are suspicious of anyone who averts their eyes during conversation.
Tuareg	North Africa	Gazing in the form of staring is the most important way to communicate besides speech because the whole body and face are covered with clothes and veils except for the eyes.
Japanese	Japan	Japanese speakers focus on the listener's neck. Little use is made of the visual–facial channels in sending and receiving messages. Communication occurs through nuances of language and body position. Japanese infants are carried on their mothers' backs most of the time and thus have less visual contact with the mother's face.

Derived from data in Argyle, 1980 (5).

Latin America cheek kiss. English and Canadians are one-cheek kissers. French and Italians are two-cheek kissers. Women often kiss each other's cheek upon greeting. To keep from smearing lipstick, sometimes an "air kiss" is blown (103).

In the Roman Catholic Church, the altar cloth or a priest's robe may be kissed as an act of reverence. Infants may be kissed at the time of their baptism. The Church passed canon laws in 1312: kissing in reverence to God was not a sin; kissing for carnal delight was a venial sin; and kissing with the intent to fornicate was a mortal sin, even if no fornication took place (103). In Judaism, when a rabbi parades the Torah around a synagogue, congregants may touch their tallis to the Torah and then touch their tallis to their lips.

It is known that adults are twice as likely to turn their heads to the right as to the left when kissing. This may be related to other types of sidedness. During late prenatal and early postnatal life, head turning occurs with a 2:1 right-to-left preference, possibly inducing later right/left asymmetries, such as favoring the right eye, the right ear, the right foot, and turning the head to the right during kissing (76).[i]

Kissing is discussed in the Indian classic the *Kama Sutra*, which is thought to have been written in the third or fourth century by Mallanaga Vatsyayana. *Kama* means *sensual pleasure* and *Sutra* means *poetic rendition*. It is not an original work, but a summary of various writings from previous centuries. The *Kama Sutra* is a book about the art of living. It is divided into seven parts and six of them deal with various relationships between men and women, ranging from furnishing a house to planting a garden to wooing a woman for marriage. Only one part deals with sensual pleasures (168). Different types of kissing found in the *Kama Sutra* are summarized in Table 7-10.

Nonhuman Primate Tongue Showing

In most species of Old World monkeys, tongue showing, called the *tongue-smacking face*, is used in greetings, in approaches for grooming, in huddling together or simply associating, in playing, and in copulating. The tongue-smacking face is exaggerated in the baboon but barely noticeable in the chimpanzee. Tongue showing occurs in greetings between gibbons. Most species of New World monkeys use less tongue showing in greetings and allogrooming than do Old World monkeys. Some lemurs also use a series of tongue protrusions in greeting behavior (3, 29, 166).

In young captive gorillas, tongue showing is found in playful cage activities requiring concentration for balancing in precarious situations.

[i]The approximate 8:1 right/left handedness is different because it has strong genetic and cultural modifying factors.

Table 7-10 Kama Sutra: On Kissing

Type	Definition
Nominal kiss	When a woman touches only the mouth of her lover
Throbbing kiss	When a woman touches the lip that is pressed into her by moving her lower lip but not her upper lip
Touching kiss	When a woman touches her lover's lip with her tongue
Straight kiss	When the lips of two lovers are brought into direct contact with each other
Bent Kiss	When the heads of two lovers are bent toward each other
Turned kiss	When one lover turns up the face of the other by holding the head and chin
Pressed kiss	When the lovers press with much force
Greatly pressed kiss	When a lover takes hold of the lower lip between two fingers and then after touching it with the tongue presses it with great force
Kiss of the upper lip	When a man kisses the upper lip of a woman while she in return kisses his lower lip
Clasping kiss	When one lover takes both lips of the other lover between his or her own lips. A woman takes this kind of kiss only from a man with no moustache. If one lover touches the teeth, tongue, and palate of the other with his or her tongue, it is called "fighting of the tongues"
Kissing of four kinds	Moderate, contracted, pressed, and soft, depending on the part of the body to be kissed
Kiss that kindles love	When a woman looks at the face of her lover when he is asleep, it shows her intention or desire
Kiss that turns away	When a woman kisses her lover when he is engaged in business, quarreling with her, or looking at something else
Kiss that awakens	When a man comes home late at night and kisses his lover while she is asleep
Kiss that shows intention	When a man kisses the reflection of his lover in the mirror, in the water, or as a shadow on the wall
Demonstrative kiss	When a man in a theater at night or in an assembly of men kisses a finger of a woman if she is standing or a toe if she is sitting or if the woman washes his body and places her face on his thigh or great toe

Based on the *Kama Sutra*, an Indian classic written by Vatsyayana (168).

Tongue showing also occurs when a gorilla is rejected by a conspecific or withdraws from social interaction. Thus, captive gorilla tongue showing is similar to certain tongue showing behaviors in humans (156) (*vide infra*).

Human Tongue Showing

Tongue protrusion by an adult triggers a tongue-protruding response in neonatal infants between 12 and 21 days of age (109) (Fig. 7-27). During infancy, the tongue may be used as a food-seeking signal or as a way of terminating feeding. The tongue may be visible with lip smacking in a suckle-like manner if the infant is hungry. The infant tongue may push food from its mouth when satiated (156).

Children may show the tongue in four basic situations: (a) when concentrating on a task or any activity that requires attention; (b) when an adverse social situation or an impending one occurs; (c) when forgoing physical control, as when a parent swings his or her child around; and (d) when teasing, mocking, or taunting another child (156). When this last situation occurs, the child literally sticks his or her tongue out at someone.

Adults may stick their tongues out slightly at the moment of being surprised or may show the tongue passively in the mouth with open-mouthed laughter. When French kissing, the tongue protrudes but is not seen. When flirting, the tongue may be shown briefly. When spitting, the tongue may appear briefly in some cases during the prespitting phase. Like children, adults may show their tongues when concentrating on various tasks or just being thoughtful. They may also show the tongue when an adverse social situation or an impending one occurs (156). Finally, the tongue protrudes in a "Bronx cheer" or "raspberry", as when baseball fans disagree with a decision by an umpire.

The tongue is used in different ways in various societies. Tibetan children have been reported to greet each other by protruding the tip of the tongue while tugging the right ear. It was also common among men, whereas men greeted women with tongue showing and a reduced ear tug. Women greeted each other as well as men by protruding the tongue while touching the tip of the nose (156). Maori men stick their tongues out with abandon during their fierce greeting challenge (Fig. 7-28).

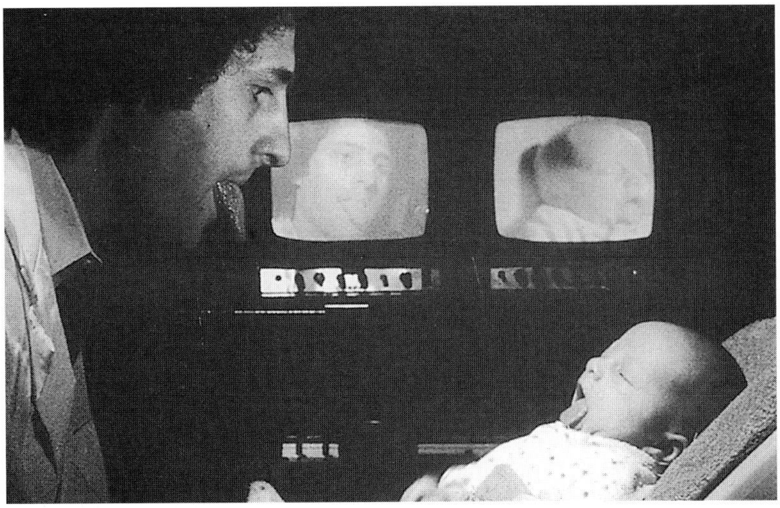

Figure 7-27 Tongue protrusion by an adult triggers a tongue-protruding response in neonatal infants between 12 and 21 days of age. Courtesy of A. Meltzoff, Seattle.

Figure 7-28 Maori men sticking their tongues out with abandon during their fierce greeting challenge. Photograph by D. Cole, Rotorua, New Zealand.

SOME STUDIES OF FACIAL PSYCHOLOGY

Facial Attractiveness

Many studies of physical attractiveness have been carried out, most frequently predicting attraction between potential romantic partners (42, 63, 73, 170, 172). The single best predictor for satisfaction with a "blind date" is facial attractiveness for both men and women (170). It has also been shown in a televised dating game that the audience could predict the dates selected by daters based on physical attractiveness (31). Cross-cultural agreement in ratings of facial attractiveness has been demonstrated in many ethnic groups from different geographical locations (37, 129).

Preference for physical attractiveness tends to be stronger for men than for women. Men are also more likely to seek someone younger than themselves (80). Some studies have shown that women change their preference for male face shape across the menstrual cycle. It has been suggested that women tend to prefer men with feminine faces that indicate warmth, honesty, and a willingness to invest in offspring (128). This preference becomes attenuated during the follicular phase of the menstrual cycle, and when conception is most likely, women tend to prefer masculine faces (125, 126).

Infancy has been the subject of several studies. In one study, infants viewed slides of pairs of faces previously rated for attractiveness by adults. Infant preference was determined by the direction and duration of gaze. It was found that they preferred to look at attractive faces (97). In another study, mothers of more attractive infants were more playful with their infants than mothers of less attractive infants. These latter mothers were more likely to be attentive to other people and to give perfunctory care to their infants (96). So-called "baby-faced adults" are discussed in the section titled "Some Avoided Psychological Terms."

Several theories have been proposed to account for facial attractiveness: symmetry, averageness, matching, and exaggerated secondary sexual characteristics. Because symmetry and averageness are related concepts, they are considered together.

Some studies have indicated that symmetric faces are considered more attractive (75, 127, 129), although other studies have found that normally asymmetric faces are preferred (94, 98, 145, 159). These results have been explained by differences in the computerized morphing techniques used: mirror-image faces, generalized facial averaging, facial averaging of attractive faces, and symmetric remapping of individual faces. Facial symmetry is discussed further in the section titled "Some Avoided Psychological Terms."

Some studies, using a matching hypothesis (173), have found that partners tend to select each other by how similar they are in terms of physical attractiveness (12, 173). In other studies, exaggerated secondary sexual characteristics of the face have been considered to enhance physical attractiveness to the opposite sex. For women, such features include a small lower face, a relatively flat midface, full lips, high eyebrows, and lack of brow ridge prominence. For men, such features include prominent cheekbones and a large jaw (75, 125, 147).

These exaggerated female and male preferred characteristics together with facial symmetry have been suggested to be biologically adaptive and related to genetic survival. According to this theory, an attractive young woman is likely to be healthy and in the prime of her reproductive years. A man with such a woman increases the probability that his genes will survive to the next generation (26, 75, 125, 163).

Some studies of sex differences in the reaction to facial resemblance have concluded that facial resemblance between self and child is more important for men than for women, which is said to have evolutionary significance because it encourages paternal investment in the child. This is considered to be important because men, unlike women, can be mistaken about parenthood (31a, 130a).

Another study of computer-generated morphed pairs of facial images were arranged as a finely graded series that spanned two original faces. Two types of morphed pairs were generated: an ethnic one between Caucasian and Japanese and one involving gender between male and female. Analysis of responses showed that choices made in a facial subset could be influenced by the after-effects of previously seen faces. For example, in the Caucasian–Japanese finely graded series, if a Japanese face at one extreme was shown, a neutral face in the middle of the series was more likely to be judged as Japanese (171a).

Critique of Studies of Facial Attractiveness

Many studies of facial attractiveness place undue emphasis on the physical aspects of dating and mating selections at the expense of other important factors. Indeed, such selections are very complex; emphasizing facial attraction is superficial and misleading. Human plasticity, both behavioral and cultural, is underplayed. The ideas that "beauty is in the eye of the beholder" and that "beauty is skin deep," although sometimes considered to be clichés, are actually important. The first idea indicates that an individual's choice of mates based on his or her idea of an attractive face may not necessarily be shared by others. Furthermore, ideas about beautiful faces change over time. For example, the earlier full round faces of feminine movie stars are different from the later occurring thin faces with the "Twiggy" look. The second idea indicates that many other factors are involved in mate selection, including personality (25), society's ideals (46), beliefs and values (27, 99), demographic factors (84, 85), age, status (42, 73, 154), and family influence (physical appearance, upbringing, conscious ideas, nonconscious expressions, religious values, and cultural background).

Material resources and occupational status may or may not be important in mate selection. A psychological survey from Peking University (36) found that many women were choosing men with promising careers and stable incomes, whereas earlier, men's age, height, and education were more important. For career women with higher economic status, selecting mates on their ability to provide financial security is less important (157, 160).

The previously discussed study finding that infants preferred to look at attractive adult faces (97) is open to question. It has been shown that infants of 6 months are *equally good* at recognizing the facial identity of both humans and macaques (122) (see section titled "Face Processing and Species Specificity in Early Infancy"). In a potential study of infants viewing slides of paired macaque faces, who would previously rate their relative attractiveness, since human adults are incapable of this?

That exaggerated female and male preferred characteristics together with facial symmetry being biologically adaptive and related to genetic survival (26, 75, 125, 163) is naive. What is important about ongoing genetic evolution is the differential leaving of offspring. A good case could be made that strongly believing Catholics may produce more offspring regardless of facial attractiveness. Furthermore, the population contains a significant number of facially attractive gay people and facially attractive career women whose choose to be single.

The studies showing that facial resemblance between self and child is more important for men than for women with its evolutionary implication of paternal investment in the child (31a, 130a) have been stated to be erroneous by several authors (43a, 63a). The computer-morphed faces used were shown to subjects as facial ovals without hair, ears, or neck. This technique is so abstract compared to natural faces that its validity is highly questionable. The other study with a finely graded series of ethnic- and gender-related images (171a) also makes use of the same highly questionable morphing technique.

Facial Prominence: Sex Differences

Visual depictions of men and women are different. A pronounced tendency to represent men with their faces and women with their bodies has been demonstrated in America, in eleven other countries, in artistic portraits over six generations, and in amateur drawings of men and women (4) (Table 7-11). Men appeared in 60.5% of American periodical photographs. In the amateur drawings of men and women, the sex of

Table 7-11 Pronounced Tendency to Represent Men with Their Faces and Women with Their Bodies

Types of Study	Sample Size	Men Represented by Faces, Women Represented by Bodies	Comments
American photographs	1,750	+++	Sampling of periodicals. Men appeared in 60.5% of the photographs
Cross-national photographs	3,500	+++	Sampling included 11 countries: United Kingdom, France, Germany, Italy, Spain, "Middle East", India, Mexico, Chile, Hong Kong, and Kenya
Portraits	920	+++	Sampling of portraits and self-portraits over a six century period
Amateur drawings	80		Drawings of men and women by men and women
		+++	Drawings by men
		+++	Drawings by women

+++, pronounced tendency.

Based on Archer et al., 1983 (4). The study design is convincing, including statistical analysis. See the reference for a fifth study of sex differences in attributions and perceptions by men and women not found in this table or the text.

the artist was unimportant. Drawings by women tended to represent men by their faces and women by their bodies. The same was true, of course, for drawings by men.

The implications of the study suggest the disturbing[j] idea that men could be perceived as being more "cerebral" and women could be perceived as being more "corporeal." The roots of the face prominence/body prominence phenomenon remain unexplained to date.

Facial Resemblance and Trust

Using computerized morphing techniques, two facial photographs can be melded together. In a study of two-person games, subjects bargained for money with a computerized face. The face of each subject was morphed with a strange face to produce a face that was different from that of the subject yet retained some resemblance to the subject. Control faces were strange faces that were not morphed. In the bargaining game, subjects did not recognize morphed faces *per se*, but trusted faces resembling their own more than they trusted strange faces. Thus, facial resemblance can modulate social behavior (43).

SOME AVOIDED PSYCHOLOGICAL TERMS

Certain common terms found in some psychological studies—facial display, caricature, and baby-faced adults—have been avoided in this chapter because I think they have been used inappropriately. Although I have to use the term "facial symmetry" to describe some studies, I think that the concept is misguided.

Facial Display

Some ethologists—biologists who study animal behavior—have used the term *facial display* instead of *facial expression* for humans and nonhuman primates, claiming that the term facial expression implies something about the underlying emotions. Certain psychologists who are not ethologists have adopted *facial display* as well. However, it is *facial expression* that remains in common use. Dorland's Medical Dictionary defines facial expression as "the facial aspect or appearance of the face, as determined by the physical *or* emotional state." The word "or" denotes that facial expression can be described as being determined by the physical state, namely, the muscles of facial expression controlled by the seventh cranial nerve. It does not necessarily refer to specific emotional states directly. In contrast, display, as used by ethologists, means a stereotyped movement or posture, such as a male peacock spreading its tail during courtship. Certainly human facial expressions are much more plastic. An anger face or a happy face can vary in degree. Human anatomists have been teaching about the muscles of facial expression to students of medicine, nursing, and dentistry for generations. To change a common and well-established usage to the "muscles of a facial display" would be completely unwarranted.

Caricature

A social science definition has been proposed: "a caricature is a symbol that exaggerates measurements relative to any measure that varies from one individual to another" (126a). Caricatures have been studied by a number of psychologists (16, 129). Such a definition, however, ignores the long-established usage of caricature in art history (62b). It misses the main types of exaggeration of good caricature—large head and concentration on commanding features only (118). Furthermore, some psychologists have blurred caricatures with cartoons (16), even though they are known to be different from each other in the field of art history (see Chapter 9).

Baby-faced Adults

Some ethologists and psychologists speak of baby-faced adults if they have a rounded face, large eyes, thin eyebrows, a small nose with a low nasal bridge, and a small chin. Although psychologists may claim that they are only studying the perceptions of *their subjects* about baby-faced adults (177), they are, after all, controlling

[j]and obnoxious

the experiments and perpetuating the idea of baby-faced adults in the literature. However, baby-faced adults are a contradiction in terms. True baby faces are entirely different from adult faces (see Figures 1-6 and 1-8, and Figures 3-21 to 3-24).

Facial Symmetry

Female preference for symmetric male characteristics has been shown in insects, such as earwigs and Japanese scorpion flies, and birds, such as swallows and Zebra finches (114, 134, 138, 158). Such biological research in animals has generated an interest in studies of "human preference for symmetric faces."

Most studies that manipulate human facial images have found that normal asymmetry is preferred to symmetry. Such studies compare normal faces with left-mirrored and right-mirrored facial images (94, 98, 145) (see Fig. 3-32). Some psychologists have questioned this technique and offered computerized alternatives to generate symmetry, such as generalized facial averaging, facial averaging of attractive faces, and symmetric remapping of individual faces, and these studies have shown subject preferences for symmetrically generated faces instead of normally asymmetric faces (75, 127, 129). However, one computerized study found that symmetrically generated faces were less attractive than normally asymmetric faces (159).

An excellent case can be made that generalized asymmetry is a normal state in human development. Asymmetry is found in the thorax, abdomen, brain, bones, teeth, and face. Asymmetric disposition of the visceral organs results in three lobes in the right lung with two in the left, right-sided liver and gallbladder, left-sided stomach and spleen, the apex of the heart pointing to the left, and the primitive gut coiling itself in a counterclockwise direction (34). Structural asymmetry of the brain is found in many studies (see Table 3-11 and Fig. 3-31). Asymmetrically paired bones include the humeri, radii, femorae, fibulae, clavicles, and ribs (34). Left–right tooth crown size also exhibits asymmetry (66).

All faces are normally asymmetric, and the mirror-image technique demonstrate this (124) (see Fig. 3-32). Furthermore, left-sided mouth asymmetry characterizes smiling and other emotional expressions—a right-sided cerebral hemisphere specialization—and right-sided mouth asymmetry characterizes speech and infant babbling (prespeech)—a left-sided cerebral hemisphere specialization (87).

SOMATOFORM DISORDERS

In somatoform disorders, emotional stress or difficult life situations are experienced as persistent physical symptoms with no physiological explanation. Somatization comprises a spectrum from stress-related exaggeration of common symptoms, such as a headache or light headedness, as might occur with a new family member, divorce, or death of a relative, to unrelenting problems that can leave a patient completely disabled and withdrawn (151).

Symptoms can affect many systems: gastrointestinal, cardiopulmonary, "pseudoneurologic," and reproductive. There are various types of pain as well as "syndromes" (14). Symptoms that affect the craniofacial region, including the central nervous system, are listed in Table 7-12.

Several vague or exaggerated symptoms, often in different organic systems, and a chronic course suggest a somatoform disorder. Although somatization does not exclude the possibility of an organic medical condition, the intensity of the symptoms is usually out of proportion to the healthy appearance of the patient. Furthermore, common psychiatric conditions frequently accompany somatization, including depression (up

Table 7-12 Symptoms Reported by Patients with Somatization Involving the Craniofacial Region Including the Central Nervous System

Nausea	Pseudodiplopia
Vomiting	Deafness
Headaches	Pseudolaryngitis
Dizziness	Bruxism
Pseudoseizures	Temporomandibular joint symptoms
Syncope	Burning tongue and oral mucosa
Amnesia	
Blurred vision	

Based, in part, on Blackwell and De Morgan, 1996 (14).

to 60%), anxiety disorders (up to 50%), personality disorders (up to 60%), and substance abuse. The risk of a psychiatric disorder increases with an increasing number of physical complaints (151).

NEUROLOGICAL AND BEHAVIORAL CORRELATES OF MINOR ANOMALIES

Minor anomalies of the face are listed in Chapter 5 (see Table 5-4). They occur in many malformation syndromes and may also be associated with neurological and behavioral abnormalities. In Down syndrome with its characteristic facial features and mental retardation, 79% of all malformations detectable by clinical examination are minor anomalies. In the fetal alcohol syndrome, developmental delay is associated with minor facial anomalies in a characteristic pattern (33, 108).

In Turner syndrome, problems are found in spatial perceptual skills, visual–motor integration, affect recognition, visual memory, attentional ability, and executive function. Features such as slow speech development, hyperactivity, learning disabilities, neuromotor deficits, and short stature can contribute to poor self-image and lowered self-esteem during adolescence. Minor anomalies occur in 73% of all individuals (74).

Substantial genetic evidence has been shown for schizophrenia, which is etiologically heterogeneous. At least two known disorders were formerly thought to occur on a behavioral basis only: metachromatic leukodystrophy and deletion 22q11 syndrome. Undoubtedly, the molecular basis of other subgroups of schizophrenia will be delineated in the future. A recent study of minor anomalies (149) divided schizophrenic subjects into four different groups by cluster analysis. Two of these subtypes had significant patterns of dysmorphogenesis and one of them had a large number of subjects with deletion 22q11 syndrome (Table 7-13).

Table 7-13 Reported Patterns of Minor Anomalies in Two Subgroups of Schizophrenics

Group 1[a]	Group 2[a]
n = 27 including 14 cases of deletion 22q11 syndrome	n = 18
Cranial	Cranial
Flat supraorbital ridges	Flat supraorbital ridges
Flat occiput	Flat occiput
Widow's peak	Abnormal hair whorl
Cheeks	Widow's peak
Flat	Low hairline
Ocular	Cheeks
Narrow/slanted palpebral fissures	Flat
Small eyes	Ocular
Nasal	Narrow/slanted palpebral fissures
Broad nasal bridge	Epicanthic folds
Large nose	Small eyes
Oral	Nasal
Lip abnormalities	Broad nasal bridge
Highly arched palate	Large nose
Narrow palate	Small nose
Small mouth	Oral
Small mandible	Lip abnormalities
Ears	Ears
Low-set	Malformed
Posteriorly angulated	
Small	
Malformed	

[a]All listed findings are significant, but not in the order of frequencies reported. Only the craniofacial findings appear in this table. There are other noncraniofacial findings not listed here.

Data from Scutt et al., 2001 (149).

Smith and Bostian (155) found that 42% of patients with mental retardation had three or more malformations of which 80% were minor anomalies. David and Osborne (40) studied scalp-hair patterns in mental subnormality. Gailey et al. (65) found minor anomalies in the offspring of epileptic mothers.[k] Miller (112) found minor anomalies associated with ataxic cerebral palsy. Quinn and Rapoport (133) studied minor anomalies and the neurological status of hyperactive boys. Accardo et al. (1), however, found that minor anomalies were not related to hyperactivity or attention deficit in a referred population. Rosenberg and Weller (140) studied the association of minor anomalies and academic performance in young school children. Waldrop et al. (169) found that minor anomalies in newborns could predict short attention span, peer aggression, and impulsivity at three years of age. LaVeck et al. (100) found that assessment of minor anomalies was not practical for screening infants to predict later aberrant behavior.

[k]It isn't clear whether or not some mothers were taking dilantin or other antiepileptic drug known to cause anomalies in fetuses.

PSYCHOLOGY OF SURGICAL RECONSTRUCTION

A surgical perspective together with illustrations of specific conditions can be found in Chapter 6.

Craniofacial Malformations

In a classic study of craniofacial anomalies, MacGregor et al. (107) found that parent–family relationships were disrupted by malformations. Initial parental responses included shock, anger, and shame followed by grieving and only partial resolution. Parents were further stressed by repeated operations, dwindling financial resources, societal reactions to their child, and emotional overinvestment in their malformed child. MacGregor et al. (107) also noted that severely malformed children seemed to have a somewhat better adjustment than children with milder facial malformations. In the former, consistent responses were elicited, whereas mixed responses, which are more confusing, were elicited from the latter.

Among body parts, the face serves as a major source of communication, reflecting the emotions, character, and personality of an individual. Self-concept is formed, in part, by one's own and other's reactions to facial features and facial expression. A person with a craniofacial malformation may receive negative social messages, resulting in lowered self-esteem. The closer a defect is to an individual's communication channels, such as the eyes or the mouth, the smaller the defect needs to be to throw a participant in the exchange off balance (70). The severity of facial disfigurement is not necessarily proportional to the degree of psychic stress it engenders in affected individuals or to the kinds of adjustments they make; factors such as parental attitudes, the individual's own personality, and the social setting play a stronger role (33, 106).

Parents need an opportunity to discuss their own ideas about the cause of their malformed infant and the meaning that they attach to the medical terms and labels employed. Counseling about the nature of the condition and its cause, when known, can clear up a number of misconceptions the parents may have. The clinician should answer any questions the parents ask, recognize that parental anger may be directed toward him or her initially, accede to parental wishes for another opinion gracefully and indicate willingness to continue working with them thereafter, be available for further information and discussion, and be prepared to repeat the information several times until it is absorbed. The parents should know that they are not alone, that craniofacial surgery, other medical services, clinics, agencies, schools, counseling services, and craniofacial support groups are available (33).

In children with craniofacial anomalies, the Draw-A-Person test may reveal a distorted body image (11, 116) (Fig. 7-29). Excessive denial is frequently used as a coping mechanism, especially among severely malformed children. In the study of Lefebvre and Munro (101), patients with craniofacial anomalies usually rated themselves as being less malformed than did either their parents or members of the craniofacial team. Following craniofacial surgery, patients rated themselves as being more attractive postoperatively than did either their parents or members of the craniofacial team. In general, a high degree of satisfaction has been reported with the results of craniofacial surgery by patients, surgeons, and psychiatrists (101, 116). Although many craniofacial malformations can be

Figure 7-29 Left, Draw-a-Person test by an 11-year-old boy with hemifacial microsomia. Right, Draw-a-Person test repeated 6 weeks later following an operation to correct facial asymmetry. Note the striking change in body image. From Murray et al., 1979 (116).

eliminated by surgery, significant residual defects may remain in some patients.

Belfer and his colleagues (11, 116) have discussed a model of body image development in malformed children based on their experience with complex craniofacial anomalies, such as Treacher Collins syndrome and Apert syndrome. Normal and distorted body image are contrasted in Figure 7-30. The malformed infant is more likely to experience disrupted parental bonding during the neonatal period, develop a sense of incompetence during infancy, become isolated from peers during school age, experience an uncertain sense of identity, have diminished career goals during adolescence, and have feelings of seclusion, incompleteness, failure, conflict, and hostility together with diminished achievement during adulthood. A distorted developmental pathway diverges from normal, becoming more constricted with each milestone. Also, a craniofacial defect resulting from trauma or a neoplasm in a normal individual tends to shift development toward distorted pathways, but often at a later point in time.

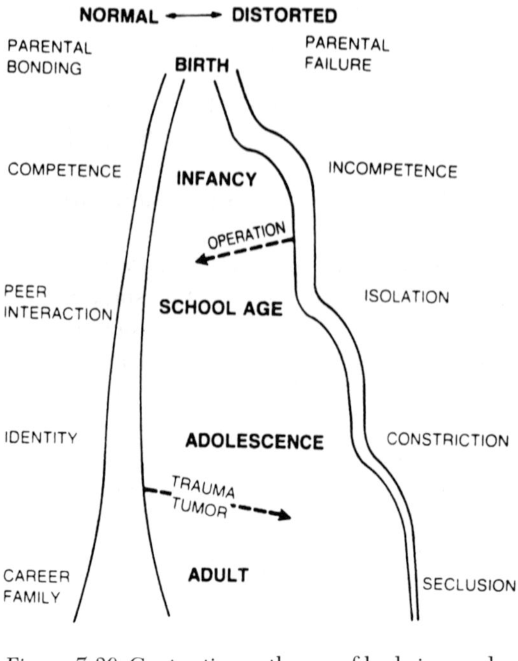

Figure 7-30 Contrasting pathways of body image development. Divergence between normal and distorted pathways with each milestone. Note the progressive constriction leading to isolation and seclusion in the malformed person. From Murray et al., 1979 (116).

Surgical treatment for a craniofacial malformation tends to shift the malformed patient's development toward normal pathways. Belfer and his coworkers (11, 116) have hypothesized four stages in the modification of body image in malformed children undergoing reconstructive surgery for complex craniofacial anomalies. The first stage, the decision to undergo surgery, represents the beginning of the dissolution of what, for many children, has been pathologic denial. The timing of the complex decision to undergo surgery is based on self-recognition, emotional state, parental and peer group pressure, social awareness, and the surgeon's assessment of reconstructive potential. During the second stage, the operative experience and the reality of physical intervention is acknowledged by the child. The third stage, the immediate postoperative period, is characterized by pain and swelling with *increased* distortion of the face. This period constitutes a psychological crisis in which the decision to undergo surgery and the hoped-for changes are reassessed. Previous psychological defenses are disrupted, leading to introspection and acknowledgement of the previously existing malformation. During the final stage, the reintegration period, the psychological defenses are reorganized, resulting in increased intellectual freedom, reordering of social priorities, and an upsurge in interpersonal relationships.

Cleft Lip and Palate

The psychosocial literature on cleft lip and palate has changed during the past 35 years from global negative impressions to relatively normal impressions to more detailed studies showing some specific subgroups at risk. Generally, educational and occupational attainments have been essentially similar to normal siblings and normal controls (130).

Three factors seem to be related to this relative normality. First, no facial disfigurement accompanies isolated cleft palate, and children with cleft lip do not have the devastating facial abnormalities found in children with severe craniofacial malformations. Second, surgical techniques for the repair of the lip and the accompanying nasal defect have greatly improved during the last 40 years. Finally, children with clefts make up the largest group of abnormalities affecting the face. They are treated in clinics in which patients and their families have exposure and access to others with similar difficulties. Such social support reduces the potential for isolation (130).

Some studies have suggested specific tendencies. Patients with clefts have been found to stay closer to home, have fewer close friendships, and report more anxiety than normal controls. Some evidence has suggested lower self-esteem, general unhappiness, behavioral inhibition in the classroom, lower expectations by parents and teachers, and underestimated intelligence of such children by teachers (91, 130).

For specific subgroups, children with isolated cleft palate tend to have more problems with anxiety and depression related to language disabilities. Children with unilateral or bilateral cleft lip tend to have more problems related to facial appearance (111, 137). Children with syndromes and associated orofacial clefting tend to have more severe problems. For example, in the velocardiofacial syndrome [del(22q.11) syndrome] in which cleft palate or submucous cleft palate occurs, patients tend to develop psychiatric conditions that may include attention deficit disorder with hyperactivity, bipolar disorder, or schizophrenia (74).

Cosmetic Surgery

Cosmetic surgical procedures like rhytidectomy (facelift) and rhinoplasty ("nose job") attempt to produce aesthetic improvements. People seeking such surgery express dissatisfaction about their body image, particularly their face. They often have feelings of self-consciousness, may avoid social situations, and may exhibit behavioral inhibition (132).

Expressing some degree of dissatisfaction with body image has increased through the years. Cash (30) showed that concern with the face had doubled from 1972 to 1985. The public has become more aware of plastic surgery and its availability.

Self-consciousness is the primary motivating factor for seeking cosmetic surgery, although fear of being considered "vain" and guilt about requesting such surgery are also expressed. Some anxiety about such a procedure is normal. The presurgical appearance is unlikely to be viewed as deviating greatly from the norm (132).

Assessing a patient's reasons for requesting cosmetic surgery should be supplemented by an evaluation of coping style, current life stresses, a history of previous cosmetic surgery, and any psychiatric problems. If the motivation is for personal reasons with a good grasp of the surgery itself and its outcome, as well as an understanding of postoperative events, such as swelling and discoloration, the patient is a good candidate for cosmetic surgery. If responses are not reasonable, the patient is a poor risk for surgery, as, for example, wanting to please someone else instead of being internally motivated (132). The presence of any psychiatric condition, such as neurosis, personality disorder, body dysmorphic disorder, or psychosis, is of great concern. However, disturbed patients are not found nearly as often as the early literature suggested. For example, in articles on rhinoplasty, the older the date of publication, the more disturbed the patient was said to be (71, 132).

Orthognathic Surgery

Patients who undergo orthognathic surgery tend to have more subtle facial alterations than patients with craniofacial anomalies.[1] They also represent a larger portion of the general population. Unlike cosmetic surgery patients who are self-referred, most orthognathic surgery patients are referred by orthodontists, dentists, or physicians (135).

Three other differences that characterize the decision to undergo orthognathic surgery contrast with those who seek cosmetic surgery. The time between defining the problem and seeking a solution is generally longer. Gaining a more attractive facial profile is a strong motivating factor in most cases, but about 30% are motivated by difficulty in chewing or other functional problems. Patients undergoing orthognathic surgery must commit themselves to extensive evaluation and long treatment, including intermaxillary fix-

[1]Except patients with repaired clefts who need secondary procedures, patients with a malunion of a maxillary or mandibular fracture, and other conditions, such as hemifacial microsomia.

ation following surgery and an extended period of time in orthodontic treatment with "braces." Similarities in the results of both cosmetic surgery and orthognathic surgery include less self-consciousness, more self-esteem, and improved social relations (135, 167).

Facial Trauma

Symptoms related to facial traumatic accidents include preoccupation with altered facial appearance, social withdrawal, depression, driving phobias, and chronic post-traumatic stress syndrome with nightmares and flashbacks. Alcohol, other substance abuse, anxiety, depression, and eating disorders are associated with increased accident risk (130).

Dysfunctional behavior includes postponing social contacts until after the completion of successive surgical procedures, withdrawal from loved ones, refusal to leave home, "doctor shopping," depression with anhedonia, suicidal thoughts, neurological abnormalities, and other problems related to head trauma (130).

With craniofacial trauma, an individual who has had a lifetime of looking normal is suddenly disfigured and stared at by strangers. Furthermore, friends and acquaintances in social situations often ask for an explanation of the altered facial appearance (130). Traumatic injuries acquired during childhood or adolescence tend to produce more psychosocial problems than those acquired in adulthood because adults may have better coping mechanisms (117).

Cancer

Malignancies affecting the craniofacial region may include rhabdomyosarcoma, squamous cell carcinoma, malignant melanoma, lymphoma, and leukemia, among others. When cancer is diagnosed in adults, universal concerns surface, including fear of death, morbidity associated with treatment, fear of recurrence, fear of dependent status, fear of abandonment, and loss of functional ability, self-esteem, economic competence, sexual function, and family role behavior. These concerns may be augmented or diminished by the particular type of neoplasm, its location, its size and extent, its degree of malignancy, the stage of life cycle during which the tumor arises, a history of previous tumors, and the presence or absence of a family history of cancer (33, 146).

Childhood cancer is a chronic life-threatening disorder in which the parents and the child must cope with a great deal of uncertainty. Problems include the type of cancer, the age of the child, the coping skills of the parents and the child, treatment-induced changes in the child's appearance, home adjustment, and school adjustment. In general, the earlier the child is told about the diagnosis, the better the adjustment tends to be. Directness and openness seem to foster better adjustment in long-term survivors. Factors favoring optimal long-term psychological adjustment include (a) having a type of cancer with a relatively short treatment time and minor permanent side effects and (b) having a recurrence-free period (95).

Childhood cancer has a profound effect on the entire family. Unlike parents with a child who has spina bifida, in which the divorce rate is high, parents with a child who has cancer tend to have a much lower divorce rate than the general population. Parents tend to draw closer together and rely on each other for support. However, about half of all childhood cancer survivors do not marry. The more marked the physical impairment is in women survivors, the less likely they are to marry, whereas men survivors with marked physical impairment are more likely to marry (95).

REFERENCES

1. Accardo PJ, Tomazic T, Morrow J, Haake C, Whitman BY: Minor malformations, hyperactivity, and learning disabilities. Am J Dis Child 145: 1184–1187, 1991.

1a. Adolphs R, Gosselin F, Buchanan TW, Tranel D, Schyms P, Damasio AR: A mechanism for impaired fear recognition after amygdala damage. Nature 433:68–72, 2005.

2. American Civil Liberties Union (online, 2004). http://archive.aclu.org/features/f110101a.html http://www.epic.org/privacy/facerecognition/ http://www.dodcounterdrug.com/facial recognition/DLs/Feret7.pdf http://www.dodcounterdrug.com/facial recognition/DLs/FRVT 2000.pdf

3. Andrew RJ: The origin and evolution of the calls and facial expressions of primates. Behavior 20: 1–109, 1963.

4. Archer D, Iritani B, Kimes DD, Barrios M: Face-ism: Five studies of sex differences in facial prominence. J Pers Soc Psych 45:725–735, 1983.
5. Argyle M. The laws of looking. In Conformity and Conflict: Readings in Cultural Anthropology. J.P. Spradley and D.W. McCurdy, Editors. Boston: Little, Brown. Chapter 5, pp. 65–74, 1980.
6. Argyle M, Cook M: Gaze and Mutual Gaze. Cambridge: Cambridge University Press, 1976.
7. Argyle M, Ingham R, Alkema F, McCallin M: The different functions of gaze. Semiotica 7:19–32, 1973.
7a. Associated Press: Honda says Domo Arigato, Mr. Roboto. December 15, 2005.
8. Bartlett JC, Searcy J: Inversion and configuration of faces. Cog Psych 25:281–316, 1993.
9. Bartlett MS, Hager JC, Ekman P, Sejnowski TJ: Measuring facial expressions by computer image analysis. Psychophysiol 36:253–263, 1999.
10. Beattie GW: Sequential temporal patterns of speech and gaze in dialogue. In Nonverbal Communication, Interaction, and Gesture: Selections from Semiotica. T.A. Sebeok, J. Uniker–Sebeok, and A. Kendon, Editors. The Hague: Mouton Publications. pp. 297–320, 1981.
11. Belfer MD, Harrison AM, Murray JE: Body image and the process of reconstructive surgery. Am J Dis Chid 133:532–535, 1979.
12. Berscheid E, Walster E: Interpersonal Attraction. Second Edition. New York: Random House, 1978.
13. Biederman I, Kalocsai P. Neural and psychophysical analysis of object and face recognition. In Face Recognition: From Theory to Applications. H. Wechsler, P.J. Phillips, V. Bruce, F.F. Soulie, and T.S. Huang, Editors. Berlin: Springer-Verlag. pp. 3–25, 1998.
14. Blackwell B, De Morgan NP: The primary care of patients who have bodily concerns. Arch Fam Med 5:457–463, 1996.
15. Breazeal CL: Designing Sociable Robots. Cambridge: MIT Press, 2002.
16. Bruce V: Recognizing Faces. London: Lawrence Erlbaum Associates, 1988.
17. Bruce V, Burton M, Dench N. What's distinctive about a distinctive face? Quart J Exp Psychol 47A:119–141, 1994.
18. Bruce V, Hancock PJB, Burton AM. Human face perception and identification. In Face Recognition: From Theory to Applications. H. Wechsler, P.J. Phillips, V. Bruce, F.F. Soulie, and T.S. Huang, Editors. Berlin: Springer-Verlag. pp. 51–72, 1998.
19. Bruce V, Young A: Understanding face recognition. Br J Psych 74:305–327, 1986.
20. Burgård W, Cremers A, Fox D, Hæhnel D, Lakemeyer G, Schulz D, Steiner W, Thrun S: The interactive museum tour-guide robot. In Proceedings of the Fifteenth National Conference on Artificial Intelligence (AAA198). Madison, WI. pp. 11–18, 1998.
21. Burson N: The Method and Apparatus for Producing an Image of a Person's Face at a Different Age, 1976.
22. Burson N: The Human Race Machine. Newtown: Wolfman Productions, 2002.
23. Burson N: Personal communication, 2004.
24. Burson N: (online, 2004). http://www.nyu.edu/greyart/exhibits/burson/
25. Buss DM: Sex differences in human mate preferences. Evolutionary hypotheses tested in 37 cultures. Behavioral Brain Sci 12:1–49, 1989.
26. Buss DM: The Evolution of Desire: Strategies of Human Nature. New York: Harper Collins, 1994.
27. Byrne D: The Attraction Paradigm. San Diego: Academic Press, 1971.
28. Calder AJ, Keane J, Manes F, Antoun N, Young AW: Impaired recognition and experience of disgust following brain injury. Nature Neurosci 3: 1077–1078, 2000.
29. Carpenter CR: A field study in Siam of the behavior and social relations of the gorilla. Comparative Psychology Monographs 16:1–212, 1940.
30. Cash TF: The psychology of physical appearance: Aesthetics, attributes, and images. In Body Images: Development, Deviance, and Change. T.F. Cash and T. Pruzinsky, Editors. New York: Guilford Press. Chapter 3, pp. 51–79, 1990.
31. Chapdelaine A, Levesque MJ, Cuadro RM: Playing the dating game: Do we know whom others would like to date? Basic Applied Soc Psychol 21:139–147, 1999.
31a. Christenfeld N, Hill E: Whose baby are you? Nature 378:669, 1995.
32. Cohen MM Jr: Perspectives on craniofacial asymmetry. III. Common and/or well-known causes of asymmetry. Int J Oral Maxillofac Surg 24:127–133, 1995.
33. Cohen MM Jr: The Child with Multiple Birth Defects. Second Edition. New York: Oxford University Press, 1997.
34. Cohen MM Jr: Asymmetry: Molecular, biologic, embryopathic, and clinical perspectives. Am J Med Genet 101:292–314, 2001.
35. Cole J: Relations between the face and the self revealed by neurological loss: The subjective experience of facial difference. Soc Res 67:187–218, 2000.
36. Cui V: Short, fat and ugly? You could be Mr. Right. Sunday Morning Post, Hong Kong, January 4, 2004.
37. Cunningham MR, Roberts AR, Barbee AP, Druen PB: Their ideas of beauty are on the whole, the same as our's: Consistency and variability in the cross-cultural perception of female attractiveness. J Personality Soc Psychol 68:261–279, 1995.
37a. Dario P, Guglielmelli E, Allotta B, Carozza MC: Robotics for medical applications. IEEE Robotics & Automation Magazine, September: 44–56, 1996.

38. Dario P, Susani G: Physical and psychological interactions between humans and robots in the home environment. In Proceedings of the First International Symposium on Humanoid Robots (HURO96). Tokyo, Japan. pp. 5–16, 1996.
39. Dautenhahn K: Design issues on interactive environments for children with autism. In Proceedings of the Third International Conference on Disability, Virtual Reality and Associated Technologies (ICDVRAT 2000). Alghero Sardinia, Italy. pp. 153–161, 2000.
40. David TJ, Osborne CM: Scalp hair patterns in mental subnormality. J Med Genet 13:123–126, 1976.
41. Davidson, RJ, Ekman P, Saron CD, Senulis JA, Friesen WV: Approach–withdrawal and cerebral asymmetry: Emotional expression and brain physiology I. J Personality Soc Psychol 58:330–341, 1990.
42. Davis S: Men as success objects and women as sex objects. A study of personal advertisements. Sex Roles 23:43–50, 1990.
43. DeBrune LM: Facial resemblance enhances trust. Proc R Soc Lond B Biol Sci 269:1307–1312, 2002.
43a. DeBruine LM: Resemblance to self increases the appeal of child faces to both men and women. Evol Hum Behav 25:142–154, 2004.
44. Dick, Philp: Do Androids Dream of Electric Sheep? New York: Ballantine Books, 1990.
45. Dimberg U, Thunberg M, Elmehed K: Unconscious facial reactions to emotional facial expressions. Psychol Sci 11:86–89, 2000.
46. Dion KK: Stereotyping based on physical attractiveness issues and conceptual perspectives. In The Ontario Symposium. Vol 3. Physical Appearance, Stigma, and Social Behavior. C.P. Herman, M.P. Zanna, E.T. Higgins, Editors. Hillsdale: Lawrence Erlbaum Associates. pp. 7–21, 1986.
47. Ekman P: Mistakes when deceiving. Ann NY Acad Sci 364:269–278, 1981.
48. Ekman P: Telling Lies: Clues to Deceit in the Marketplace, Marriage, and Politics. New York: Norton, 1986.
49. Ekman P: The argument and evidence about universals in facial expressions of emotion. In Handbook of Social Psychophysiology. H. Wagner and A. Manstead, Editors. Chichester: Wiley. pp. 143–164, 1989.
50. Ekman P: Facial expression of emotion: New findings, new questions. Psych Sci 3:34–38, 1992.
51. Ekman P: Facial expression and emotion. Am Psychol 48:384–392, 1993.
52. Ekman P: Strong evidence for universals in facial expressions: A reply to Russell's mistaken critique. Psychol Bull 115:268–287, 1994.
53. Ekman P, Friesen WV: Nonverbal leakage and clues to deception. Psychiatry 32:88–105, 1969.
54. Ekman P, Friesen W: Unmasking the Face. Englewood Cliffs: Prentice-Hall, 1975.
55. Ekman P, Friesen WV: Facial Action Coding System. Palo Alto: Consulting Psychologists Press, 1978.
56. Ekman P, Friesen WV: Felt, false and miserable smiles. J Nonverbal Behavior 6:238–252, 1982.
57. Ekman P, Friesen WV, Hager JC: New Version of the Facial Action Coding System, 2002 online, 2004. http://dataface.nirc.com/Expression/FACS/New_Version/new_version.html
58. Ekman P, Friesen WV, O'Sullivan M: Smiles when lying. J Personality Soc Psychol 54:414–420, 1988.
59. Ekman P, O'Sullivan M: Who can catch a liar? Am Psychol 46:913–920, 1991.
60. Ellis HD: Processes underlying face recognition. In R. Bruyer, Editor. The Neuropsychology of Face Perception and Facial Expression. Hillsdale: Lawrence Erlbaum Associates, 1986.
61. Exline RV: Visual interaction: The glances of power and preference. In Nebraska Symposium on Motivation. J. Cole, Editor. Lincoln: University of Nebraska Press. Chapter 4, pp. 163–206, 1972.
62. Exline RV, Yellin A: Eye contact as a sign between man and monkey. Symposium on Nonverbal Communication. Nineteenth International Congress of Psychology, London, 1969.
62a. EXPO 2005: Robot exhibition. Aichi, Japan, June 9–19, 2005.
62b. Feaver W: Masters of Caricature from Hogarth and Gillray to Scarfe and Levine. New York: Alfred A. Knopf, 1981.
62c. Fellous J-M, Arbib MA: Who Needs Emotions? The Brain Meets the Robot. New York: Oxford University Press, 2005.
63. Folkes VS: Forming relationships and the matching hypothesis. Personality Soc Psychol Bull 8: 631–636, 1982.
63a. French RM, Brédart S, Huart J, Labiouse C: The resemblance of one-year-old infants to their fathers: Refuting Christenfeld & Hill (1995). Proceedings of the 22nd Annual Cognitive Science Society Conference, 2000, NJ:LEA.
64. Fridlund AJ: Human Facial Expression: An Evolutionary View. New York: Academic Press, 1994.
65. Gaily E, Granstrom M-L, Hiilesmaa V, Bardy A: Minor anomalies in offspring of epileptic mothers. J Pediatr 112:520–529, 1988.
66. Garn SM, Lewis AB, Kerewsky RS: The meaning of bilateral asymmetry in the permanent dentition. Angle Orthodont 36:55–62, 1966.
67. Geschwind N: Specialization of the human brain. In The Brain: A Scientific American Book. San Francisco: W.H. Freeman. pp. 108–117, 1979.
68. Ghazanfar AA, Logothetis NK: Facial expressions linked to monkey calls. Nature 423:937–938, 2003.

69. Ginsburg AG: Visual information processing based on spatial filters constrained by biological data. AMRL Tech Rep 78–129, 1978.
70. Goffman E: Encounter: Two Studies in the Sociology of Interaction. Indianapolis: Bobbs-Merrill, 1963.
71. Goin MK, Goin JM: Psychological effects of aesthetic facial surgery. Adv Psychosom Med 15:84–108, 1986.
72. Goldman–Eisler F: Psycholinguistics: Experiments in Spontaneous Speech. London: Academic Press, 1968.
73. Goode E: Gender and courtship entitlement. Responses to personal ads. Sex Roles 34:141–168, 1996.
74. Gorlin RJ, Cohen MM Jr, Hennekam RCM: Syndromes of the Head and Neck. Fourth Edition. New York: Oxford University Press, 2001.
75. Grammer K, Thornhill R: Human (*Homo sapiens*) facial attractiveness and sexual selection: The role of symmetry and averageness. J Comp Psychol 108:233–242, 1994.
76. Güntürkün O: Adult persistence of head-turning asymmetry. Nature 421:711, 2003.
77. Haber J: MEDUSA–A Facial Modeling and Animation System, Heinz-Billing-Preisträger 2001: Forschung und wissenschaftliches Rechnen. (online, 2004). http://www.billingpreis.mpg.de/hbp01/haber.pdf
78. Haidt J, Keltner D: Culture and facial expression: Open-ended methods find more expressions and a gradient of recognition. Cogn Emot 13:225–266, 1999.
78a. Harlow J: The smile that says where you're from. The Sunday Times—Britain, February 20, 2005.
79. Harmon LD: The recognition of faces. Sci Am 229:70–83, 1973.
80. Harrison AA, Saeed L: Let's make a deal: An analysis of revelations and stipulations in lonely hearts advertisements. J Personality Soc Psychol 35:257–264, 1977.
81. Harrison R: Controversy on the face. Brochure: Spirit of the Age, 1977.
82. Hatfield E, Cacioppo JT, Rapson RL: Primitive emotional contagion. In Emotion and Social Behavior. M.S. Clark, Editor. London: Sage. pp. 151–177, 1992.
83. Hay DC, Young AW: The human face. In A.W. Ellis, Editor. Normality and Pathology in Cognitive Functions. London: Academic Press, 1982.
84. Hendrick C: Roles and gender relationships. In S. Duck, Editor. Handbook of Personal Relationships. New York: Wiley. pp. 429–448, 1988.
85. Hill CT, Rubin Z, Peplau LA: Breakups before marriage. The end of 103 affairs. J Soc Issues 32: 147–168, 1976.
86. Hill H, Bruce V. Effects of lighting on matching facial surfaces. J Exp Psych: Human Percept Perform 22:986–1004, 1996.
86a. Holden C: Random samples. Science 302:1500, 2003.
87. Holowka S, Petitto LA: Left hemisphere cerebral specialization for babies while babbling. Science 297:1515, 2002.
88. Iverson JM, Goldin–Meadow S: Why people gesture when they speak. Nature 396:228, 1998.
89. Izard CE: Innate and universal facial expressions: Evidence from developmental and cross-cultural research. Psychol Bull 115:288–299, 1994.
90. Johnston A, Hill H, Carman N. Recognizing faces: Effects of lighting direction, inversion and brightness reversal. Cognition 40:1–19, 1992.
91. Kapp–Simon K: Self-concept of primary-school-age children with cleft lip, cleft palate, or both. Cleft Palate J 23:24–27, 1986.
92. Kendon A: Some functions of gaze–direction in social interaction. Acta Psychol 26:22–63, 1967.
93. Knight B, Johnston A: The role of movement in face recognition. Vis Cog 4:265–274, 1997.
93a. Korb W, Marmulla R, Raczkowsky J, Mühling J, Hassfeld S: Robots in the operating theatre—chances and challenges. Int J Oral Maxillofac Surg 33:721–732, 2004.
94. Kowner R: Facial asymmetry and attractiveness judgment in developmental perspective. J Exp Psych Hum Percept Perform 22:662–675, 1996.
95. Koocher GP, O'Malley JE: The Damocles Syndrome: Psychosocial Consequences of Surviving Childhood Cancer. New York: McGraw-Hill Book Company, 1981.
96. Langlois JH, Ritter JM, Casey RJ, Sawin DB: Infant attractiveness predicts maternal behaviors and attitudes. Dev Psychol 31:464–472, 1995.
97. Langlois JH, Ritter JR, Roggman LA, Vaughn LS: Facial diversity and infant preferences for attractive faces. Dev Psychol 27:79–84, 1991.
98. Langlois JH, Roggman LA, Musselman L: What is average and what is not average about attractive faces. Psychol Sci 5:214–220, 1994.
99. LaPrelle J, Hoyle RH, Insko CA, Bernthal P: Interpersonal attraction and descriptions of the traits of others. Ideal similarity, self similarity, and liking. J Res Personality 24:216–240, 1990.
100. LaVeck B, Hammond MA, LaVeck GD: Minor congenital anomalies and behavior in different home environments. J Pediatr 96:940–943, 1980.
101. Lefebvre A, Munro I: The role of psychiatry in a craniofacial team. Plast Reconstr Surg 61:564–569, 1978.
102. Le Grand R, Mondloch CJ, Maurer D, Brent HP: Early visual experience and face processing. Nature 410:890, 2001.
103. Lesem J: Loving history: Did a nose rubbing miss become our first kiss? Seattle Post-Intelligencer, February 3, 1981.
104. Levenson RW, Ekman P, Friesen WV: Voluntary facial action generates emotion-specific autonomic nervous system activity. Psychophysiol 27: 363–384, 1990.
105. Levine JA, Pavlidis I, Cooper M: The face of fear. Lancet 357:1757, 2001.

105a. Lewis L: Sony aims to give pint-size humanoid ability to "think." The Times of London, August 7, 2004.
105b. Livingstone MS: Is it warm? Is it real? Or just a low spatial frequency? Science 290:1299, 2000.
106. MacGregor FC: Some psycho-social problems associated with facial deformities. Am Soc Rev 16:629–638, 1951.
107. MacGregor FC, Abel TM, Brgt A, Laver E, Weissman S: Facial Deformities and Plastic Surgery. Springfield: C.C. Thomas, 1953.
108. Marden PM, Smith DW, McDonald MJ: Congenital anomalies in the newborn infant, including minor variations. J Pediatr 64:358–371, 1964.
109. Meltzoff AN, Moore MK: Imitation of facial and manual gestures by human neonates. Science 198:75–78, 1977.
110. Menzel P, D'Aluisio F: *Robo sapiens*: Evolution of a New Species. Cambridge: MIT Press, 2000.
111. Millard T, Richman LC: Different cleft conditions, facial appearance, and speech: relationship to psychological variables. Cleft Palate-Craniofac J 38:68–74, 2001.
112. Miller G: Minor congenital anomalies and ataxic cerebral palsy. Arch Dis Child 64:557–562, 1989.
113. MIRA Lab, University of Geneva, 2004.
114. Møller AP: Female swallow preference for symmetrical male sexual ornaments. Nature 257: 238–240, 1992.
115. Motley MT, Camden CT: Facial expression of emotion: A comparison of posed expressions versus spontaneous expressions in an interpersonal communication setting. West J Speech Commun 52:1–22, 1988.
116. Murray JE, Mulliken JB, Kaban LB, Belfer M: Twenty years experience in maxillocraniofacial surgery: An evaluation of early surgery on growth, function and body image. Ann Surg 190: 320–331, 1979.
117. Murray JE, Swanson LT, Strand RD, Hricko GH: Evaluation of craniofacial surgery in the treatment of facial deformities. Ann Surg 182: 240–265, 1975.
118. Nelson RP: Comic Art and Caricature. Pickering: Beaverbook, 1978.
119. Noh J-y, Neumann U: A survey of facial modeling and animation techniques. USC Technical Report 99–705. (online, 2004). http://graphics.usc.edu/cgit/pdf/papers/survey.pdf
120. O'Sullivan M, Ekman P, Friesen W, Scherer K: What you say and how you say it: The contribution of speech content and voice quality to judgments of others. J Personality Soc Psychol 48:54–62, 1985.
121. Parade Magazine, December 7, 1980, p. 10.
122. Pascalis O, de Haan M, Nelson CA: Is face processing species–specific during the first year of life? Science 296:1321–1323, 2002.
123. Pavlidis I, Eberhardt NL, Levine JA: Seeing through the face of deception. Nature 415:35, 2002.
124. Peck H, Peck S: A concept of facial esthetics. Angle Orthodont 40:284–318, 1970.
125. Penton–Voak LS, Perrett DI: Consistency and individual differences in facial attractiveness Judgments: An evolutionary perspective. Soc Res 67:219–244, 2000.
126. Penton–Voak LS, Perrett DI, Castles D, Burt M, Kobayashi T, Murray LK: Female preferences for male faces change cyclically. Nature 399:741–742, 1999.
126a. Perkins D: A definition of caricature, and caricature and recognition. Studies Anthropol Visual Commun 2:1–24, 1975.
127. Perrett DI, Burt M, Penton–Voak IS, Lee KJ, Rowland DA, Edwards R: Symmetry and human facial attractiveness. Evol Hum Behav 20:295–307, 1999.
128. Perrett DI, Lee KJ, Penton–Voak IS, Rowland DR, Yoshikawa S, Burt DM, Henzi SP, Castles DL, Akamatsu S: Effects of sexual dimorphism on facial attractiveness. Nature 394:884–887, 1998.
129. Perrett DI, May KA, Yoshikawa S: Facial shape and judgments of female attractiveness. Nature 368:239–242, 1994.
130. Pertschuk MJ: Reconstructive surgery. In Body Images: Development, Deviance, and Change. T.F. Cash and T. Pruzinsky, Editors. New York: Guilford Press. Chapter 11, pp. 237–252, 1990.
130a. Platek SM, Raines DM, Gallup GG Jr, Mohamed FB, Thomson JW, Myers TE, Panyavin IS, Levin SL, Davis JA, Fonteyn LCM, Arigo DR: Reactions to children's faces: Males are more affected by resemblance than females are, and so are their brains. Evol Hum Behav 25:394–405, 2004.
131. Preuschoft S: Primate faces and facial expressions. Soc Res 67:245–271, 2000.
132. Pruzinsky T: Collaboration of plastic surgeon and medical psychotherapist: Elective cosmetic surgery. Med Psychotherapy 1:1–13, 1988.
133. Quinn PQ, Rapoport JL: Minor physical anomalies and neurologic status in hyperactive boy. Pediatrics 53:742–746, 1974.
134. Radesäater T, Halldórsdóttir H: Fluctuating asymmetry and forceps size in earwigs, *Forficula anricuria*. Animal Behaviour 45:626–628, 1993.
135. Redmond AC, Donner L, Tilghman D: Psychological evaluation of facial change following orthognathic surgery. Monograph no. 11. Craniofacial Growth Series. G.W. Lucker, K.A. Ribbens, and J.A. McNamara, Jr., Editors. Ann Arbor: University of Michigan pp. 81–117, 1980.
136. Rhodes G: Lateralized processes in face recognition. Brit J Psych 76:249–271, 1985.
137. Richman LC, Eliason MJ, Lindgren SD: Reading disability in children with clefts. Cleft Palate J 25:21–25, 1988.
138. Ridley M: Swallows and scorpionflies find symmetry is beautiful. Science 257:327–328, 1992.
139. Rosenberg EL, Ekman P, Jiang W, Babyak M,

Coleman E, Hanson M, O'Connor C, Waugh R, Blumenthal JA: Linkages between facial expressions of anger and transient myocardial ischemia in men with coronary artery disease. Emotion 1: 107–115, 2001.

140. Rosenberg JB, Weller GM: Minor physical anomalies and academic performance in young school age children. Dev Med Child Neurol 15: 131–135, 1973.

141. Russell JA: Is there universal recognition of emotion from facial expression? A review of cross-cultural studies. Psychol Bull 115:102–141, 1994.

142. Russell JA: Facial expressions of emotion: What lies beyond minimal universality? Psychol Bull 118:379–399, 1995.

143. Russell JA, Bachorowski J-A, Fernández–Dols J-M: Facial and vocal expressions of emotion. Annu Rev Psychol 54:329–349, 2003.

144. Russell JA, Fernández–Dols JM, Editors. The Psychology of Facial Expression. New York: Cambridge University Press, 1997.

145. Samuels CA, Butterworth G, Roberts T, Graupner L, Hoyle G: Facial aesthetics: Babies prefer attractiveness to symmetry. Perception 23:823–831, 1994.

146. Schain WS: Sexual functioning, self-esteem and cancer care. Front Radiat Ther Onc 14:12–19, 1980.

147. Scheib JE, Gangestad SW, Thornhill R: Facial attractiveness, symmetry, and cues to good genes. Proc R Soc London B Biol Sci 266:1913–1917, 1999.

148. Schwartz EI: A face of one's own. Discovery 16: 78–87, 1995.

149. Scutt LE, Chow EWC, Weksberg R, Honer WG, Bassett AS: Patterns of dysmorphic features in schizophrenia. Am J Med Genet 105:713–723, 2001.

150. Sergent J: Microgenesis of face perception. In Aspects of Face Processing. H.D. Ellis, M.A. Jeeves, F. Newcombe, and A. Young, Editors. Nijhoff : Dordrecht. 1986.

151. Servan–Schreiber D, Kolb NR, Tabas G: Somatizing patients: Part I. Practical diagnosis. Am Fam Physician 61:1073–1078, 2000.

152. Shenk D: Watching you. National Geographic 204:2–29, 2003.

153. Shepherd JW, Davies GM, Ellis HD. Studies of cue saliency. In Perceiving and Remembering Faces. G.M. Davies, H.D. Ellis, and J.W. Shepherd, Editors. London: Academic Press, 1981.

154. Sitton S, Rippee ET: Women still want marriage. Sex differences in lonely hearts advertisements. Psychol Rep 58:257–258, 1986.

155. Smith DW, Bostian KE: Congenital anomalies associated with idiopathic mental retardation. J Pediatr 65:189–196, 1964.

156. Smith WJ, Chase J, Lieblich AK: Tongue showing: A facial display of humans and other primate species. Semiotica 11:201–246, 1974.

157. Spitze G: Women's employment and family relations. A review. J Marriage Family 50:585–618, 1988.

158. Swaddle, JP, Cuthill IC: Preference for symmetric males by Zebra finches. Nature 367:165–166, 1994.

159. Swaddle JP, Cuthill IC: Asymmetry and human facial attractiveness: Symmetry may not always be beautiful. Proc RSoc London Series B Biol Sci 261:111–116, 1995.

160. Taeuber CM: Statistical Handbook on Women in America. Phoenix: Oryx, 1991.

161. Taister MA, Holliday SD, Borrman HIM: Comments on facial aging in law enforcement investigation. Forensic Sci Commun. 2:1–11, 2000.

162. Thiessen T: 'Big brother' is watching: Cameras keep close tabs on Britons. South China Morning Post, January 1, 2004.

163. Thornhill R, Grammer K: The body and face of women: One ornament that signals quality? Evol Hum Behav 20:105–120, 1999.

164. Thrun S, Bennewitz M, Burgård W, Cremers A, Dellært F, Fox D, Hæhnel D, Rosenberg C, Roy N, Schulte J, Schulz D. MINERVA: A second generation mobile tour-guide robot. In Proceedings of the IEEE International Conference on Robotics and Automation (ICRA'99), Detroit, MI, 1999.

165. Trumble A: A Brief History of the Smile. New York: Basic Books, 2004.

166. van Hooff JARAM: The facial display of catarrhine monkeys and apes. In Primate Ethology. D. Morris, Editor. New York: Anchor Books. pp. 9–88, 1969.

167. Vargo JK, Gladwin M, Ngan P: Association between ratings of facial attractiveness and patients' motivation for orthognathic surgery. Orthod Craniofac Res 6:63–71, 2002.

168. Vatsyayana, Mallanaga: Kama Sutra. New Delhi: Lustre Press Put. Ltd., 1996 (assumed to have been written in the third or fourth century).

168a. Vuilleumier P: Staring fear in the face. Nature 433:22–23, 2005.

169. Waldrop MF, Bell RQ, McLaughlin B, Halverson CF: Newborn minor physical anomalies predict short attention span, peer aggression, and impulsivity at age 3. Science 199:563–565, 1978.

170. Walster E, Aronson V, Abrahams D, Rottman L: Importance of physical attractiveness in dating behavior. J Personality Soc Psychol 4:508–516, 1966.

171. Waters K: A muscle model for animating three-dimensional facial expression. Computer Graphics 21:17–24, 1987.

171a. Webster MA, Kaping D, Mizokami Y, Duhamel P: Adaptation to natural facial categories. Nature 428:557–561, 2004.

171b. Whalen PJ, Kagan J, Cook RG, Davis FC, Kim H, Polis S, McLaren DG, Somerville LH, Mc-

Lean AA, Maxwell JS, Johnstone T: Human amygdala responsivity to masked fearful eye whites. Science 306: 2061, 2004.

172. Woll S: So many to choose from. Decision strategies in video dating. J Soc Pers Relat 3:43–52, 1986.

173. Wong FY, McCreary DR, Bowden CC, Jenner SM: The matching hypothesis. Factors influencing dating preferences. J Hum Behavior 28:27–31, 1991.

174. Wu Y, Beylot P, Thalmann NM: Skin aging estimation by facial simulation. IEEE Computer Animation. Geneva, Switzerland, May 26–28, 1999. pp. 210–219.

175. Wu Y, Thalmann NM, Thalmann D: A dynamic wrinkle model in facial animation and skin aging. J Visualization Computer Animation 6:195–205, 1995.

176. Yik MSM, Meng Z, Russell JA: Adults' freely produced emotion labels for babies' spontaneous facial expressions. Cogn Emot 12:723–730, 1998.

177. Zebrowitz LA, Montepare JM: Impressions of babyfaced individuals across the life span. Dev Psychol 28:1143–1152, 1992.

178. Zhao W, Chellappa R, Phillips PJ, Rosenfeld A: Face recognition: A literature survey. ACM Computing Surveys 35:399–458, 2003.

179. Zuker RM, Goldberg CS, Manktelow RT: Facial animation in children with Moebius syndrome after segmental gracilis muscle transplant. Plast Reconstr Surg 16:1–8, 2000.

8

Sociocultural and Symbolic Perspectives

CRANIOFACIAL DIVINATION: DETERMINATION OF CHARACTER

Terminology

Divination is defined as *revealing knowledge by supernatural means*. *Mancy* is the suffix for divination. Litho*mancy* is divination by using stones. *Physiognomy* is defined as *judging character and personality traits from facial features*. Lessa (36), an anthropologist, proposed the term *somatomancy* for *divination from the human body*. For example, neomancy is divination by using the number, type, and position of moles on the body. According to doctrine, moles on the face duplicate the position of moles on the body, making it unnecessary for the neomancer to examine the whole body, thus avoiding embarrassment to the client (36).

Somatomancy

There are two main types of somatomancy. *Astral somatomancy* established a link between the features of the human body and astrology. By these means, a person's temperament and character were determined. Also, the future course of events could be predicted, thus determining the person's destiny. In *natural somatomancy*, divination was directly from the human body. Neomancy, discussed above, is an example (36).

There are four general types of natural somatomancy. In *theriologic physiognomy*, divination employed human facial comparison with animals. By comparing the likeness of a specific face with that of a specific animal, a person's character could be likened to the behavioral characteristics of that animal. A second type was the *doctrine of the average person*. The basic tenet was that the mental, moral, and physical attributes of persons were most favorable when they approached the average in the population. People who departed from the norm were, in proportion to their deviation, inferior in mentality, morality, and appearance. Third, the *pseudo-inductive type* was characterized by the single-case method with overgeneralization. If a woman with a dimple in her chin had a personality that pleased everyone, then all women with chin dimples had the same personality that pleased everyone. Fourth, the *intuitive method* had no logical basis whatsoever. Because physiognomists had quasi-mystical powers, they could determine character and personality traits by examining the face (36).

Chinese Siang Mien: Reading Faces

In ancient China, body balance was based on the *yin*, as a negative force, and the *yang*, as a positive force. Equilibrium between these two forces was necessary for good health. Siang Mien—the reading of faces—is more than 2000 years old; it is still practiced today. The yin has feminine and maternal influences on the right side of the face. The yang has masculine and

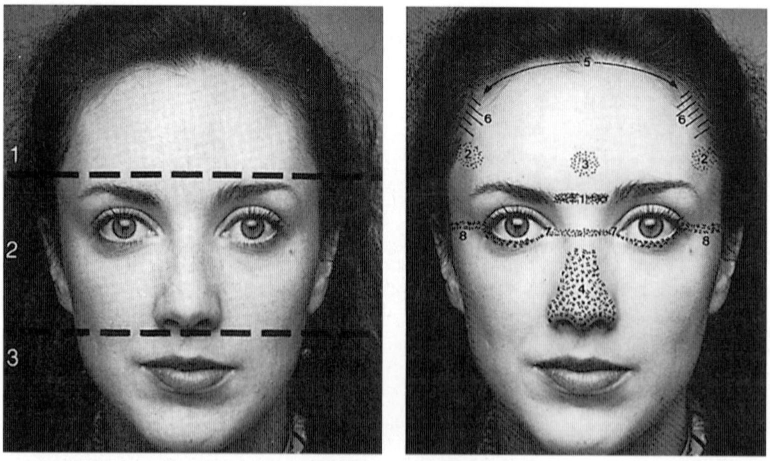

Figure 8-1 Left, Siang Mien divides the face into three *zones*: (1) forehead, which reveals mental capacity; (2) eyebrows to tip of the nose, which reveals luck or fortune; and (3) tip of the nose to chin, which reveals suitability for old age. Right, Eight facial *regions* of Siang Mien: (1) life region, (2) pulse points, (3) career region, (4) wealth region, (5) friendship region, (6) parental region, (7) health and energy region, and (8) love region. From Young, 1984 (65).

paternal influences on the left side of the face. Both sides should be in equilibrium. If the yin is dominant on the right side of the face, the person may be an underachiever, who is submissive and overemotional. If the yang is dominant on the left side of the face, the person may be a high achiever, who strives for power and success (33).

Chinese face reading is based on detailed consideration of facial components: facial shape; three facial zones, including the forehead zone, eyebrow-to-nosetip zone, and nosetip-to-chin zone; eyebrows; eyes; nose; mouth; ear; cheek and cheekbones; chin; and moles. Of the three zones (Fig. 8-1), the forehead indicates a person's mental capacity. The eyebrow-to-nosetip zone indicates a person's fortune and ability to overcome obstacles. The nosetip-to-chin zone indicates the suitability for old age. In addition to the three zones, the face is also divided into eight regions: life region, pulse points, career region, wealth region, friendship region, parental region, health and energy region, and love region (65) (Fig. 8-1).

Physiognomy in Ancient Greece

Niquetius, a seventeenth century Jesuit, named more than 100 classical Greek and Roman scholars who had written on physiognomy. Aristotle wrote two treatises on the subject: *Physiognomonia* and *Historia Animalium*. He noted earlier attempts to read faces by the Egyptians, Thracians, and Sythians (1, 2).

In his *Historia Animalium*, Aristotle thought that people who resembled particular animals had analogous temperaments. Fat ears, like those of an ox, indicated laziness. A fat nose, like that of a pig, signified stupidity. A nose, like that of a hawk, indicated magnanimity. Fine hair, like that of a stag, meant timidity (1).

Pythagoras was said to have selected his pupils on the basis of their facial features. An early physiognomist thought that the face of Socrates indicated brutality and excessive love for women. Hippocrates, considered the father of medicine, used the term *facies Hippocratica* to describe the face in impending death: sharp nose, hollow eyes, tense facial skin, and yellowish color. Galen, a Greek anatomist and physician, thought that personality was a reflection of four humors or fluids that make up the human body (28, 51, 59, 64) (see Chapter 1).

Facial Astrophysiognomy

In 1272, *De Hominis Physiognomia* was written by Scott (53), an astrologer. Although the volume contained some physiological explanations of facial muscles and nerves, the causes were said to be astrological. In 1654, Belot (4) showed the relationship of the face to the planets (Fig. 8-2). In 1658, *Metoposcopia*, and in 1659, *Fisionomia Astrologica* were written by Cardano (10, 11), an Italian mathematician, philosopher, and physician. He founded *metoposcopy*, in which the forehead was mapped into zones, each one denoting a planet (Fig. 8-2). Practically every ex-

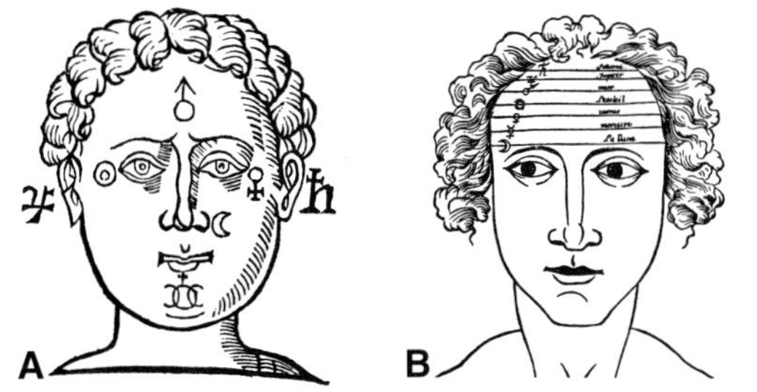

Figure 8-2 Left, Facial astrophysiognomy; the relationship of the face to the planets. From Belot, 1654 (4). Right, Metoposcopy; the relationship of the forehead to the planets. From Cardano, 1658 (10).

ternal part of the body had been the subject of astrophysiognomists (36).

The Physiognomy of Della Porta

Although divination by comparison of humans to animals had been employed by the ancient Greeks (*vide supra*), the concept was advanced by Della Porta during the Renaissance. In 1627, he wrote *Of Celestial Physiognomy* (17), in which he rejected any astrological influence on human facial features, character, or personality. However, in his *De Humana Physiognomonia* of 1601 (16), he compared humans to animals to determine character. Men whose faces looked like lions were courageous, whereas those who looked like goats were stupid (Fig. 8-3).

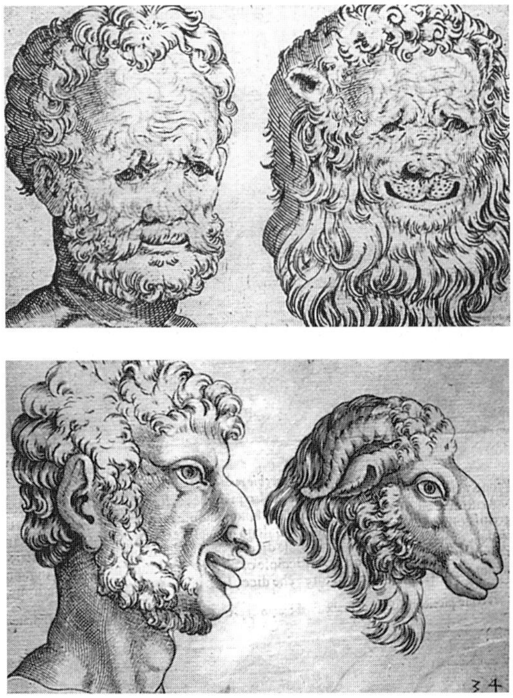

Figure 8-3 Top, Men whose faces looked like lions were courageous. Bottom, Men whose faces looked like goats were stupid. From Della Porta, 1601 (16).

The Physiognomy of Lavater

Johann Kaspar Lavater,[a] a pastor and teacher from Zürich, wrote a famous four volume treatise *Essays in Physiognomy* with his own illustrations. Appearing first in 1772 (35), the volumes became so popular that they were translated into many languages. By the 1940s, 151 editions had appeared (51).

Lavater's objective was "to promote the love of man for his neighbor through the realization of the beauty of the mind and the spirit—a beauty so clearly made by God and so clearly visible in the face, if only one took the trouble to look" (35).

Lavater distinguished between *physiognomy*—static features of the face—and *pathognomy*—the face in motion. Every part of the face was analyzed. The nose was an indicator of taste, sensibility, and feeling. The lips were indicators of mildness, anger, love, and hatred. The chin was an indicator of the degree of sensuality. Women with brown hairy warts on the chin were industrious but amorous to folly or even to frenzy (35, 36, 51, 59).

Lavater thought that by the shape and proportion of facial parts, he could reveal the secrets of the mind. He knew nothing about facial bone structure or the muscles of facial expression. He generalized metaphoricaly. Straight facial features meant straightness of character. Rough hair, eyebrows, and skin meant roughness of character. He eschewed scientific method: "What folly it is to make physiognomy into a science . . ." (24, 35, 36).

Phrenology

Phrenologists argued that the human personality consisted of a number of independent mental

[a]Often Anglicized as John Caspar Lavater.

faculties, each of which was located in a specific region of the brain. The larger each region, the stronger the faculty. By examining the bumps on the skull, a person's character could be determined (25).

Franz Josef Gall received his MD in Vienna in 1785. After developing the principles of phrenology, he left Vienna on a speaking tour accompanied by his pupil and colleague Johann Gaspar Spurzheim. Their work generated great public interest, and hundreds of phrenological societies sprang up in Europe, England, and in America, where numerous phrenological journals also flourished. In Boston, Dr. John C. Warren, Professor at Harvard Medical School, incorporated phrenology into his lectures. He also made phrenology the subject of an annual address to the Massachusetts Medical Society (15). By today's standards, phrenology represented the union of localized brain functions with a primitive behavioristic psychology.

The underpinnings of phrenology were based on four postulates. First, humans were divided into four psychological types: (a) nervous, distinguished by a large brain, delicate health, and emaciation; (b) bilious, marked by harsh features and firm muscles; (c) sanguine, characterized by large lung capacity and moderate plumpness; and (d) lymphatic, with rounded form and heavy countenance. Phrenologists later refined the original four types into three: mental, motive, and vital (15).

Second, the mind was composed of 37 independent and ascertainable faculties. Examples included, among others, benevolence, veneration, secretiveness, combativeness, and amativeness. Third, these faculties were located in different regions of the brain. Finally, these 37 faculties affected the size and contour of the cranium, which could be examined directly (Fig. 8-4). Added to these four postulates of phrenological doctrine was a social philosophy that appealed to reformers because it was possible to cultivate socially desirable and diminish socially undesirable propensities (15).

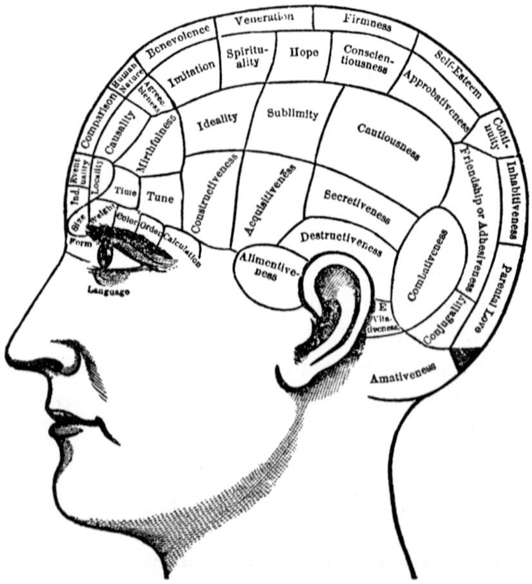

Figure 8-4 In phrenology, independent mental faculties were thought to be located in different regions of the brain. From Davies, 1955 (15).

Criminal Anthropology

Criminal anthropology was established by an Italian physician Cesare Lombroso. On the basis of anthropometric data, he claimed that criminals were evolutionary throwbacks. Their atavism was both physical and mental. Their physical signs were "decisive," based on anatomical grounds. Lombroso's *Criminal Man* was published in 1887 (37) and another book and two articles appeared later (38–40). He discussed what he called "criminal behavior" in animals. Two examples suffice. An ant driven by rage may kill and dismember an aphid! An adulterous stork with her lover killed her husband! (37). He also compared criminals with what he called "inferior groups," such as the Dinka of the Upper Nile (39).

Lombroso (37) listed the following apish features as characteristic of criminals: greater skull thickness, simplicity of cranial sutures, large jaw, dominance of the face compared to the cranium, precocious wrinkles, low and narrow forehead, large ears, darker skin, greater visual acuity, inability to blush, and absence of baldness. Other criminal features were derived from animals. For example, Lombroso (37) compared prominent canine teeth and a flat palate with the anatomy of lemurs and rodents.

On the frontispiece of Lombroso's *Criminal Man* (37) is a montage showing different groups of criminal faces, including German murderers, burglars, purse snatchers, shoplifters, swindlers, and men who had declared bankruptcy fraudulently.

Earnest A. Hooton, Professor of Physical Anthropology at Harvard University, investigated the relationship of crime to physical characteristics in 1939 (31, 32). He found that robbers tended to have heavy beards, diffuse iris pigmentation, attached earlobes, and six other body traits. He asserted "whatever the crime may be, it ordinarily arises from a deteriorated organism . . . You may say that this is tantamount to a declaration that the primary cause of crime is biological inferiority—and that is exactly what I mean."

Human Mismeasurement and Racism

Human mismeasurement has already been discussed in phrenology and criminal anthropology (*vide supra*). Early nineteenth century quantitative approaches were also used to reinforce pre-existing racial prejudice. Blacks and Indians were considered to be separate and inferior species. Nott and Glidden in their *Types of Mankind* in 1854 (48) compared Blacks with apes. The book was not considered to be a fringe document of the time, but the leading American text on human racial differences (26).

Paul Broca, Professor of Clinical Surgery, founded the Anthropological Society of Paris in 1859. He made an exhaustive study of various methods used in determining the cranial capacity of human skulls, although he preferred weighing brains during autopsies that he himself performed (26). Two amazing conclusions from his studies are as follows:

> In general, the brain is larger in mature adults than in the elderly, in men than in women, in eminent men than in men of mediocre talent, in superior races than inferior races . . . Other things being equal, there is a remarkable relationship between the development of intelligence and the volume of the brain (7).

> A prognathous[b] face, more or less black color of the skin, woolly hair and intellectual and social inferiority are often associated, while more or less white skin, straight hair and an orthognathous[c] face are the ordinary equipment of the highest groups of the human series . . . A group with black skin, woolly hair and a prognathous face has never been able to raise itself spontaneously to civilization (9).

Broca (8) also tried to demonstrate a steady increase in brain size with the advance of European civilization from medieval to modern times.

Samuel George Morton, a scientist and physician from Philadelphia, studied a large number of Indian skulls of different tribes and published his results in 1839 (44). He also studied other races in 1844 (45) and 1849 (46). His hypothesis was that ranking various races could be established objectively by the physical characteristics of the brain, particularly its size, which could be determined by measuring skulls. His conclusion was that whites were the highest, Indians were in the middle, and blacks were the lowest. Among whites, Teutons and Anglo-Saxons were the highest, Jews were in the middle, and Hindus were the lowest (26).

Dr. John Langdon Haydon Down described Down syndrome or trisomy 21 syndrome in 1866. The diagnosis was established by the facial features together with mental deficiency. He used the term *idiot*, which was the quasi-technical term of his day. He noted that "A very large number of congenital idiots . . . [were] . . . typical Mongols." He found other idiots of the "Ethiopian variety," "white negroes of European descent," "idiots of the Malay type," and "analogues of the people with shortened foreheads, prominent cheeks, deep-set eyes, and slightly apish nose, originally inhabiting the American continent" (20).

There are many other examples of human mismeasurement. Sheldon (54, 55) in his *Varieties of Human Physique and Varieties of Human Temperament*, classified body types, which were thought to be determinants of personality. Because the subject dealt primarily with the body rather than the face, no further discussion is provided.

Figure 8-5 shows the facial profiles of various humans and animals. In this nineteenth century typology, the Caucasian profile is the highest type with the largest brain and most upright face. Also shown are progressively smaller brains with correspondingly more slanted profiles. In descending order, these types include enlightened,

[b]Forward-jutting.
[c]Straight.

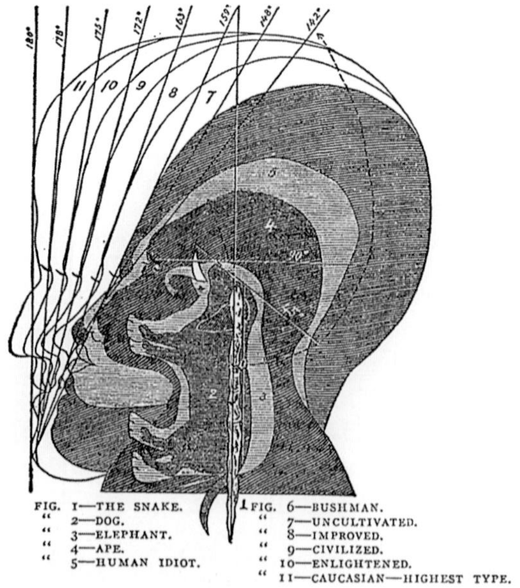

Figure 8-5 Nineteenth century typology showing the Caucasian profile as the highest type with the largest brain and most upright face. Also shown are progressively smaller brains with correspondingly more slanted profiles. In descending order, these types include enlightened, civilized, improved, uncultivated, Bushmen, human idiot, and ape. From Davies, 1955 (15).

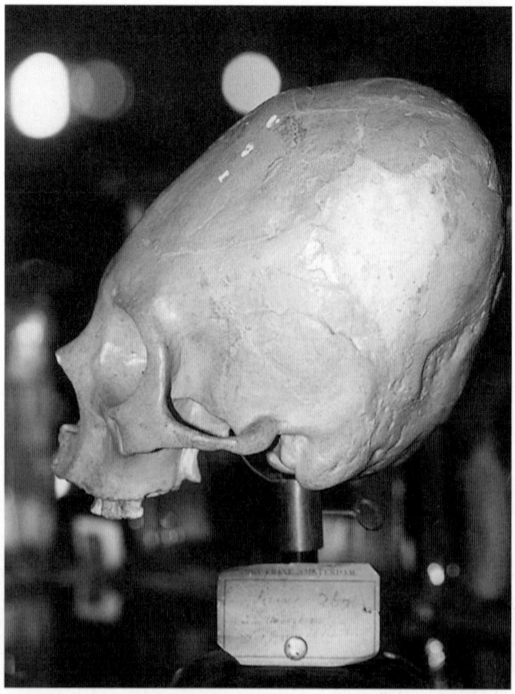

Figure 8-6 Culturally sanctioned skull deformation. From the Vrolik Museum, Amsterdam.

civilized, improved, uncultivated, Bushman, human idiot, and ape (15).

CULTURAL ALTERATIONS OF THE SKULL AND DENTITION

Skull Deformation

Culturally sanctioned skull deformation is both very ancient and widespread and includes hyperbrachycephaly, occipital flattening, and elongation of the skull (Fig. 8-6) or some combination of these. A variety of techniques have been used, such as flat cradle boards, head wrappings, and tight-fitting caps. Both Hippocrates and Pliny noted the popularity of head deformation in high-born Greeks and Romans. Elongated heads are found in Egyptian sculpture, as for example, in Princess Nefertiti and in one of Akhanaten's daughters. Shaping of the skull was widely practiced in the Americas and reached its zenith in the pointed heads of high-born Mayan Indians. Skull deformation is also found in some African peoples (18, 19, 23, 62).

Dental Customs

Dental customs are widespread (Table 8-1). Filing front teeth to a point and knocking out front teeth have been practiced as initiation rites. Teeth have been stained dark brown or black by betel nut chewing or by the application of a black lacquer. Gold, semiprecious stones, and brass studs have been inserted into the front teeth in various groups (14). In North America, moving teeth into ideal positions is a common practice for aesthetic purposes (Fig. 8-7).

FACIAL ADORNMENT

Facial adornment is culturally ubiquitous and varied. It may indicate status either collectively or individually, such as social class, warrior

Table 8-1 Some Dental Customs

Customs	Comments	Geographical Areas Where Practiced[a]
Filing	Front teeth filed to a point in boys and girls for initiation rites	Parts of Africa, parts of Southeast Asia, Central America
Avulsion	Knocking out front teeth, usually lower incisors, but sometimes upper and lower incisors; practiced on boys for initiation rites	Borneo, Malaysia, scattered parts of Africa, Central Australia
Betel nut	Chewed to produce a dark brown or black stain	Asia, Malaysia, Tibet
Black lacquer	Produces black teeth, usually among women only	Thailand, South Vietnam
Gold caps	Gold-capped front teeth	Pre-Colombian Pacific Coast (canines only), Borneo
Semiprecious stones or brass studs	Inserted into the front teeth	Borneo

[a]In many instances, these practices took place aboriginally or, more recently, practices took place before extensive white contact. However, some of these practices still occur, e.g., Borneo.

Data from Davies, 1972 (14).

status, a punk-style haircut, or clown-face makeup. It may indicate style, beauty, sex appeal, or hiding one's age, such as with cosmetics, a coiffure, earrings, electrolysis (facial hair removal), hair dyeing, wearing a wig, hair transplantation, having a rhytidectomy (face-lift) or a rhinoplasty (changing the shape of the nose), or having orthodontic treatment or some types of orthognathic surgery. In other societies, facial adornment may include tattoos, scarification, piercing, cutting, mutilation, and the insertion of various objects into the nose, lips, or ears. Facial adornment may have religious and magical significance, as with Hopi Indian Katchina masks during religious ceremonies or with shamen's masks during curing ceremonies. Facial adornment may also be used to hide one's identity from public view, as with hair covering and face covering in some Moslem countries or with green, silver, or blue sunshades that interfere with eye contact during social intercourse.

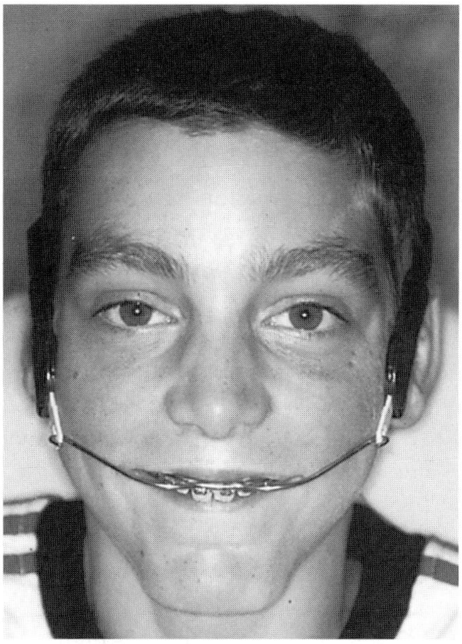

Figure 8-7 Orthodontic treatment with headgear use. Courtesy of K. Russell, Halifax.

During the fifteenth century, it was fashionable for women to shave their foreheads to produce a higher hairline. This can be seen in the profiles of Italian women painted by Pollaiuollo (see Figure 9-15). The eyebrows were also shaved, as can be seen in the famous Netherlandish painting *Portrait of a Lady* by Rogier Van Der Weyden at the National Gallery of Art in Washington, DC.

Adornment in Tribal Societies

Facial tattooing is of great antiquity, having been practiced by the Thracians, Assyrians, Egyptians, Ainu, and Britons. *Tattoo* comes from the Tahitian *tatau*, which means *to strike*—an appropriate term for the delivery of a blow to a pointed instrument that deposits color under the skin. Figure 8-8 shows a New Zealand Maori chief with an elaborate facial tattoo that indicates his rank, stature, and distinction in Maori society. Tattoos in modern society are common, but facial tattooing is rare (61).

Figure 8-8 also shows a young Kayan woman

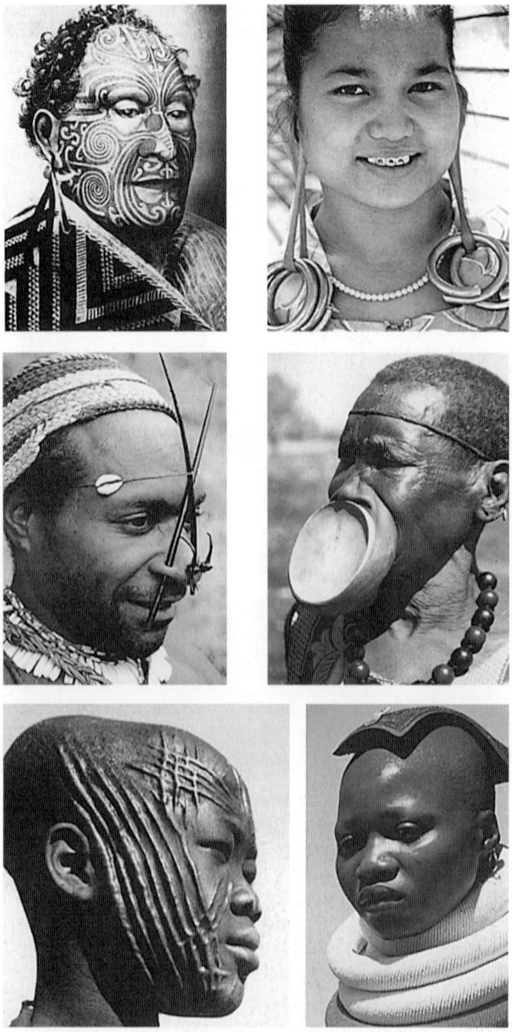

Figure 8-8 Top left, Elaborately tattooed New Zealand Maori chief. From the Bernice P. Bishop Museum, Hawaii. Top right, Kayan woman from Borneo with gold inlaid teeth containing emeralds and rubies. Her pierced ears hold heavy rings that have distended her earlobes below her neck. From Virel, 1980 (60). Middle left, New Guinea man with a bandeau containing a shell, two long points through his nose, and the horns of two scarab beetles also inserted into his nose to make him look fierce. From Ebin, 1979 (21). Middle right, Large wooden lip plug or labret worn by the Chad people of North Central Africa. From Ebin, 1979 (21). Bottom left, Facial scarification in a young man from Chad. From Brain, 1979 (6). Bottom right, Young Bakoba bride from the Transvaal with a thick beaded choker. From Ebin, 1979 (21).

from Borneo with gold inlaid teeth containing emeralds and rubies. Her pierced ears hold heavy rings that have distended her earlobes below her neck (60). Also shown is a New Guinea man with a bandeau containing a shell, two long points through his nose, and the horns of two scarab beetles also inserted into his nose to make him look fierce (21) (Fig. 8-8).

A large wooden lip plug or *labret* is worn by the Chad people of North Central Africa (Fig. 8-8). A hole in the upper lip is gradually enlarged by increasing the size of the lip plugs. Labrets may be worn in the upper lip, in the lower lip, or in both lips. They have been used in West Africa where they are made of stone, bone, clay, or quartz (21). Lip plugs have also been used by the Botocudo of Eastern Brazil.

Facial scarification is common in Africa, serves as tribal markings, and plays an important role during initiation rites. Scarification is shown in a young man from Chad in Figure 8-8 (6). A young Bakoba bride from the Transvaal with a thick beaded choker is also shown (21).

Several customs involve disembodied heads. During the eighteenth century, the Maoris of New Zealand mummified the highly tattooed heads of famous chieftains. Among the Jivaro of the Ecuadorian highlands, the heads of dead enemies were cut off and their facial skin and scalp were removed from their underlying bony attachments and shrunken through a process of mummification. The miniature head, known as a *tsantsa*, retained its facial features. The lips were sealed with three long strings. During the nineteenth century in the Trobriand Islands, widows carried the skulls of their husbands around for the rest of their lives.

Modern Facial Makeup

Facial makeup among women is almost universal in modern society. Often used are moisturizers, creams, liquids, gels, rouge, powders, mascara, eyeshadow, false eyelashes, eyelash tinting, eyebrow tinting, color contact lenses, and lipstick. Various applicators and treatments include cotton balls, swabs, sponges, brushes, pencils, tissue paper, eyelash combs, and eyelash curlers (3, 13). Figure 8-9 shows a woman before and after facial makeup (52). In this advertisement, her assessment of her own "imperfections" appear

Figure 8-9 Woman before (left) and after (right) makeup. From Redbook, 1979 (52).

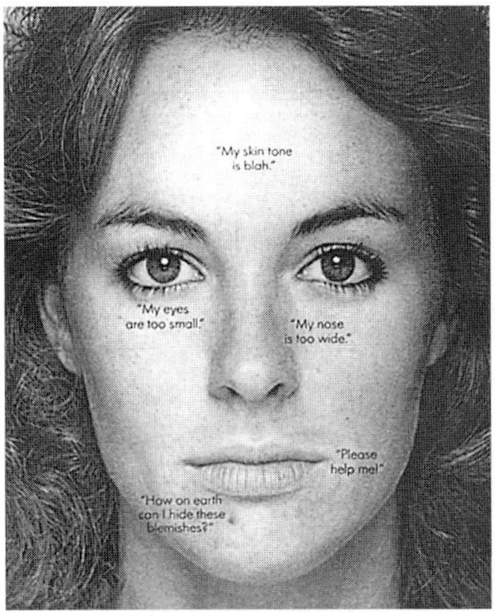

on her premakeup face: "My skin tone is blah"; "My eyes are too small!"; "My nose is too wide"; "How on earth can I hide these blemishes?"; and "Please help me!"

Aoi Festival of Kyoto

Figure 8-10 shows several beautiful Japanese women with elegant dress and makeup at the Aoi Festival in Kyoto held at the Kamigama Shrine and the Shimagamo Shrine in the city of Kyoto during the month of May. The festival, held every year since 544, was originated to pray for a good harvest following frequent yearly storms and floods. Except for musicians and priests, the participants are women dressed in beautiful silk costumes. They use white facial makeup and wear wigs, representing the hair style of the Heian period (794–1185). All women hold in their hands a small paper representation of a human, which is cross-shaped with a long wide base. Women touch the paper to their bodies or gently breathe on them to rid themselves of ill fortune and diseases. The paper human figures are then cast into the river (56).

Figure 8-10 Aoi Festival of Kyoto. Women with white facial makeup, wigs, and silk costumes.

Children's Face Painting

Face painting for children is common for parties, for plays, and just for fun at home. Child makeup kits are available from toy stores, novelty stores, and theatrical supply stores. Water-based face paints are applied with brushes of different sizes.

Figure 8-11 Children's face painting. Left, Girl made up as a mouse. Right, Boy made up as a panda. From Boase and Graham, 1995 (5).

Also used are sponges for applying dry powder and special pencils for fine details (5). Figure 8-11 shows a girl made up as a mouse and a boy made up as a panda.

Circus Clowns

Circus clowns use many different types of facial makeup. Figure 8-12 shows two famous clowns. Paul Jung played with the Ringling Brothers and Barnum & Bailey Circus. He was also a producer of amusing mechanical devices used in circuses. Annie Fratellini is a French female clown and Director of the Ecole National du Cirque (57).

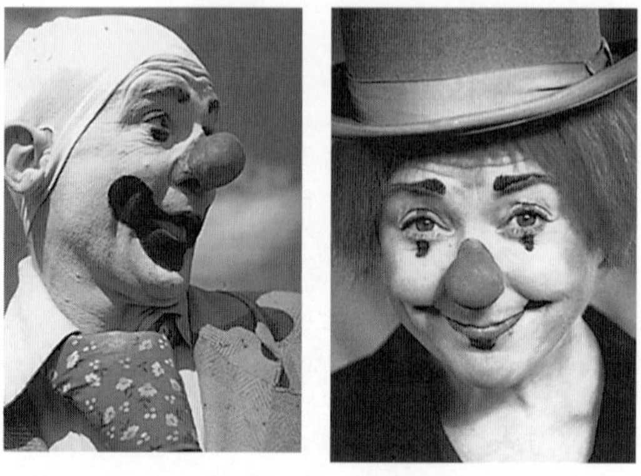

Figure 8-12 Circus clowns with different types of facial makeup. Left, Paul Jung of Ringling Brothers and Barnum & Bailey Circus. Photograph by E. Kennedy. Right, Annie Fratellini, a French female clown and Director of the Ecole National du Cirque. Photograph by M. Fielding.

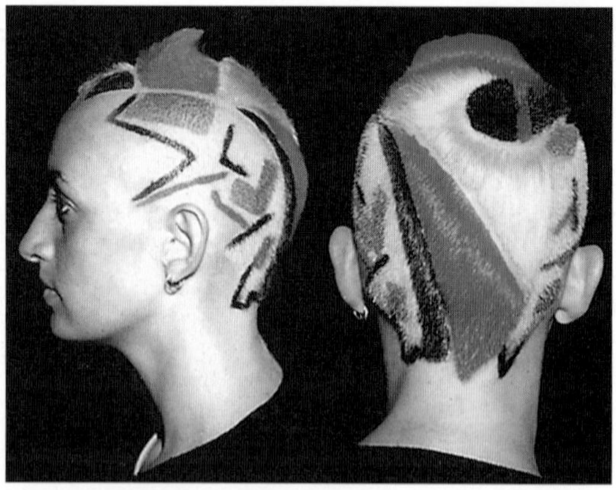

Figure 8-13 Man with punk hair styling and an earring. Hair art by J.-P. Pagès and photograph by S. Richter, Art Unlimited, Amsterdam.

Punk Hair Styling and Facial Piercing

A man with remarkable punk hair styling is shown in Figure 8-13. He also wears an earring. Facial piercing may also involve the eyebrows, nose, mouth, and tongue.

Humages

Humages, the English version of Mario Mariotti's Italian book *Ignani* appeared in 1985 (42). Humages is a contraction of *human* and *images*. The book has an introductory *PreFace* instead of *Preface*. By using his own makeup, Mariotti can transform his face into unexpected and surprising appearances. In Figure 8-14, his face becomes six different men in profile.

FACIAL MASKS

Noh is a highly stylized Japanese masked song and drama, dating back to the fourteenth century. Subject matter is simple and deals with a single emotion such as filial piety, love, jealousy, revenge, and Samuri spirit. Noh plays usually lack a coherent plot and avoid dramatic contrasts and development. Characteristics include (a) a special stage; (b) the role of the *shite* (pronounced shtay), who wears a mask; (c) dance as a means of expression; (d) a special type of vo-

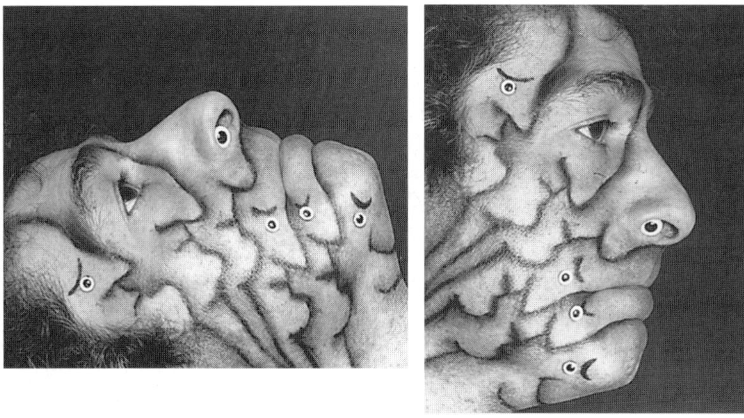

Figure 8-14 Makeup transforms a face into six different men in profile. Left, Side view of the face showing six upright profiles. Right, Upright facial view showing the eye, nose, and chin together with six different profiles facing sideways. From Mariotti, 1985 (42).

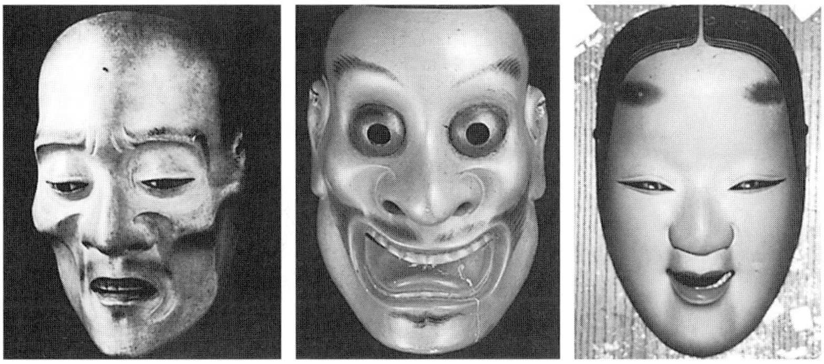

Figure 8-15 Japanese Noh play masks. Left, Shunkan, a man exiled to a deserted island for life. Mask has a sad facial expression with furrowed brow, lowered eyelids, and downturned mouth. Mask from the seventeenth century. From Nakanishi and Komma, 1983 (47). Middle, Ō-tobide, a mask used for role of violent demons. From the sixteenth century. From Nakanishi and Komma, 1983 (47). Right, Waka-onna. Used to play a young woman; however, all Noh play actors are males.

calization; and (e) a group of musicians playing specific instruments: a large hand drum (*ōtsuzumi*), a small hand drum (*kotsuzumi*), a transverse flute (*fue*), and sometimes a drum played with drumsticks (*taiko*) (43, 47, 49, 50). Three Japanese Noh play masks are shown in Figure 8-15.

Figures 8-16 and 8-17 show masks from the Northwest Coast of the North American continent.[d] Although somewhat distinctive art styles flourished, general characteristics include (a) black as a primary color with red as a secondary color; (b) stylistic rather than realistic appearances; (c) highly abstract designs involving distortions, rearrangements, ornamentation, and accentuation of certain features; (d) the use of rounded rectangles known as ovoids and also the use of U forms and S forms; (e) the representation of animals and other creatures often with two profiles; and (f) symmetry but with some exceptions. Typical structural forms are found among the Kwakiutl, but mask paintings in the north among the Bella Coola and the Tlingit are independent of structural form (27, 30, 34, 58).

[d]The Northwest Coast Culture Area in anthropology extends from Northern Oregon to Southern Alaska.

Figure 8-16 Left, Kwakiutl sun mask with open rays. From Hawthorn, 1979 (27). Right, Shaman's mask, perhaps used by the Tlingit. From King, 1979 (34).

Among the Kwakiutl, the sun is shown as a round human-like face surrounded by a halo or fringe of pointed pieces of wood, denoting rays. Figure 8-16 shows a Kwakiutl sun mask with open rays that can also close (27). In contrast to the Kwakiutl sun mask, a shaman's mask, perhaps used by the Tlingit, is less highly ornamented and does not have structural form. Such masks are frequently adorned with inset hair (34) (Fig. 8-16). Figure 8-17 illustrates *Medicine Mask Dance* painted by Paul Kane in 1847. It is his rendition of an imaginary shaman's ritual with Indians wearing masks that he had previously seen on Vancouver Island (34).

Figure 8-18 shows a masked exorcism dance in Sri Lanka (60). Three unusual masks are shown in Figure 8-19. On the left is a mask with asymmetric interruption of the lips and teeth used in an attempt to cure stammering and stuttering in Sri Lanka. In the center is a demon mask with a central cyclopean eye used for the exorcism of demons in Sri Lanka (41). On the right is a wooden mask from the Fon tribe of Benin in West Africa. The mask has a cleft lip in an adult, which was used for public curiosity and even amusement because an adult with a cleft lip did not reflect the traditional facial image of the Fon people (12).

Figure 8-17 Medicine Man Dance, painted by Paul Kane in 1847. From the Royal Ontario Museum, Toronto.

Figure 8-18 Masked exorcism dance in Sri Lanka. From Virel, 1980 (60).

THE FACE IN SYMBOLIC PERSPECTIVE

The word *symbol* has different meanings and is used in various fields, such as philosophy, mathematics, psychology, and art history. Here, I use the term *symbol* to mean the *ambiguous representation of two things by association or resemblance*. In this section, symbolic transformations of faces are shown in Oscar Wilde's *Portrait of Dorian Gray*, in visual illusions, in carved vegetables, in an inverted coffee percolator, and, by imagination, even in multicolored pansies.

The Picture of Dorian Gray by Oscar Wilde

The *Picture of Dorian Gray* was written by Oscar Wilde in 1891 (63). In the book, Basil Hallward painted a portrait of Dorian, who when he saw it, was struck by his own beauty. His portrait takes on symbolic meaning and becomes more revealing than his own face.

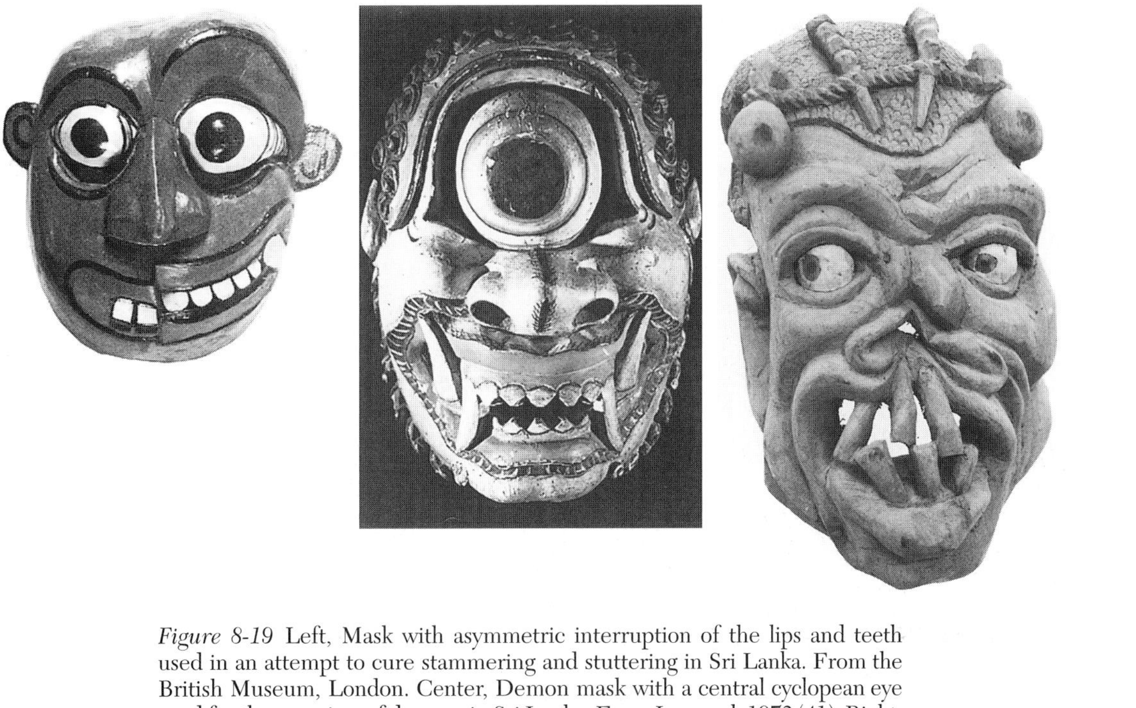

Figure 8-19 Left, Mask with asymmetric interruption of the lips and teeth used in an attempt to cure stammering and stuttering in Sri Lanka. From the British Museum, London. Center, Demon mask with a central cyclopean eye used for the exorcism of demons in Sri Lanka. From Lommel, 1972 (41). Right, Wooden mask with cleft lip in an adult from the Fon tribe of Benin in West Africa. From Cervenka, 1984 (12).

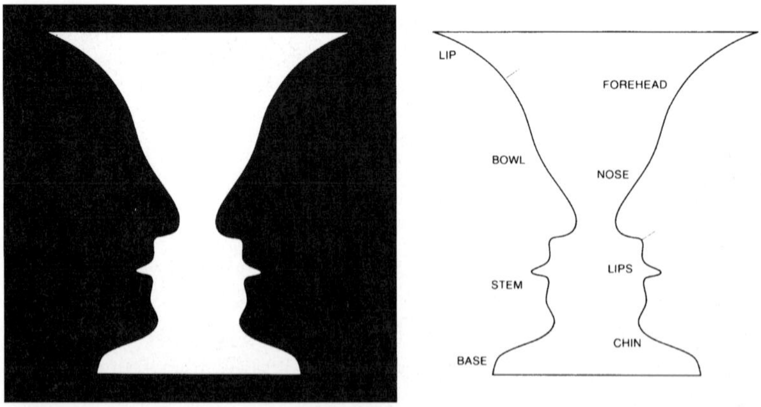

Figure 8-20 Facial goblet illusion. The picture can be seen as a goblet or as a pair of facial profiles. From Hoffman, 1983 (29).

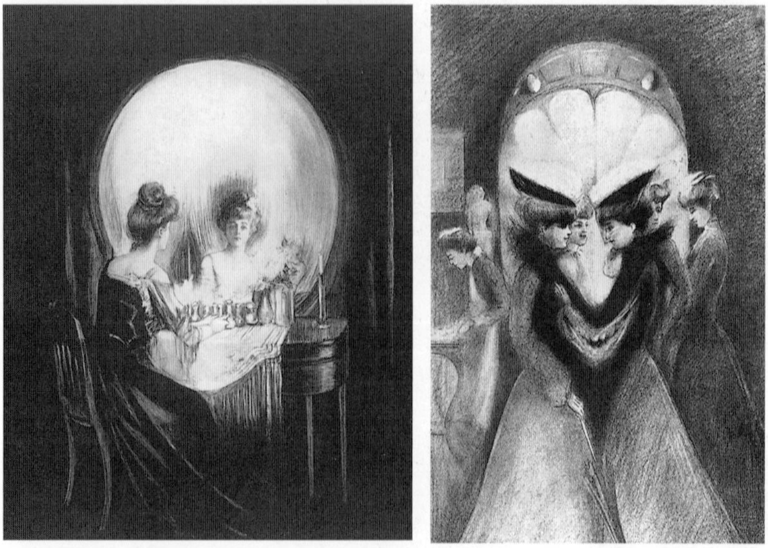

Figure 8-21 Excessive vanity and gossiping are considered vices in these pictures, the vanity doubling as a skull (left) and gossiping ladies doubling as the devil (right). Left, *All is Vanity*, by C.A. Gilbert, House of Art, N.Y. Right, *Gossip and Satan Came Also*, by G.A. Wotherspoosy, House of Art, N.Y.

> "How sad it is!" murmured Dorian Gray, with eyes still fixed upon his own portrait. 'How sad it is' I shall grow old, and horrible, and dreadful. But this picture will remain always young. It will never be older than this particular day of June . . . If it were only the other way! If it were I who was to be always young, and the picture that was to grow old! For that–for that–I would give everything! Yes, there is nothing in the whole world I would not give! I would give my soul for that!" (63)

Abandoning himself to every sin his profligate mind could devise, Dorian brought misery and disgrace to everyone who accepted his companionship. He remained youthful and to his mind "beautiful," while his portrait, now in his attic, became, by degrees, old, ugly, cruel looking, satanic, and satyr-like. The story unfolds as vividly as an Edgar Allan Poe narrative, involving the murder of Basil Hallward by Dorian.

> He looked round, and saw the knife that had stabbed Basil Hallward . . . As it had killed the painter, so it would kill the painter's work, and all that it meant. It would kill the past, and when that was dead he would be free. It would kill the monstrous soul-life, and without its hideous warnings, he would be at peace. He seized the thing, and stabbed the picture with it. (63)

A piercing cry was heard, followed by a crash. Dorian Gray was found dead with a knife in his chest. He looked as old, ugly, cruel looking, satanic, and satyr-like as his portrait, which had now assumed its original youthful appearance.

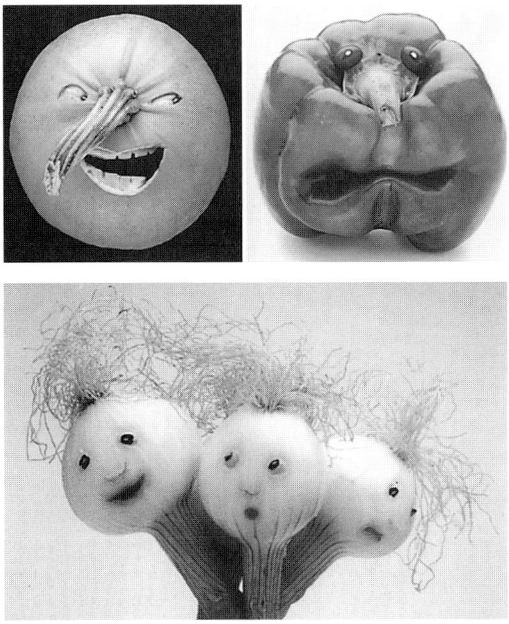

Figure 8-22 Amusing vegetable transformations. Top left, Pumpkin face. Top right, Green pepper goblin. Bottom, Gossiping scallions. From Elffers, 1997 (22).

Visual Illusions

Figure 8-20 shows a famous illusion devised by Edgar Rubin around 1915 (29). The picture can be seen as a goblet or as a pair of facial profiles. When the faces are considered, the curvature of the goblet—bowl, stem, and base—becomes two foreheads, two noses, two pairs of lips, and two chins. Figure 8-21 illustrates a lady putting on makeup in front of her vanity and two ladies talking at some type of gathering. Excessive vanity and gossiping are considered vices in these pictures, the vanity doubling as a skull and the gossiping ladies doubling as the devil.

Fruit and Vegetable Faces

With imaginative alterations, fruits and vegetables can suggest faces that heighten the food's appeal and make them more appetizing. Three amusing vegetable transformations are shown in Figure 8-22. In the pumpkin face, the stem serves as a comic nose and the lines that converge around it suggest facial wrinkles. Shallow carving of the eyes and teeth provide a lighter color contrast to the outer orange color. Small deep holes that completely penetrate the pumpkin serve as pupils, which look off to the side. Also shown is a green pepper goblin with two beans for eyes perched on top of the pepper. The face is suggested by the pepper's shape alone and no alterations have been made. Three gossiping scallions have black-eyed peas for eyes and beet-juice lipstick (22).

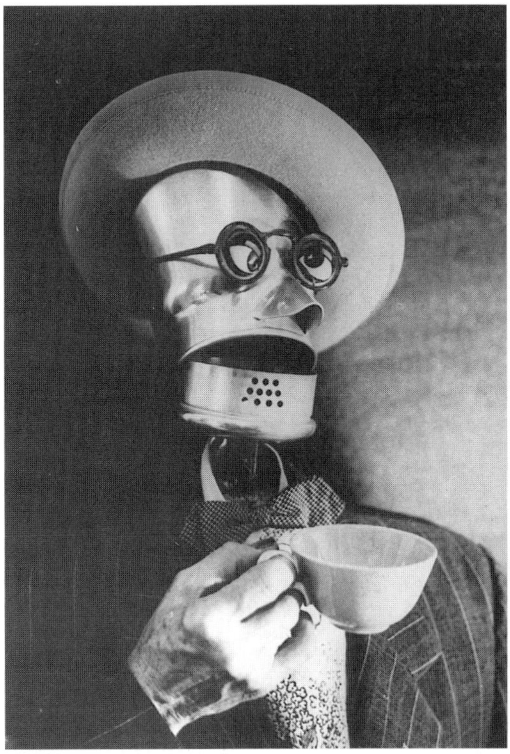

Figure 8-23 *Just Too Much Coffee* is an upside down coffee percolator with eyeglasses transformed into a man's face. From Flashcards, Fort Lauderdale.

Coffee Percolator Face

Americans love coffee. "I must have my morning coffee," but coffee after dinner is also common. Some people drink coffee regularly during the work day and others claim that they need coffee to accomplish their work. But coffee also has a social function. "It's time for a coffee break" or "Let's go for coffee." Although at gatherings and

Table 8-2 Idiomatic Expressions Using Facial Components

Head

To go head on
To have a tête-à-tête[a]
To be headstrong
To keep one's head
To lose one's head
To have a good head on one's shoulders
Get your head straight
Use your head
What's your heading? Which way are you headed?
Heads up!
Head first
A strong headwind
Head-on collision
Headline
Headlight
Department head
Head cheese
Put a head on it[b]

Face

To save face
To lose face
To face up to something, to face the music
To face down
To face the future
On the face of it, at face value
In the face of
To put on one's face, to prepare a face[c]
To show one's face
To come face-to-face, vis-à-vis[d]
To have a long face
Let's face it
About face!
Face-off[e]

Eye

To eye something, to have one's eye on
To see eye to eye
An eye for an eye
With an eye to
Evil eye
In a twinkling of an eye[f]
It's an eyesore[g]
To cause raised eyebrows
To have shifty eyes
To have eyes for
He's all eyes
Bright-eyed and bushy-tailed
I see!
In the eye of the wind
The eye of a hurricane
The eye of a needle
Potato eyes
My eye!
In the public eye
In a pig's eye!
Deja-vu[h]
Eyeball to eyeball
Eyecatcher
Eye opener
Eyeful
Eye shot
Eye tooth

Ear

To grin from ear to ear
To keep one's ear to the ground
To be all ears
His ears are burning
Lend me your ears[i]
I hear!

Nose[j]

To be nosy, to stick one's nose into, to nose around
To put one's nose in the air, to turn one's nose up at
To follow your nose
To lead one around by the nose
To take a nosedive
To thumb one's nose at
A nosegay of flowers[k]

Mouth

To be down in the mouth
He has a big mouth
He has a dirty mouth
To wash one's mouth out with soap
Put your money where your mouth is
Keep a stiff upper lip
Don't give me any lip!
Grin and bear it
To be all smiles
Let a smile be your umbrella

Tongue

To speak tongue-in-cheek
To be tongue-tied
To speak with forked tongue
To stick one's tongue out at

Teeth

To sink one's teeth into
To gnash one's teeth, to clench one's teeth, to grind one's teeth
To have a sweet tooth
To have a toothy grin
To bare one's teeth
By the skin of one's teeth

Other

To be cheeky
Keep your chin up

[a]Although French, the term is used in English to mean a *head-to-head private conversation between two people*.
[b]Meaning to refill a beer glass or mug.
[c]*To prepare a face* was used by T.S. Eliot in *The Love Song of J. Alfred Prufrock*.
[d]Although French, it is used in English, meaning *face-to-face with*.
[e]Method of starting play in ice hockey, lacrosse, and other games in which an official drops the puck or the ball between two opposing players who contend for its control.
[f]In German, the term *Augenblick* means *eyeblink*, literally *in the blink of an eye*. It is used on occasion in English.
[g]Something that is unpleasant or offensive to view.
[h]Although French, it is used in English, meaning *the impression of having been seen before*.
[i]Used in Shakespeare's Julius Caesar in Marcus Antonius's famous speech.
[j]Many well-known people and characters are famous for their noses: *Jimmy Durante, W.C. Fields, Cyrano de Bergerac,* and *Pinocchio*. In *Sesame Street,* the elephant, *Snuffleupagus–Snuffy* for short–has a trunk-nose and a name suggestive of this fact.
[k]Nosegay means a *small bunch of flowers*.

home parties, women traditionally made the coffee, one well-known coffee maker has been called, ironically, "Mr. Coffee." A picture titled *Just Too Much Coffee* is illustrated in Figure 8-23. An upside down coffee percolator with eyeglasses is transformed into a man's face.

Pansy Faces

Pansies are flowers that vary from monochromatic to multicolored. With imagination, pansies can suggest faces of different kinds. The left pansy face in Figure 8-24 appears "happy,"

Figure 8-24 Pansies can suggest faces. Left, "Happy" pansy face. Right, "Surprised" pansy face.

whereas the one on the right appears "surprised."

IDIOMATIC EXPRESSIONS WITH FACIAL COMPONENTS

Americans use a great many idiomatic expressions with various facial components. Table 8-2 lists those involving the head, face, eye, ear, mouth, and teeth. Several selected entries are discussed here.

The word *head* is used in many ways, such as *to keep one's head, to lose one's head, to have a good head on one's shoulders, to get one's head straight*, or *to use one's head. Put a head on it* means to refill a beer glass or beer mug. Although *tête-a-tête* is French, it is also used in English to mean a head-to-head private conversation between two people.

The phrase *to prepare a face* was used by T.S. Eliot in *The Love Song of J. Alfred Prufrock* (22a):

> There will be time, there will be time
> To prepare a face to meet the faces that you meet

A *face-off* is a method starting play in ice hockey, lacrosse, and other games in which an official drops the puck or the ball between two opposing players who contend for its control. A French term, also used in English, *vis-a-vis* means face to face with.

The *evil eye* is a look or stare that is believed to harm others by magical or supernatural means. The English phrase *in a twinkling of an eye* is similar to the German *Augenblick*, meaning eyeblink, literally in the blink of an eye. A French term, also used in English, *déjà-vu* means having the impression of having seen before. The term *eye* has some common but perhaps unusual usages when considered: *the eye of a hurricane, the eye of a needle*, and *the eyes of a potato*. Two expressions–one using the eye, the other using the face–have different kinds of emphasis. *Eyeball to eyeball* suggests confrontation, whereas *face-to-face* suggests frank discussion.

Lend me your ears is an expression used by Shakespeare in Julius Caesar. Marcus Antonius opens his famous speech with

> Friends, Romans, Countrymen, lend me your ears. (26a).

Many well-known people and characters are famous for their noses: Jimmy Durante, W.C. Fields, Cyrano de Bergerac, and Pinocchio. The term *nosegay* means a small bunch of flowers. It is often used redundantly as a *nosegay of flowers*.

REFERENCES

1. Aristotle: Historia Animalium, approximately 340 B.C.
2. Aristotle: Physiognomonia, approximately 340 BC.
3. Aucoin K: Making Faces. Boston: Little, Brown, 1997.
4. Belot: Oeuvres, 1654.
5. Boase P, Graham B: Step-by-Step Face Painting and Dressing Up for Kids. New York: SMITHMARK Publishers, 1995.
6. Brain R: The Decorated Body. New York: Harper & Row, 1979.
7. Broca P: Sur le volume et la forme du cerveau suivant les individus et suivant les races. Bulletin Société d'Anthropologie Paris 2:139–207, 301–321, 441–446, 1861.
8. Broca P: Sur la capacité des crânes parisiens des

diverses époques. Bulletin Société d'Anthropologie Paris 3:102–116, 1862.
9. Broca P: Anthropologie. In Dictionnaire Encyclopédique des Sciences Médicales. A. Dechambre, Editeur. Paris: Masson, pp. 276–300, 1866.
10. Cardano: Metoposcopia, 1658.
11. Cardano: Fisionomia Astrologica, 1659.
12. Cervenka J: African mask with cleft lip and palate. Cleft Palate J 21:38–40, 1984.
13. Clark F: Vogue Guide to Make-up. New York: Penguin Books, 1981.
14. Davies DM: Influences of Teeth, Diet and Habits on the Human Face. London: Heinemann Medical Books, 1972.
15. Davies JD: Phrenology: Fad and Science. New Haven: Yale University Press, 1955.
16. Della Porta GB: De Humana Physiognomonia, 1601.
17. Della Porta GB: Of Celestial Physiognomy, 1627.
18. Dembo A, Imbelloni J: Deformaciones Intencionales del Cuerpo Humano de Carácter Étuico, Buenos Aires, 1938.
19. Dingwall EJ: Artificial Cranial Deformation: A Contribution to the Study of Ethnic Mutilations. London: John Bale, Sons and Danielson, 1931.
20. Down JLH: Observations on an ethnic classification of idiots. London Hospital Reports, pp. 259–262, 1866.
21. Ebin V: The Body Decorated. New York: Thames and Hudson, 1979.
22. Elffers J: Play with Your Food. New York: Stewart, Tabori & Chang, 1997.
22a. Eliot T.S.: The Love Song of J. Alfred Prufrock, 1917. In T.S. Eliot, The Complete Poems and Plays, 1909–1950. New York: Harcourt Brace, pp. 3–7, 1952.
23. Ewing JF: Hyperbrachycephaly as Influenced by Cultural Conditioning. Papers of the Peabody Museum of American Archaeology and Ethnology. Cambridge: Harvard University. Volume XXIII, Number 2, 1950.
24. Fridlund AJ: Human Facial Expression: An Evolutionary View. San Diego: Academic Press, 1994.
25. Gardner M: Fads and Fallacies in the Name of Science. New York: Dover Publications, 1957.
26. Gould SJ: The Mismeasure of Man. New York: W.W. Norton, 1981.
26a. Harrison GB, Editor: Shakespeare: The Complete Works. New York: Harcourt, Brace, 1952.
27. Hawthorn A: Kwakiutl Art. Seattle: University of Washington Press, 1979.
28. Hill T: The Contemplation of Mankind, 1588.
29. Hoffman DD: The interpretation of visual illusions. Sci Am 249:154–162, 1983.
30. Holm B: Northwest Coast Indian Art: An Analysis of Form. Vancouver: J.J. Douglas, 1965.
31. Hooton EA: Crime and the Man. Cambridge: Harvard University Press, 1939.
32. Hooton EA: The American Criminal. Cambridge: Harvard University Press, 1939.
33. Hu H-C: The Chinese concept of 'face.' Am Anthropol 1:45–65, 1944.
34. King JCH: Portrait Masks from the Northwest Coast of America. London: Blacker Calmann Cooper, 1979.
35. Lavater JC: Essays in Physiognomy, 1772.
36. Lessa A: Somatomancy–precursor of the science of human constitution. In Reader in Primitive Religion: An Anthropological Approach. W.A. Lessa and E.Z. Vogt, Editors. Evanston: Row, Peterson and Company. pp. 355–365, 1958.
37. Lombroso C: L'homme criminel. Paris: F. Alcan, 1887.
38. Lombroso C: Criminal anthropology applied to pedagogy. Monist 6:50–59, 1895.
39. Lombroso C: Histoire des progrès de l'Anthropologie et de la Sociologie criminelles pendant les années 1895–1896. Trav 4th Cong Int d'Anthrop. Crim. Geneva. pp. 187–199, 1896.
40. Lombroso C: Crime: Its Causes and Remedies. Boston: Little, Brown, 1911.
41. Lommel A: Masks: Their Meaning and Function. New York: McGraw-Hill Book Company, 1972.
42. Mariotti M: Humages. Florence: La Nuova Italia Editrice S.p.A., 1985.
43. Maruoka D, Yoshikoshi T: Noh. Osaka: Hoikusha, 1996.
44. Morton SG: Crania American or, A Comparative View of the Skulls of Various Aboriginal Nations of North and South America. Philadelphia: John Pennington, 1839.
45. Morton SG: Observations on Egyptian ethnography, derived from anatomy, history, and the monuments [separately reprinted subsequently as Crania Aegyptiaca, with title above as subtitle]. Transactions of the American Philosophical Society 9:93–159, 1844.
46. Morton SG: Observations on the size of the brain in various races and families of man. Proceedings of the Academy of Natural Sciences Philadelphia 4:221–224, 1849.
47. Nakanisha T, Komma K: Noh Masks. Osaka: Hoikusha, 1983.
48. Nott JC, Gliddon GR: Types of Mankind. Philadelphia: Lippincott, Grambo and Company, 1854.
49. O'Neill PG: A Guide to Nō. Tokyo: Hinoki Shoten, 1954.
50. Ortolani B: The Japanese Theatre. Princeton: Princeton University Press, 1990.
51. Piderit T: Mimik und Physiognomik. Detmold: Meyer, 1886.
52. Redbook: The Magazine for Young Women. Volume 153, Number 6, October, 1979.
53. Scott M: De Hominis Physiognomia, 1272 (printed in 1477).
54. Sheldon WH: The Varieties of Human Physique. New York: Harper, 1940.
55. Sheldon WH: The Varieties of Human Temperament. New York: Harper, 1942.
56. Shiota K: Personal communication, 2003.

57. Speaight G: The Book of Clowns. New York: MacMillan Publishing, 1980.
58. Stewart H: Looking at Indian Art of the Northwest Coast. Seattle: University of Washington Press, 1979.
59. Tytler G: Physiognomy in the European Novel: Faces and Fortunes. Princeton: Princeton University Press, 1982.
60. Virel A, Lenars C, Lenars J: Decorated Man: The Human Body as Art. New York: Harry N. Abrams, 1980.
61. Webb S: Pushing Ink: The Fine Art of Tattooing. New York: Simon and Schuster, 1979.
62. Weinman JP, Sicher H: Bone and Bones. Second Edition. St. Louis: C.V. Mosby, 1955.
63. Wilde, Oscar: The Picture of Dorian Gray. New York: Penguin Books, 1982.
64. Wilson JQ, Herrnstein RJ: Crime and Human Nature. New York: Simon & Schuster, 1985.
65. Young L: Secrets of the Face: The Chinese Art of Reading Character through Facial Structure and Features. Boston: Little, Brown, 1984.

9

Artistic Perspective

PHYSIOGNOMIC LIKENESS, ART, AND BEAUTY

Physiognomic Likeness

There is often a tendency to think that a portrait should be as lifelike as possible. Although this has been true during certain periods of art history, it cannot be said to represent a general tendency (*vide infra*). Two arguments have been used against true likeness.

First, a portrait is timeless long after the sitter dies. What the sitter really looked like is not remembered, but the artist's rendition. When it was remarked that Michelangelo's Medici portraits in the Sagrestia Nuova were not good likenesses, Michelangelo said that in a thousand years, it won't matter what these men looked like (5). It was said that the 1906 portrait of Gertrude Stein by Picasso was not a true likeness when it was painted. Picasso observed that, "in time, she will come to look like my painting." Gertrude herself liked the portrait from the beginning, stating "For me, it is I" (17).

Second, during the Renaissance, Filippino Lippi is said to have painted a portrait that was more real than the sitter himself (31), suggesting that the artist can penetrate through the veil of mere physical appearance and reveal some deeper psychological meaning, value, or truth (11).

Art and Beauty

With respect to facial art, the meanings of art and beauty need to be addressed briefly, although these are complex philosophical topics (19, 25, 30, 32).

Art is often defined as an attempt to create pleasing forms (25). Generally, every work of art has structure, patterns of some kind, and even distortions. A typical Greek statue is distorted in an ideal way with an aqualine nose that lacks a nasofrontal angle. There can be various degrees of distortion. Many of Picasso's works are quite distorted and yet are great works of art (3). Although structure, pattern, distortion, or abstraction characterize various art forms, works of art cannot be reduced to simple or explicable formulas. Indeed, some of the most interesting works may have ambiguities, may have meanings at several different levels, or may suggest multiple interpretations.

The identification of art with beauty is very problematic. A good Renaissance portrait of an ugly man can be "beautiful." Picasso often combined "beauty" and "ugliness" in his paintings, which gave them great power (3). Concepts of beauty have differed during various periods in art history. The classic concept of beauty arose in ancient Greece, was perpetuated in Rome, and was revived during the Renaissance. It differs from the ideals of beauty in primitive art, Byzantine art, Oriental art, and modern art.

The school of aesthetics based on beauty has been superseded by an illuminating theory of aesthetics derived from the philosophy of

Benedetto Croce. Its basic tenet is that art can be defined simply as *intuition* and that *expression* is a basic creative act. Although Croce's theory has been met with a flood of criticism, it does circumvent any concept of beauty (25).

THEORIES OF FACIAL PROPORTIONS

Theories of human proportions reflect the history of art styles. There are fundamental differences between the methods used in ancient Egypt, ancient Greece, the Middle Ages, the Renaissance, and modern times (20). Several are discussed here.

Byzantine Proportions

Byzantine facial proportions are geometric and based on the length of the nose as a basic unit of measure. The Byzantine three-circle scheme is illustrated in Figure 9-1, which shows a close-up of Christ's head from the mosaic ceiling of the Baptistry associated with the Cathedral in Florence. Three concentric circles give the external measurements for the face and head, using nose-lengths as the radius. A circle of one nose-length outlines the forehead, cheeks, and nasal tip. A circle of two nose-lengths gives the external measurements of the head. A circle of about three nose-lengths passes through the neck and outlines the halo (20).

Dürer's Facial Proportions

Albrecht Dürer was a great German portraitist of the Renaissance. Basing his theories on Piero della Francesca's earlier work with perspective, he expressed his own ideas about human proportions in his *Four Books on Human Proportion* published posthumously in 1528 (7). Using a coordinate system, Dürer could alter facial proportions and thus create various disproportions (Fig. 9-2).

Golden Ratio

The Golden Ratio is based on the Fibonacci series:

$$0, 1, 1, 2, 3, 5, 8, 13, 21, 34, 55, 89, 144, \ldots \infty$$

The ratio of any two sequential numbers in the series falls into an increasingly narrow range. The ratio oscillates around 1.618, known by the Greek letter Φ (phi). However, before its mathematical refinement by Leonardo Pisano (better known by his nickname Fibonacci) in the early part of the thirteenth century, the Golden Ratio was used by the ancient Egyptians and Greeks. Historically, it has been variously

Figure 9-1 Close-up of Christ's head from the mosaic ceiling of the Baptistry associated with the Cathedral in Florence. Byzantine facial proportions are based on nose length as a basic unit of measure. Three concentric circles give the external measurements of the face and head. A circle with a radius of one nose-length outlines the forehead, cheeks, and nasal tip. A circle of two nose-lengths gives the external measurements of the head. A circle of about three nose-lengths passes through the neck and outlines the halo.

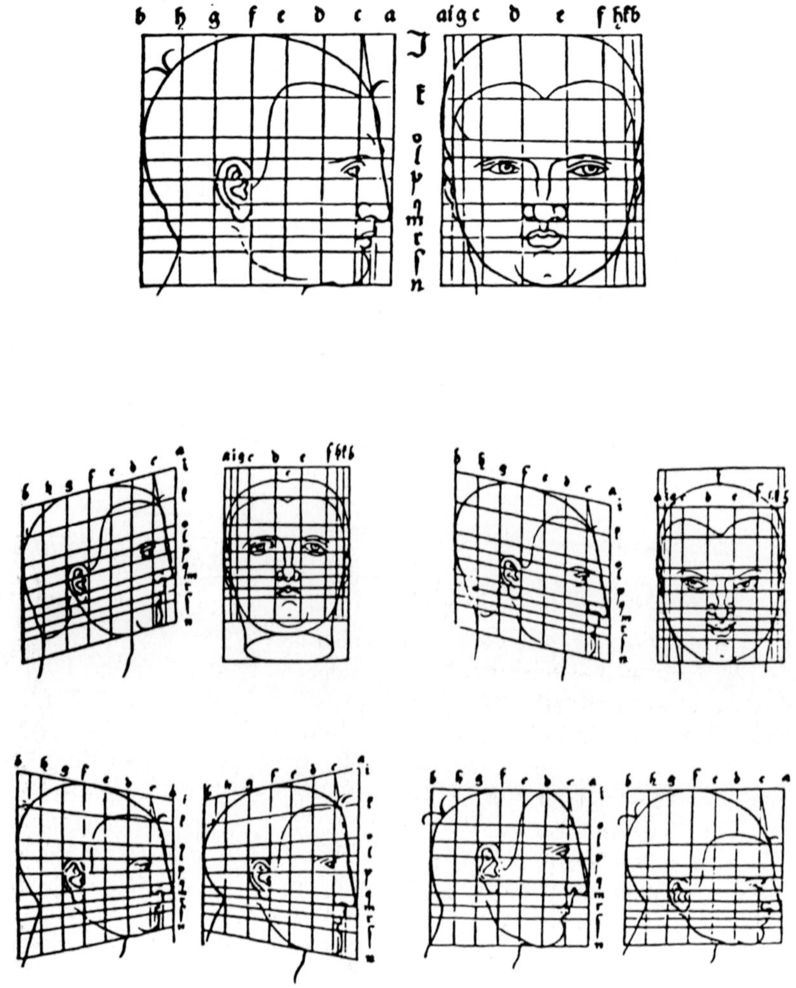

Figure 9-2 Dürer's use of a coordinate system to alter facial proportions, creating various disproportions. Top, Proportionate face. Middle, Bottom, Disproportionate faces. From Dürer, 1587 (7).

termed the *Golden Section*, the *Golden Mean*, and the *Divine Proportion*. It has been used in mathematics, architecture, art, botany, and by clinicians, such as plastic surgeons, to determine the most ideal or "pleasing" facial proportions.

Figure 9-3 demonstrates ideal facial proportions using Φ. If the distance between the hairline and the eye is set at 1, then the distance between the eye and the chin is Φ. If the distance between the wing of the nose and the chin is set at 1, then the distance between the hairline and the wing of the nose is Φ. If the distance between the eye and the wing of the nose is set at 1, then the distance between the wing of the nose and the chin is Φ. If the distance between the lip and the chin is set at 1, then the distance between the eye and the lip is Φ. Finally, if the distance between the wing of the nose and the lip is set at 1, then the distance between the lip and the chin is Φ (26).

Face Shape and Eye Widths

The basic face shape may be rounded, ovoid, squared, or pear-shaped (Fig. 9-4). The face can

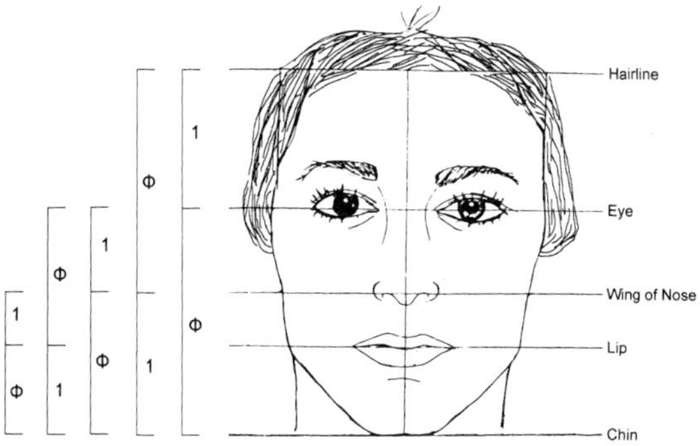

Figure 9-3 Use of ϕ to demonstrate ideal or "pleasing" facial proportions. See text. Modified from Ricketts, 1982 (26).

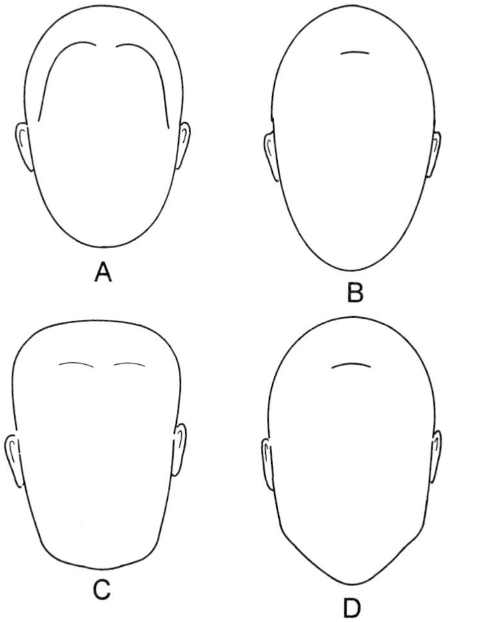

Figure 9-4 Basic face shapes. A, Rounded. B, Ovoid. C, Squared. D, Pear-shaped (inverted) From Powell and Humphreys, 1984 (24).

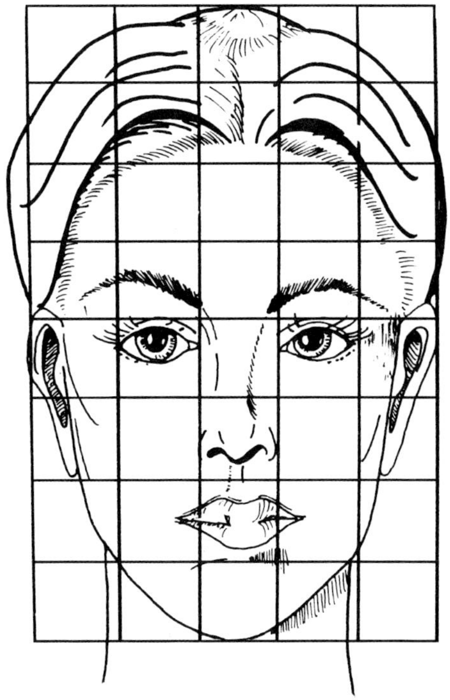

Figure 9-5 The face can be divided equally into five eye widths vertically and approximately eight eye widths horizontally. Modified from Powell and Humphreys, 1984 (24).

be divided equally into five eye widths vertically and approximately eight eye widths horizontally (24) (Fig. 9-5).

DRAWING FACES

The face can be divided into three equal parts by four horizontal lines at trichion, nasion, subnasale, and menton (Fig. 9-6A). Trichion is lo-

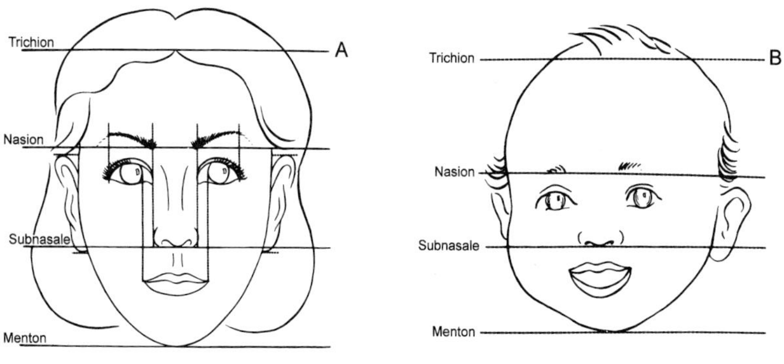

Figure 9-6 A, Drawing of an adult woman's face. Horizontal lines through trichion, nasion, subnasale, and menton divide the face into three equal parts. B, Drawing of an infant's face. The forehead between trichion and nasion occupies the greatest proportion of the face. From Lee and Lee, 1979 (15).

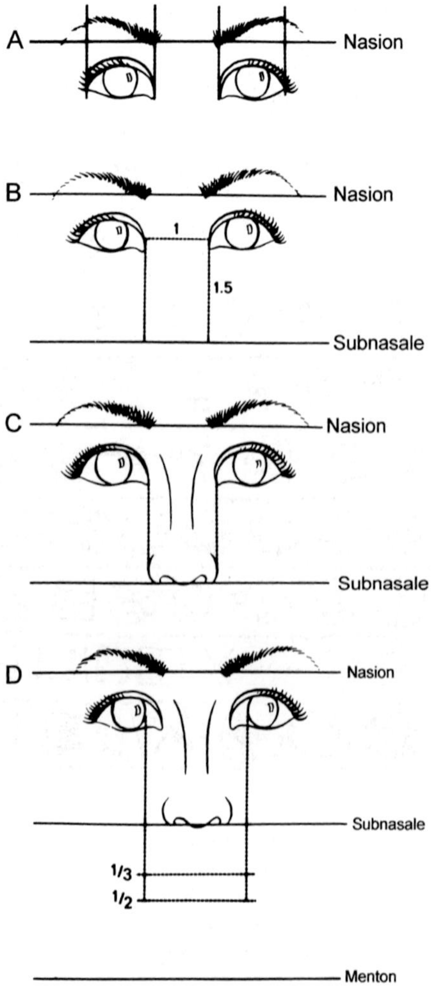

Figure 9-7 A, B, The width of each eye equals the distance between the eyes. Nose length is 1.5 times one eye width. C, Nose width is the same as one eye width and vertical lines of such length dropped from the inner canthi determine the alar rims of the nose. D, The approximate width of the mouth is determined from vertical lines dropped from the medial edge of each limbus. The mouth should be placed between one-third and one-half the distance between subnasale and menton. All facial measurements are reviewed in Figure 9-6A; note that the ears are positioned from slightly below nasion to slightly below subnasale. From Lee and Lee, 1979 (15).

cated at the hairline[a]; nasion corresponds to the frontonasal suture; subnasale corresponds to the anterior nasal spine; and menton corresponds to the symphysis of the mandible. In the infant, the head and the eyes are prominent and the face is diminutive (see Figs. 1-6, 3-16). In drawing an infant face, the head is disproportionately large (15) (Fig. 9-6B).

Figure 9-7(A, B, C) shows that nose width is the same as eye width and nose length is one and a half times eye width. The mouth should be placed between one-third and one-half the distance between subnasale and menton. The approximate width of the mouth can be determined from vertical lines dropped from the medial edge of each limbus (Fig. 9-7D). The ears should be positioned from slightly below nasion to slightly below subnasale (15) (Fig. 9-6A).

CHILDREN'S ART

Young children's art depends on neuromuscular control, previously drawn structures, and self-teaching. Humans are not drawn from life by stumbling efforts. Children know that arms and legs are not attached to the face even though they draw them that way. The perceptual process in children's art is entirely different than their obvious recognition of normal facial and body parts (13).

Figure 9-8 shows the artistic developmental stages that children go through. About 20 basic scribbles are learned. They depend on muscular tension but do not require visual guidance; blind children can make the same basic scribbles. The sequence of events consists of scribbles at age 2 and combines, aggregates, suns, and sun faces at ages 3–4 years. Humans with head-top markings and arms and legs attached to the face are drawn at age 4. Head-top markings become simplified or removed and the arms drop out also at age 4 (Fig. 9-9). Children begin to draw figures with bodies, including dresses and pants by 4–5 years of age (13) (Fig. 9-8). Some drawings of faces by older children are illustrated in Figure 9-10.

[a]For sketching a bald man, trichion can be estimated by having him wrinkle his forehead to form a triangular area from the point where the wrinkling stops to where the aponeurosis of the scalp begins.

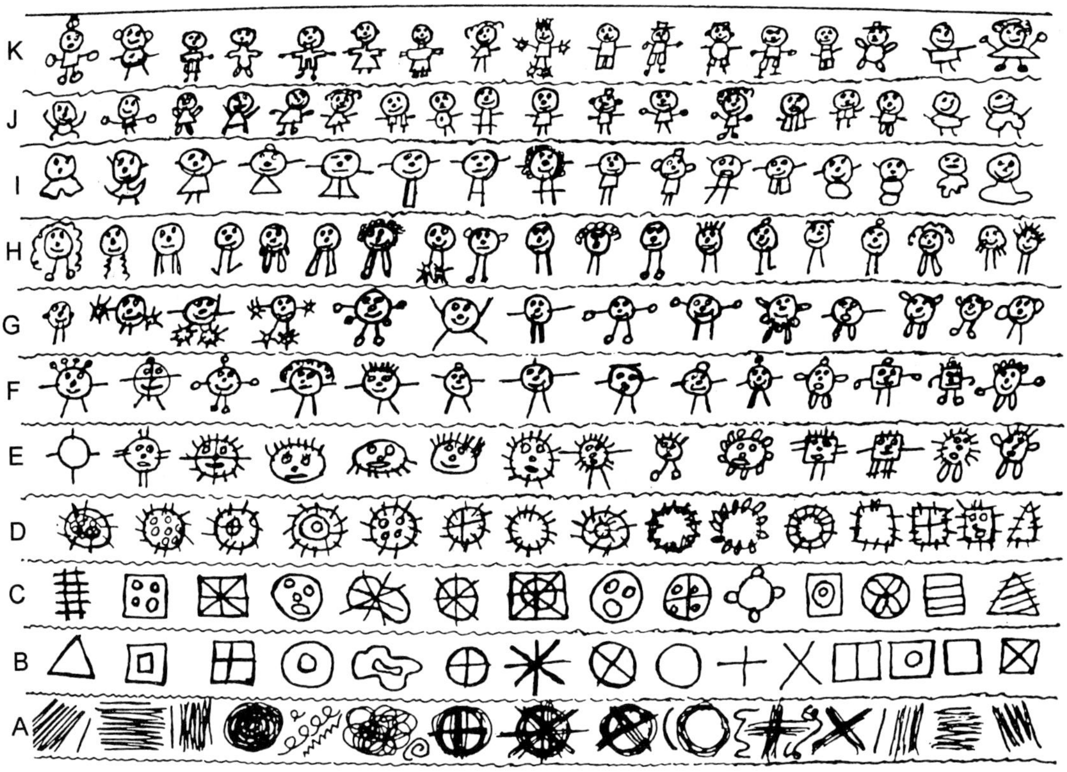

Figure 9-8 The probable evolution of human drawings from the earlier scribbling of children. Basic scribbles (A) lead to diagrams and combines (B), aggregates (C), suns (D), sun faces and figures (E), humans with head-top markings and with arms attached to the head (F), humans without head-top markings (G), armless humans (H), humans with varied torsos (I), humans with arms attached to the torso (J), and relatively complete human images (K). From Kellogg, 1970 (13).

Figure 9-9 Children draw humans with arms and legs attached to the face at age 4. Armless humans are drawn *after* drawing faces with attached arms.

Although artistic development in children is sequential (Fig. 9-8), the path is not always straight. In some instances, simplifications follow more complex structures, as with head-top markings and arms discussed above. Unfortunately, parents sometimes look at children's art in terms of mistakes and forgetfulness, as, for example, when a child draws an armless human after drawing earlier faces with arms. One reason is that, proportionately, facial figures without arms may begin to look better to children. Stick figures are really not part of children's natural artistic development; they are taught to children by adults or by older children (13).

Figure 9-10 Some drawings of faces by older children.

Figure 9-11 A series of caricatures of King Louis Philippe showing that he looked like a pear. Pear in French is *poire* but it has a double meaning as a *fathead.* Drawn by newspaper editor Charles Philipon, Daumier's employer (12).

CARICATURES

The word *caricature*, derived from the Italian *caricare*, was defined in Samuel Johnson's first *Dictionary of the English Language: To hold to ridicule . . . A drawing . . . intended as humour, satire and comment.* Good caricature depends on enlargement of the head, simplification, and exaggeration. Concentration on a few commanding features is emphasized instead of too many features (9, 12, 18).

Figure 9-11 illustrates a series of caricatures of King Louis Philippe drawn by newspaper editor Charles Philipon, Daumier's employer. He had transformed the king's head (Fig. 9-11, upper left) directly into a pear (Fig. 9-11, lower right). Pear in French is *poire* but it has a double meaning of *fathead.* Philipon was summoned

Figure 9-12 Groucho Marx and two different caricatures of him. They both illustrate his distinctive hair shape, bushy eyebrows, thick mustache, and large cigar. Left, Photograph from A. Devaney, Inc., New York. Center, Caricature by D. Levine, New York Review of Books. Right, Caricature by H. Montiel.

and a 6,000 franc fine was levied. In his own defense, he drew the now famous sequence of four faces, showing how the king could be transformed into a pear (12, 22).

Figure 9-12 shows a photograph of Groucho Marx and two caricatures of him. Although very different in style, they both illustrate his distinctive hair shape, bushy eyebrows, thick mustache, and large cigar.

Differences Between Caricatures and Cartoons

Caricatures can be differentiated from cartoons. Caricatures are commonly distorted representations of well-known particular people rather than general types of people. In contrast, cartoons are commonly based on imaginary beings. Some editorial cartoons in newspapers and magazines may deal with real people and, although the facial features may be exaggerated, the head is usually not enlarged as in caricatures. Problems in the definition and classification of caricatures and cartoons also arise because many caricaturists also draw cartoons (9, 18).

CARTOONS

Figure 9-13 illustrates two cartoons of Tweety and Sylvester (1) and Jerry, Tuffy, and Tom (34). Well-liked animals like Tweety, Jerry, and Tuffy have large infantile heads in relation to their bodies. Such heads make them cute in the same way that enlarged heads, prominent eyes, and diminutive faces make human infants cute (see Figs. 1-6 right, 1-8 left). In contrast, "villains" like Sylvester and Tom have adult head sizes in relation to their bodies.

PORTRAITURE

Portrait Styles

There are many portrait styles that represent different historical time periods, individual artistic approaches, and many possible facial expressions of each individual sitter. Two different types of portraits are illustrated in Figure 9-14.

One of these is an 1832 portrait by Jean Auguste Dominique Ingres, *Louis-François Bertin,* that hangs in the Louvre (Fig. 9-14A). Bertin was a powerful newspaper owner. The portrait suggests a large predatory bird. Bertin is a substantial man lurching forward, as if to spring from his seat. He has feather-like hair, a hawk-like nose, and talon-like hands (8).

The other is a 1940 portrait titled *Linda* by Chuck Close, the reigning portrait painter of the Information Age (Fig. 9-14B). His face-forward heads are monumental in size, about 9 feet tall; they are derived from photographs. His portraits

Figure 9-13 Left, Tweety and Sylvester. Right, Tom, Jerry, and Tuffy. Well-liked animals like Tweety, Jerry, and Tuffy have large infantile heads that make them cute in the same way that enlarged heads, prominent eyes, and diminutive faces make human infants cute. In contrast, "villains" like Sylvester and Tom have adult head sizes in relation to their bodies. Left, From Tweety and Sylvester. Courtesy of Warner Bros. Right, Tom and Jerry, Number 322, September, 1979. Courtesy of Warner Bros.

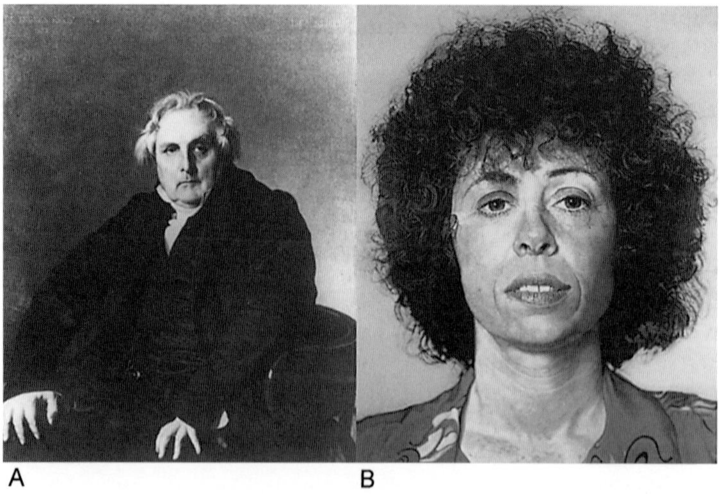

Figure 9-14 Two portrait styles. A, *Louis-François Bertin*, 1832 portrait by Jean Auguste Dominique Ingres, Louvre, Paris. B, *Linda*, 1940 portrait by Chuck Close, Pace Gallery, New York.

have frozen immobile expressions. The technique used is a system of incremental dots and other tiny markings arranged on an enlarged photographic grid. The dots ultimately coalesce in a pointillistic Seurat-like manner into a recognizable image (27, 33).

Renaissance Portraits

The Renaissance was a revival and recapitulation of the classical forms of Greek and Roman art. Human character, human emotions, and human motivation became important. A person could be commemorated for all time. The first modern portraits were painted during the Renaissance. Three factors were important for developing portraiture: (a) the study of the facial profiles on Romain coins and medallions; (b) the inclusion of wealthy patrons in religious narrative art of the churches; and (c) the painting of whole families, both living and dead, in the same work of art (23).

Profiles were used initially for several reasons. A profile casts a shadow on the wall, forming an image that can be traced and cut with scissors. A profile view is easy to paint. Also, a profile is

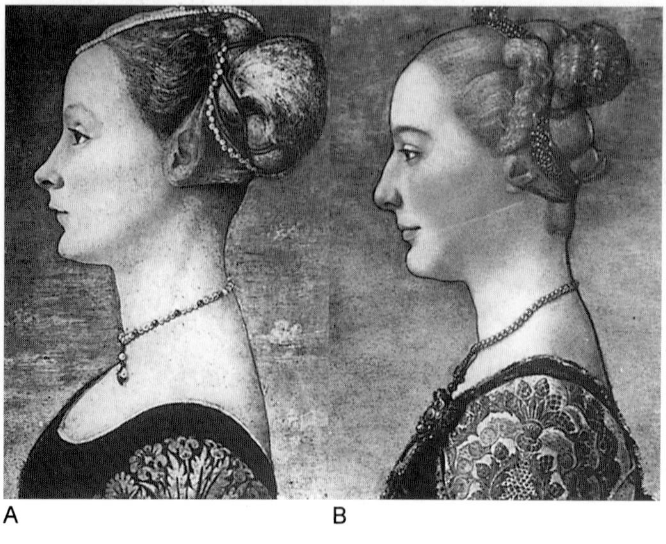

Figure 9-15 Renaissance profile portraits by Antonio del Pollaiuolo. A, *Portrait of a Woman*, Poldi Pezzoli Museum, Milan. B, *Portrait of a Young Woman*, Uffizi Gallery, Florence.

often the most flattening view (23). Two profile paintings of women by Pollaiuolo are illustrated in Figure 9-15.

In religious narrative painting, patrons were usually placed in the foreground at the left and right lower corners of the painting, where they could view the religious scene; they commonly knelt in prayer. In religious paintings and in family portraits, three-quarter views characterize most faces with profile views relegated to secondary people (23). Paintings in frontal view by Botticelli and Tintoretto are found in Chapter 1 (see Fig. 1-20).

Giuseppe Arcimbaldo: Unusual Facial Paintings

Giuseppe Arcimbaldo, who lived from 1527 to 1593, was born in Milan, where his Italian family name was actually Acimbaldi. He and his father, both artists, designed several stained glass windows for the cathedral in Milan, and Giuseppe also designed a religious mural tapestry for the Como Cathedral (14).

In 1562, Arcimbaldo left Milan to become a court artist in Vienna and Prague. He worked under Ferdinand I and his successors Maximil-

Figure 9-16 Unusual facial paintings of Giuseppe Arcimbaldo. A, *Summer*, Kunsthistorisches Museum, Vienna. B, *Autumn*, Louvre, Paris. C, *The Librarian*, Skoklosters Slott, Styrelsen, Stockholm. D, *The Jurist*, Statens Konstsamlingar Gripsholm Slott, Stockholm.

ian II and Rudolph I. As a court painter, he developed a technique of painting human heads composed of fruits, vegetables, trees, vines, leaves, and other items. Profiles of the four elements—water, fire, earth, and air—were painted in 1566, as were two other paintings—*The Jurist* and *The Librarian*. Four sets of the *four seasons* were produced—one in 1563, one in 1572, and two in 1573 (14). Four of these unusual works are illustrated in Figure 9-16.

As a court painter in Vienna and Prague, Arcimbaldo signed his name in various ways: Giuseppe, Josephus, Joseph, Josepho, Arcimbaldi, Arcimbaldo, or Archimbaldus. Most books and articles written about him use the name Arcimbaldo (14).

For the heads of the four element series, water is composed of fish, crustaceans, and other marine life; fire—flames, fire-making equipment, and firearms; earth—various terrestrial animals; and air—different kinds of birds. For the heads of the four seasons, *Spring* is composed of flowers; *Summer*—vegetables and fruit (Fig. 9-16A); *Autumn*—mostly fruits and some vegetables (Fig. 9-16B); and *Winter*—mostly arboreal forms (2, 4, 14). Also shown in Figure 9-16 (C,D) are *The Librarian*, composed of books, and *The Jurist*, composed of chicken and fish.

SURREALISM, SEXUAL SYMBOLISM, AND THE FACE

The object of *Surrealism* was to free the artist from the normal association of pictorial ideas and accepted means of expression so that art could be created according to the irrational dictates of the unconscious mind (35). Familiar objects were placed in unfamiliar contexts with bizarre juxtapositioning of known images. Together with De Chirico and Duchamp (10), Picasso was one of three major influences on the development of Surrealistic art.[b]

Picasso's relationship to Surrealism is complex. He did influence the Surrealists who, in turn, influenced him and he exhibited his paintings with them. However, Picasso did not consider himself a Surrealist and although the early Surrealists considered him one of them, ultimately they did not. He never became a true Surrealist because he was unable to approach external reality with his eyes closed (29). He rejected the supremacy of the dream world over the conscious, waking, visual world (10).

The Surrealists were interested in symbolism, in the interchangeability of images, and in Freudian sexual metamorphosis. Picasso shared these same ideas with Miro, Dali, and Magritte. Sexual themes had been recurrent in Picasso's art and appeared early, but became predominant in the 1920s and 1930s (10, 28, 35). Picasso frequently confounded nose with penis, mouth with vulva, and breasts with testicles (28) for aesthetic purposes.

Two surrealistic paintings with sexual puns of the face are illustrated in Figure 9-17. In René Magritte's 1934 painting *The Rape*, the face is made up of the woman's body. The breasts are the eyes, the nose in the navel, and the vulva with pubic hair is the mouth. In the painting by Barbara Millett, the nose is the man's penis and his beard is substituted by his scrotal sac.

PICASSO AS A RESEARCHER AND EXPERIMENTALIST

The work of Picasso shows that he was a researcher and an experimentalist. To illustrate one of Picasso's many facial experiments, four drawings of a woman's head, three done on June 9 and one completed on June 13, 1938, are shown in Figure 9-18. In Figure 9-18A, the cheek is composed of a burrette-like structure with the ear as a stopcock and the eye as a keylike stopper at the top. Most striking is the fact that the cheek appears to arise from the lower jaw but seems somewhat independent from the rest of the face. In Figure 9-18B, the cheek is now an integral part of the face but the ear has been transformed into a bugle or musical horn. A somewhat tubular ear on the opposite side is confounded positionally with the eye on that side, suggesting telescopic powers. In Figure 9-18C, the ear has become a flower and the other ear appears to be a more open blossom. Inter-

[b]Surrealism began as a literary movement and continued as one.

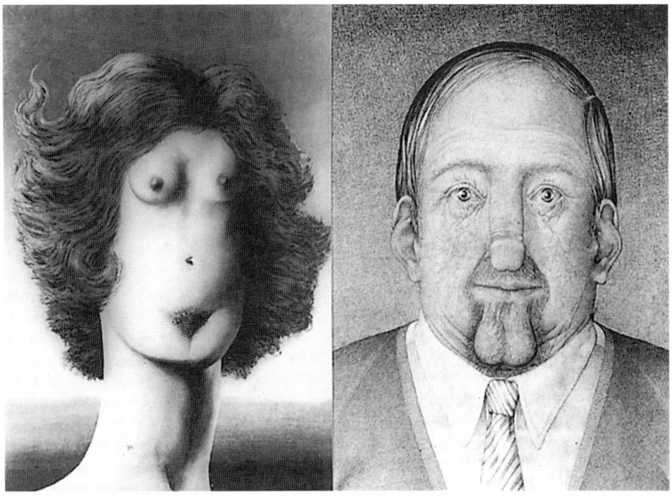

Figure 9-17 Surrealism, sexual symbolism, and the face. A, *The Rape*, by Réne Magritte, George Melly Collection, London. B, *Man with Penis-Nose and Scrotal Beard*, by Barbara Millett, Menil Foundation, Texas.

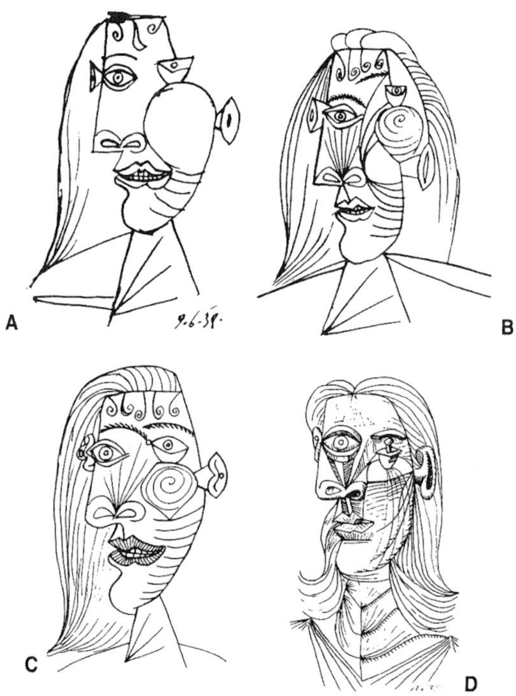

Figure 9-18 Four drawings of a woman's head, by Picasso. A, Zervos IX, 153, June 9, 1938. B, Zervos IX, 154, June 9, 1938. C, Zervos IX, 155, June 9, 1938. D, Zervos IX, 156, June 13, 1938. India ink, 45.1 × 24.1 cm (36).

Figure 9-19 Head of a Woman, by Picasso. 1925. Lithograph, 12.7 × 11.4 cm, Mourlot XX.

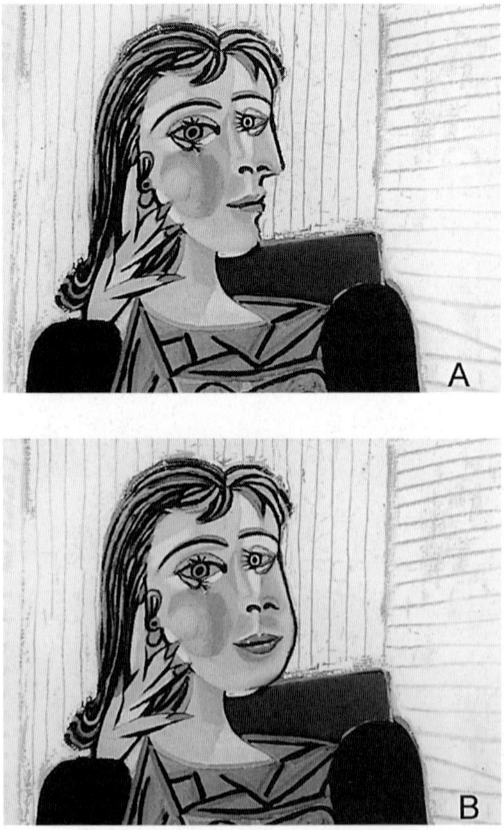

Figure 9-20 A, *Portrait of Dora Maar*, by Picasso. 1937. Oil on canvas, 92.1 × 64 cm. Musée Picasso, Paris. B, Reconstruction of three-quarter view by the author (3).

estingly, the neck now resembles a twisted stalk whereas in the two earlier versions, the head was balanced on the neck. In Figure 9-18D, a section of the face around the eye is cut away so that the workings of an inner light bulb become visible. The ear suggests a handle that upon turning might allow the face-cover to swing open on a hinge, revealing other underlying structures. Further dissection is suggested by the vertically oriented muscle fiber-like bands associated with the other eye and the stitched-up neck that looks like it had been previously cut open (3).

Many of Picasso's experiments explore a variety of possible solutions to problems. Within bounded time frames, works may be viewed as logical sequences (3). In 1956, Picasso asserted:

> Paintings are but research and experiment . . . I never do a painting as a work of art. All of them are research. I search incessantly and there is a logical sequence in all this. . . . (16)

That Picasso wanted his art to be understood sequentially is borne out by the fact that he dated his sketches, drawings, and paintings beginning in the 1930s, indicating that they were experiments in time.[c]

In *Head of a Woman*, a lithograph of 1925 (Fig. 9-19), frontal and profile views coexist in the same work by accommodation. Because Picasso often began with realistic drawings or paintings, he could start afresh from a naturalistic perspective to find a new way of tackling an old problem. Figure 9-20 shows one of Picasso's lovers—Dora Maar. In 1937, he painted a portrait of her in frontal view with a profile inserted into the cheek (Fig. 9-20A). Some art critics believe these paintings represent both frontal and profile views equally; others believe they represent a profile view primarily with some degree of frontality. My own experiment on Picasso's painting is shown in Figure 9-20B; it shows that neither interpretation is likely. In Figure 9-20B, the profile has been eradicated by painting a rounded line from the cheek to the chin, erasing the profile line, and completing the lip. It is extremely easy to do and results in a reasonably realistic facial image which Picasso must have started with; he painted a face in frontal view and then simply inserted a profile on the cheek (3).

The implications of profile insertion for experimental simultaneity were far reaching. Picasso evolved a series of stages in which the nose began to acquire a life of its own. Two nostrils and one eye became transformed into a snout. The next stage consisted of snout elevation, in

[c]It should be realized, however, that Picasso was quite capable of ambiguities, paradoxes, and even falsehoods (3). For example, in the 1920s, he vigorously denied being a researcher and experimentalist (6). And as for experiments in time, almost all of his work between 1907 and 1914 was unsigned as well as undated, and there are known reasons for this. Differences in what Picasso said should be perceived not so much as contrary statements, but as a change of opinion with time or as conversations with different emphasis addressing different questions; the person to whom he was talking was also a factor (3). He once told Penrose (21), "You mustn't always believe what I say. Questions tempt you to tell lies, particularly when there is no answer."

Figure 9-21 Snout formation, snout elevation, and snout rotation. A, *Marie-Thérèse Walter*, March 4, 1939. Zervos IX, 261. B, October 12, 1939. Zervos IX, 352, 1939. C, April 14, 1940. Zervos X, 368. D, *Woman With Hat*, 1941. Oil on canvas, 51 × 38 cm. Zervos XI, 374. Emanuel Hoffmann Foundation, Oeffentliche Museum, Basel. E, June 13, 1941. Zervos XI, 151. F, June 16, 1940. Zervos X, 553 (36).

which the snout began to acquire independence from the rest of the face. In the final stages—snout rotation—the detached snout is rotated to the opposite side of the face (3).

The montage in Figure 9-21 reviews snout formation, snout elevation, and snout rotation. Figure 9-21A shows that the idea for the snout arises from deep profile insertion. In Figure 9-21B, the snout with two nostrils and one eye make up a separate compartment of the face, and in Figure 9-21C, the snout is elevated freely from the face. In Figure 9-21 (D,E), the freely movable snout has been lifted and swung to the opposite side of the face. Snouts are associated with animals, and it is of interest that in Figure 9-21E, the lower part of the face is Dora Maar, but the snout belongs to Kasbek, Picasso's Afghan hound; the dog's ears make up part of Dora's hat. In Figure 9-21F, the snout is elevated. Two interpretations are possible. The snout may have swung from the opposite side of the face and then simply elevated itself, as is suggested by its origin from the concave edge of the left side of the face. Alternately, it may represent simple snout elevation from the same side of the face, as in Figure 9-21C, the only difference being the direction the subject faces (3).

REFERENCES

1. Arneson DJ: Stories: Tweety and Sylvester. Middleton: Xerox Education Publications, 1971. Courtesy of Warner Bros.
2. Barthes R: Arcimbaldo. Milano: Franco Maria Ricci, 1980.
3. Cohen MM Jr: Picasso and the art of distortion and dislocation: The artist as researcher and experimentalist. J Craniofac Genet Dev Biol 11:61–95, 1991.
4. Cohen MM Jr: The unusual facial paintings of Giuseppe Arcimbaldi. Unpublished manuscript.
5. De Tolnay C: Michelangelo. Princeton: Princeton University Press, 1948.
6. De Zayas M: Picasso speaks. The Arts (New York) 3:315–326, 1923.
7. Dürer, Albrecht: Vier Bücher von menschlichen Proportion. Nürnberg, 1528.
8. Elsen AE: Purposes of Art. Fourth Edition. New York: Holt, Rinehart and Winston, 1981.
9. Feaver W: Masters of Caricature from Hogarth and Gillray to Scarfe and Levine. New York: Alfred A. Knopf, 1981.
10. Golding J: Picasso and Surrealism. In Picasso in Retrospect. R. Penrose and J. Golding, Editors. New York: Praeger. pp. 77–121, 1973.
11. Gombrich EH: The mask and the face: The perception of physiognomic likeness in life and art. In Art, Perception, and Reality. E.H. Gombrich, J. Hochberg, and M. Black, Editors. Baltimore: Johns Hopkins University Press. pp. 1–46, 1972.
12. Gombrich EH: Art and Illusion: A Study in the Psychology of Pictorial Representation. Princeton: Princeton University Press, 1972.
13. Kellogg R: Analyzing Children's Art. Palo Alto: Mayfield Publishing, 1970.
14. Kriegeskorte W: Giuseppe Arcimbaldo. Cologne: Benedikt Tascher Verlag, 1988.
15. Lee L, Lee K-J: A study of facial proportions and sketching of facial contours. Ear, Nose, and Throat 58:14–19, 1979.
16. Liberman A: Picasso. Vogue, New York, November 1, pp. 132–134, 1956.
17. Mellow JR: Charmed Circle: Gertrude Stein & Company. New York: Avon Books, 1974.
18. Nelson RP: Comic Art and Caricature. Pickering: Beaverbooks, 1978.
19. Osborne H: Aesthetics and Art History. New York: E.P. Dutton, 1970.
20. Panofsky E: Meaning in the Visual Arts. Garden City: Doubleday Anchor Books, 1955.
21. Penrose R: Picasso: His Life and Work. New York: Schocken Books, 1962.
22. Philipon C: Les Poires, Le Charivari, 1834.
23. Pope–Hennessy J: The Portrait in the Renaissance. Princeton: Princeton University Press, 1979.
24. Powell N, Humphreys B: Proportions of the Aesthetic Face. New York: Thieme-Stratton, 1984.
25. Read H: The Meaning of Art. New York: Praeger Publishers, 1972.
26. Ricketts RM: Divine proportions in facial aesthetics. Clin Plast Surg 9:401–422, 1982.
27. Rose B: American Painting: The Twentieth Century. New York: Rizzoli, 1986.
28. Rosenblum R: Picasso and the anatomy of eroticism. In Studies in Erotic Art. T. Bowie and L.V. Christenson, Editors. New York: Basic Books. pp. 337–350, 1970.
29. Rubin WS: Dada and Surrealist Art. New York: Harry N. Abrams, 1968.
30. Santayana G: The Sense of Beauty. New York: Dover Publications, 1955.
31. Scharf A: Filippino Lippi, Vienna, 1935.
32. Sircello G: A New Theory of Beauty. Princeton: Princeton University Press, 1975.
33. Solomon D: Portraitist for the Information Age. New York Times Magazine, Section 6, pp. 24–73, 1998.
34. Tom and Jerry, Number 322, September, 1979. Courtesy of Warner Bros.
35. Wilson S: The Surrealists. Uxbridge: Hillingdon Press, 1974.
36. Zervos C: Pablo Picasso: Oeuvres. Paris: Editions Cahiers d'Art, 33 Volumes, 1932–1978.

10

Conclusion

The world has billions of faces, each one unique, except for identical twins. Yet each person is capable of remembering hundreds of faces. Both known and unknown faces are essential in providing the fabric for social interaction. In this book, a series of facial perspectives has been developed. Each perspective can be conceived of as a self-contained, harmonious, and systematic body of knowledge with a network of interconnected elements and concepts. These perspectives are stratified in Table 10-1.

Within each self-contained perspective, phenomena can be explained in their own terms. Two examples suffice. First, within the embryonic perspective, the brain is derived from the neural tube formed by fusion of the neural folds. Second, within the artistic perspective, three factors are important for the development of the first modern portraits during the Renaissance: (a) the study of facial profiles on Roman coins and medallions; (b) the inclusion of wealthy patrons in religious narrative art of the churches; and (c) the painting of whole families, both living and dead, in the same work of art (13).

Each perspective also bears an interactive relationship to its adjacent perspectives and sometimes to more distant perspectives. For example, the embryonic, dysmorphic, surgical, and psychological levels can be combined. If the left maxillary swelling fails to fuse with its corresponding medial nasal swelling, left-sided cleft lip results and this can be surgically repaired, resulting in alterations in the family dynamics.

Stratified facial perspectives can also be grouped under generic categories, such as biological, applied biological, social, and symbolic (Table 10-2). In this context reductionism is possible. This means that elements in one or more perspectives can be explained at a more basic level. For example, some biological phenomena at the evolutionary, embryonic, anatomic–physiologic, and dysmorphic levels have a genetic basis (Table 10-3). Certain conditions at the surgical and psychological–psychiatric levels also have a genetic basis. The craniofacial malformations in Apert syndrome, Crouzon syndrome, and Treacher Collins syndrome can be surgically repaired. These disorders are caused by known genetic mutations. At the psychological–psychiatric level, a *FOXP1* mutation has been found in a speech and language disorder, and an *OX2R* mutation has been identified as one of several possible causes in Tourette syndrome. However, reductionism is not always possible for some perspectives. For example, in the sociocultural phenomena discussed in Chapter 8, the diverse societies considered in their historical contexts and their cultural changes over time are not susceptible to reductionism.

Broader comparisons are also possible. Both art (Chapter 9) and science (Chapters 2–4) attempt to extend our understanding of experience. In some ways art and science are different. Art tends to deal with universals in terms of particulars. For example, the dove is commonly used to symbolize peace. Science, on the other hand, tends to deal with particulars in terms of universals. Thus, the dove is subsumed within the order *Columbiformes* together with the pi-

Table 10-1 Stratified Facial Perspectives

Evolutionary (Chapter 2)
Embryonic and Developmental (Chapter 3)
Anatomical and Physiological (Chapter 4)
Dysmorphic and Genetic (Chapter 5)
Surgical (Chapter 6)
Psychological (Chapter 7)
Sociocultural and Symbolic (Chapter 8)
Artistic (Chapter 9)

Table 10-2 Generic Categories of Stratified Perspectives

Categories	Perspectives
Biological	Evolutionary Embryonic–Developmental Anatomic–Physiologic Dysmorphic–Genetic
Applied Biological	Surgical
Social	Psychological Sociocultural[a]
Symbolic	Symbolic[a] Artistic

[a]Chapter 8 is titled Sociocultural Sociocultural and Symbolic Perspectives. In the grouped stratified perspectives in this table, sociocultural and symbolic have been disjoined and appear under separate groups. Oscar Wilde's *The Picture of Dorian Gray* is both symbolic and artistic (see Chapter 8).

geon and the sandgrouse. But art and science need not be relegated to C.P. Snow's two cultures (15) because in some respects they are similar. Both explore the unknown; both seek truth; both use intuition; both use creative imagination; and both use rational as well as irrational faculties. Although the scientist is usually envisioned as a researcher and experimentalist, the artist sometimes works in a similar fashion (1).

Popper has characterized an underlying pattern of continuous development that is equally applicable to the activities of both scientists and artists (11). The essence of his formulation is expressed as

$$P_1 \rightarrow TS \rightarrow EE \rightarrow P_2,$$

where P_1, stands for an initial problem, TS stands for a proposed trial solution, EE stands for the process of error elimination applied to the trial solution, and P_2 stands for the result, which contains the seeds of its own problem to solve. Popper's formulation is a noncyclical feedback process in which P_2, resulting from a critical series of successive adjustments, is always different from P_1. P_2 then undergoes its own trial solution and error elimination, producing still another result, and so on *ad infinitum* as the history of art and the development of scientific thought through the ages have shown (1). The work of Picasso, for example, shows that he was a researcher and experimentalist. His four drawings of a woman's head shown in Figure 9-18 (see Chapter 9) serve as examples.

Quantitative measurements are used within every perspective (Table 10-4). For example, in evolutionary perspective, measurements and statistical analyses are commonly applied to fossil skulls. In artistic perspective, facial proportions require measurements, such as the Byzantine method of using nose length or Dürer's method of using a coordinate system (Chapter 9).

Both symmetric and asymmetric phenomena occur within each perspective. Although symmetry is common, asymmetry is found with surprising frequency (Table 10-5). Three examples follow. First, speech and language are functions of the left cerebral hemisphere, whereas tonal

Table 10-3 Genetic Reductionism

Categories	Perspectives	Genetic Examples
Biological	Evolutionary Embryonic–Developmental Anatomic–Physiologic Dysmorphic–Genetic	Molecular homologues of head-specific genes in the hydra Molecular determinants of craniofacial development Olfactory genes, gustatory receptors, color vision genes Genetic disorders of craniofacial development
Applied Biological	Surgical	Craniofacial surgery for Apert syndrome, Crouzon syndrome, and Treacher Collins syndrome, all of which involve mutant genes
Social	Psychological–Psychiatric	*FOXP1* mutation in speech and language disorder, *OX2R* mutation in Tourette syndrome and in narcolepsy, metachromatic leukodystrophy, del22q11 syndrome

Data from Chapters 1–7.

Table 10-4 Quantitative Measurements

Categories	Perspectives	Examples
Biological	Evolutionary	Measurements and statistical analyses of fossil skulls
	Embryonic–Developmental	Crown–rump length measurements, embryonic measurements of the head and face, anthropometric and cephalometric studies of the craniofacial region
	Anatomic–Physiologic	Vision: perimetry, relative index of refraction Hearing: measurement of Hertz and decibel
	Dysmorphic–Genetic	Molecular tools for genetic diagnosis
Applied Biological	Surgical	Inner and outer canthal distance measurements, telorbitism measurements
Social	Psychological	Measurements and statistical analyses for face processing, face recognition, and facial expression
	Sociocultural	Craniofacial measurements showing a pseudo-relationship to various psychological characteristics, crime, or inferiority
Symbolic	Symbolic	Measurements for creating facial illusions
	Artistic	Measurements of facial proportions (Byzantine proportions by nose length, Dürer's measurements using a coordinate system, Golden Ratio)

Examples from Chapters 2–9.

Table 10-5 Asymmetry

Categories	Perspectives	Examples
Biological	Evolutionary	Asexual budding from the lateral column of the hydra (see Fig. 2-1, left)
	Embryonic–Developmental	Flounder: one eye translocates to the opposite side of the head; brain and skull become asymmetrically shaped Narwhal: Of the two teeth that develop, the left one grows outward into a 3 meter twisted tusk Unilateral cleft lip (2:1 left-to-right)
	Anatomic–Physiologic	Left hemisphere: speech and language Right hemisphere: tonal stimuli and music Face: normally asymmetric
	Dysmorphic–Pathologic	Various types of synostotic and deformational plagiocephaly, hemifacial microsomia, unilateral facial paralysis
Social	Psychological	Prenatal and postnatal head turning, 2:1 right-to-left. Kissing with 2:1 right-to-left head turning
	Sociocultural	Asymmetric facial piercing (one ear, one eyebrow, side of nose), women's one-sided hairstyles, women's hats with asymmetric ornaments
Symbolic	Artistic	Facial painting of Picasso, Soutine, and Modigliani

Examples from Chapters 2–5 and 7–9 and from Cohen, 1995, 2001 (1–8); Lipschitz, 1954 (10); Narwhal, 2004 (12); Schreiber 2004 (14); Weiss, 2004 (17); and Wright, 2004 (18).

stimuli and music are functions of the right cerebral hemisphere. Second, kissing occurs with a 2:1 ratio of right-to-left head turning. Third, for Picasso, although the eyes, nose, and mouth could be distributed in various ways, a face still remained a face (16). The left and right eyes in his portrait of Marie-Thérèse Walter are asymmetrically placed and of different colors as well (9).

This volume has presented eight facial perspectives that can be differentiated from one another as separate yet related fields of knowledge. Faces, like images seen in a kaleidoscope, are infinitely complex and beautiful. They are woven into our daily lives, our dreams, and our memories.

REFERENCES

1. Cohen MM Jr: Picasso and the art of distortion and dislocation: The artist as researcher and experimentalist. J Craniofac Genet Dev Biol 11:61–95, 1991.
2. Cohen MM Jr: Perspectives on craniofacial asym-

metry. I. The biology of asymmetry. Int J Oral Maxillofac Surg 24:2–7, 1995.

3. Cohen MM Jr: Perspectives on craniofacial asymmetry. II. Asymmetric embryopathies. Int J Oral Maxillofac Surg 24:8–12, 1995.
4. Cohen MM Jr: Perspectives on craniofacial asymmetry. III. Common and/or well-known causes of asymmetry. Int J Oral Maxillofac Surg 24:127–133, 1995.
5. Cohen MM Jr: Perspectives on craniofacial asymmetry. IV. Hemi-asymmetries. Int J Oral Maxillofac Surg 24:134–141, 1995.
6. Cohen MM Jr: Perspectives on craniofacial asymmetry. V. The craniosynostoses. Int J Oral Maxillofac Surg 24:191–194, 1995.
7. Cohen MM Jr: Perspectives on craniofacial asymmetry. VI. The hamartoses. Int J Oral Maxillofac Surg 24:195–200, 1995.
8. Cohen MM Jr: Asymmetry: Molecular, biologic, embryopathic, and clinical perspectives. Am J Med Genet 101:292–314, 2001.
9. Gilot F, Lake C: Life With Picasso. New York: McGraw Hill, 1964.
10. Lipschitz J: Modigliani. New York: Harry Abrams, 1954.
11. Magee B: *Karl Popper*. New York: Viking Press, pp. 60–69, 1973.
12. Narwhal. Online, 2004. www.enchantedlearning.com/subjects/whales/species/Narwhal.shtml.
13. Pope–Hennessy J: The Portrait in the Renaissance. Princeton: Princeton University Press, 1979.
14. Schreiber A: Morphogenesis, thyroid hormone, asymmetry, and craniofacial remodeling. Online, 2004. http://www.ciwemb.edu/labs/schreiber/.
15. Snow CP: The Two Cultures. London: Cambridge University Press, 1959.
16. Stein L: Appreciation, Painting, Poetry, and Prose. New York: Crown, p. 177, 1947.
17. Weiss B: Chaim Soutine. Online 2004. www.jewishvirtuallibrary.org/jsource/biography/Soutine.html.
18. Wright J: Hydra. Online 2004. www.northern.edu/natsource/INVERT/Hydra1.htm.

Index

Page numbers followed by f *indicate figures; those followed by* t *indicate tables.*